LTE SIGNALING, TROUBLESHOOTING AND PERFORMANCE MEASUREMENT

LTE SIGNALING, TROUBLESHOOTING AND PERFORMANCE MEASUREMENT

SECOND EDITION

Ralf Kreher and Karsten Gaenger

NETSCOUT – Mobile Access, Germany

Library of Congress Cataloging-in-Publication Data

Kreher, Ralf, author.
  LTE signaling, troubleshooting and performance measurement / Ralf Kreher and Karsten Gaenger, NETSCOUT – Mobile Access, Germany. – Second edition.
    pages cm
  Includes bibliographical references and index.
  ISBN 978-1-118-72510-8 (cloth : alk. paper) 1. Long-Term Evolution (Telecommunications) I. Gaenger, Karsten, author. II. Title.
  TK5103.48325.K74 2016
  621.3845′6–dc23
                              2015034555

A catalogue record for this book is available from the British Library.

Set in 9/11pt, TimesLTStd by SPi Global, Chennai, India

1   2016

Ralf Kreher

I dedicate this book to my grandmother Emilie (*1914, †2014). She managed to raise four children in the aftermath of World War II after losing her husband, home and all her valuables.

Karsten Gaenger

To my lovely lady for her care and patience. To my parents and my sister for being with me and for their continuous support.

As a global standard, LTE might connect more people than ever before. It is my hope that as we increase our ability to communicate we increase our ability to live peaceably together.

Contents

Author Biography

Karsten Gaenger

Karsten Gaenger received a Dipl.-Ing. degree in electrical engineering from the Berlin University of Technology. He worked for the Fraunhofer HHI research institute from 2004 to 2006. During this time, he has published several IEEE papers on the development of a reliable real-time streaming system and protocol for mobile ad-hoc networks. His research interests are mobile communications, geolocation, and robust real-time media streaming.

From 2006 to 2015, he worked for Tektronix Communications as a Solution Architect and Product Line Manager with focus on RAN testing, monitoring, and optimization. One of his major projects was the development of a passive real-time LTE air interface monitoring probe.

Currently, he works for NETSCOUT Mobile Access product line in the field of 3G and LTE radio access networks (RANs). He is a Product Line Manager with focus on RAN and geolocation in today's and next generation mobile networks. His current projects include the network wide passive geolocation of all devices (UEs).

He currently resides in Germany.

Ralf Kreher

Ralf Kreher works as a Principal Engineer/Senior Solution Architect for NETSCOUT Mobile Access product line where he specializes in Performance Measurement and Key Performance Indicator (KPI) implementation. Previously, he was head of the Tektronix Communications Test and Optimization Customer Training Department for almost 4 years and was responsible for a world-class seminar portfolio for mobile technologies and measurement products.

Tektronix Communications and NETSCOUT have combined and merged their businesses in July 2015. Before joining Tektronix, Kreher held a trainer assignment for switching equipment at Teles AG, Berlin. Kreher holds a Communication Engineering Degree from the University of Applied Science, Deutsche Telekom Leipzig. He is internationally recognized as an author of the following books: *UMTS Signaling* (Wiley) and *UMTS Performance Measurement: A Practical Guide to KPIs for the UTRAN Environment* (Wiley). He currently resides in Germany.

Foreword

Over the past 30 years, society has undergone a profound and fundamental change. This change has been driven by radical advancements in communications technology. At the heart of this change has been cellular or mobile communications. It is the ability to instantly connect to anyone, anywhere, from the palm of your hand.

Today, there are more connected devices than there are people on the planet. It is forecasted that by 2019 over 50% of the mobile phones on the planet will be smartphones – devices that provide far more than just the ability for voice communications, but devices that allow internet browsing, video streaming, and unified communications. They are the platform for innovation and economic growth, and it is a platform that governments around the globe view as vitally important to economic success – regionally, nationally, and internationally.

As mobile communications have developed, the networks have become increasingly more complex. Nowhere is this more apparent than in the radio access network (RAN). Today, RAN provides voice, data, and video services to a countless array of devices with varying capabilities moving through different terrains at different speeds. All the while, the user is expecting a seamless, uninterrupted user experience.

That is why it is vitally important for engineers, planners, managers, and regulators to understand the complexity of the modern RAN, and the importance of having monitoring capabilities in the network. From understanding capacity utilization to user experience, the mobile network is the platform for tomorrow's innovation. It is incumbent upon all of us to ensure that this platform is ready and available to fuel the breakthroughs of tomorrow.

Richard Kenedi, President NETSCOUT Service Provider Business Unit

Acknowledgments

We would like to take the chance to acknowledge the effort of all who participated directly or indirectly in creating and publishing this book.

First of all a special "thank you" goes to Ralf Kreher's sister Brit who created and formatted all figures you will find in this book. Another one goes to our family members and all who supported and encouraged us to get this work done.

Eiko Seidel and his team at Nomor Research have not just created some excellent primers about LTE radio interface procedures and set up the 3GPP LTE Standards Group at www.linkedin.com, but also gave us deep insight into their scheduling simulator, a tool used to design scheduling algorithms for eNodeB vendors.

Antonio Bovo who used to work as a System Architect for Tektronix Communications Padova contributed a very detailed research on E-UTRAN protocols and functions. From his work, we have derived the major part of the S1AP chapter of this book.

Karsten Gienskey and Marcus Garin working for Tektronix Berlin shared with us their earliest prototypes and design specifications for RLC reassembly and radio interface tracing. Without their great job, we would have been "blind" on the radio interface.

Ulrich Jeczawitz, freelancer and ex-colleague of Tektronix Berlin, and the former development team of the Tektronix G35 protocol simulator led by Dirk-Holger Lenz generated traces of E-UTRAN and Enhance Packet Core signaling procedures long before they would occur in any live network field trial.

Lars Chudzinsky, working on LTE call trace and call analysis modules for Tektronix Berlin, contributed design specifications of protocol failure events that became the raw material for chapter 2.10.

This book would not exist without the ideas, questions, and requirements contributed by customers, colleagues, and subcontractors. Besides all others that cannot be personally named, we would like to express thanks especially to the following people listed in an alphabetical order:

Jürgen Forsbach
Andre Huge
Steffen Hülpüsch
Armin Klopfer

In addition, thanks goes to the Management of the Tektronix Communications Mobile Access product line, in particular the Human Resources Department represented by Nadine Eckert and R&D Berlin Director Jens Dittrich who supported the idea to write this book and approved usage of Tektronix material in the contents.

Maïssa Bahsoun, Jeanne Lancry-Gulino, and Stéphanie Langlois have been our prime contacts in 3GPP/ETSI to get copyright permissions and, last but not least, we also would like to express our thanks to the team at Wiley, especially Mark Hammond, Sarah Tilley, Sophia Travis, Liz Wingett, and Teresa Netzler for their strong support.

                                                                                    Berlin, 1 July, 2015
                                                                              Ralf Kreher and Karsten Gaenger

1

Standards, Protocols, and Functions

Long-Term Evolution (LTE) of Universal Mobile Telecommunications Service (UMTS) is one of the latest steps in an advancing series of mobile telecommunication systems. The standards body behind the paperwork is the 3rd Generation Partnership Project (3GPP).

Along with the term LTE, the acronyms EPS (Evolved Packet System), EPC (Evolved Packet Core), and SAE (System Architecture Evolution) are often heard. Figure 1.1 shows how these terms are related to each other: EPS is the umbrella that covers both the LTE of the Evolved Universal Terrestrial Radio Access Network (E-UTRAN) and the SAE of the EPC network.

LTE was and is standardized in parallel to other radio access network technologies like EDGE (Enhanced Data Rates for GSM Evolution) and HSPA (High-Speed Packet Access). This means that LTE is not a simple replacement of existing technologies. Rather it is expected that different kinds of radio access will coexist in operator networks.

From this background, it emerges that understanding LTE also requires understanding alternative and coexisting technologies. Indeed, one of the major challenges of LTE signaling analysis will concern the analysis of handover procedures. Especially, the options for possible inter-RAT (Radio Access Technology) handovers have multiplied compared to what was possible in UMTS Release 99. However, also intra-system handover and dynamic allocation of radio resources to particular subscribers will play an important role.

The main drivers for LTE development are:

- reduced delay for connection establishment;
- reduced transmission latency for user plane data;
- increased bandwidth and bit rate per cell, also at the cell edge;
- reduced costs per bit for radio transmission;
- greater flexibility of spectrum usage;
- simplified network architecture;
- seamless mobility, including between different radio access technologies;
- reasonable power consumption for the mobile terminal.

It must be said that LTE as a RAT is flanked by a couple of significant improvements in the core network known as the EPS. Simplifying things a little, it is not wrong to state that EPS is an all-IP (Internet Protocol) transport network for mobile operators. IP will also become the physical transport layer on the wired interfaces of the E-UTRAN. This all-IP architecture is also one of the facts behind the bullet point on simplified network architecture. However, to assume that to be familiar with the TCP/IP world is

---

LTE Signaling, Troubleshooting and Performance Measurement, Second Edition. Ralf Kreher and Karsten Gaenger.
© 2016 John Wiley & Sons, Ltd. Published 2016 by John Wiley & Sons, Ltd.

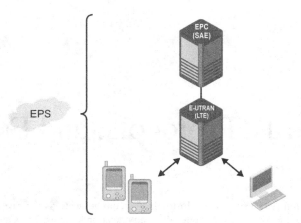

Figure 1.1   EPC and LTE under the umbrella of EPS

enough to understand and measure LTE would be a fatal error. While the network architecture and even the basic signaling procedures (except the handovers) become simpler, the understanding and tracking of radio parameters require more knowledge and deeper investigation than they did before. Conditions on the radio interface will change rapidly and with a time granularity of 1 ms, the radio resources assigned to a particular connection can be adjusted accordingly.

For instance, the radio quality that is impacted by the distance between the User Equipment (UE) and base station can determine the modulation scheme and, hence, the maximum bandwidth of a particular connection. Simultaneously, the cell load and neighbor cell interference – mostly depending on the number of active subscribers in that cell – will trigger fast handover procedures due to changing the best serving cell in city center areas, while in rural areas, macro cells will ensure the best possible coverage.

The typical footprint of an LTE cell is expected by 3GPP experts to be in the range from approximately 700 m up to 100 km. Surely, due to the wave propagation laws, such macro cells cannot cover all services over their entire footprint. Rather, the service coverage within a single cell will vary, for example, from the inner to the outer areas and the maximum possible bit rates will decline. Thus, service optimization will be another challenge too.

1.1   LTE Standards and Standard Roadmap

To understand LTE, it is necessary to look back at its predecessors and follow its path of evolution for packet switched services in mobile networks.

The first stage of the General Packet Radio Service (GPRS), which is often referred to as the 2.5G network, was deployed in live networks starting after the year 2000. It was basically a system that offered a model of how radio resources (in this case, GSM time slots) that had not been used by Circuit Switched (CS) voice calls could be used for data transmission and, hence, profitability of the network could be enhanced. At the beginning, there was no pre-emption for PS (Packet Switched) services, which meant that the packet data needed to wait to be transmitted until CS calls had been finished.

In contrast to the GSM CS calls that had a Dedicated Traffic Channel (DTCH) assigned on the radio interface, the PS data had no access to dedicated radio resources and PS signaling, and the payload was transmitted in unidirectional Temporary Block Flows (TBFs) as shown in Figure 1.2.

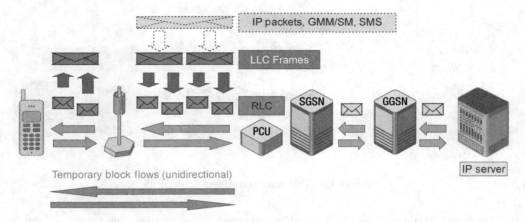

Figure 1.2　Packet data transfer in 2.5G GPRS across Radio and Abis interfaces

These TBFs were short and the size of data blocks was small due to the fact that the blocks must fit the transported data into the frame structure of a 52-multiframe, which is the GSM radio transmission format on the physical layer. Larger Logical Link Control (LLC) frames that contain already segmented IP packets needed to be segmented into smaller Radio Link Control (RLC) blocks.

The following tasks are handled by the RLC protocol in 2.5G:

- Segmentation and reassembly of LLC packets → segmentation results in RLC blocks.
- Provision of reliable links on the air interface → control information is added to each RLC block to allow Backward Error Correction (BEC).
- Performing sub-multiplexing to support more than one MS (Mobile Station) by one physical channel.

The Medium Access Control (MAC) protocol is responsible for:

- point-to-point transfer of signaling and user data within a cell;
- channel combining to provide up to eight physical channels to one MS;
- mapping RLC blocks onto physical channels (time slots).

As several subscribers can be multiplexed on one physical channel, each connection has to be (temporarily) uniquely identified. Each TBF is identified by a Temporary Flow Identifier (TFI). The TBF is unidirectional (uplink (UL) and downlink (DL)) and is maintained only for the duration of the data transfer.

Toward the core network in 2.5G GPRS, the Gb interface is used to transport the IP payload as well as GPRS Mobility Management/Session Management (GMM/SM) signaling messages and short messages (Short Message Service, SMS) between SGSN and the PCU (Packet Control Unit) – see Figure 1.3. The LLC protocol is used for peer-to-peer communication between SGSN and the MS and provides acknowledged and unacknowledged transport services. Due to different transmission conditions on physical layers (E1/T1 on the Gb and Abis interfaces, 52-multiframe on the Air interface), the size of IP packets needs to be adapted. The maximum size of the LLC payload field is 1540 octets (bytes) while IP packets can have up to 65 535 octets (bytes). So the IP frame is segmented on SGSN before transmission via LLC and reassembled on the receiver side. All in all, the multiple segmentation/reassembly of IP

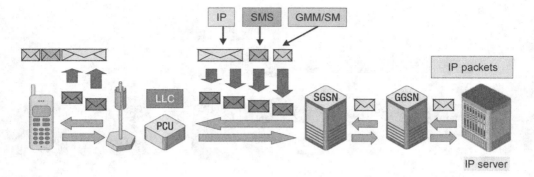

Figure 1.3   Packet data transfer in 2.5G GPRS

payload frames generates a fair overhead of transport header information that limits the chargeable data throughput. In addition, the availability of radio resources for PS data transport has not been guaranteed. So this system was only designed for non-real-time services like web browsing or e-mail.

To overcome these limitations, the standards organizations proposed a set of enhancements that led to the parallel development of UMTS and EGPRS (Enhanced GPRS) standards. The most successful EGPRS standard that is found today in operators' networks is the EDGE standard. From the American Code Division Multiple Access (CDMA) technology family, another branch of evolution led to the CDMA2000 standards (defined by the 3GGP2 standards organization), but since the authors have not seen any interworking between CDMA2000 and Universal Terrestrial Radio Access Network (UTRAN) or GSM/EDGE Radio Access Network (GERAN) so far, this technology will not be discussed further in this book.

The most significant enhancements of EGPRS compared to GSM/GPRS are shown in Figures 1.4 and 1.5. On the one hand, a new modulation technique, 8-Phase Shift Keying (8PSK), was introduced to allow transmission of 8 bits per symbol across the air interface and, thus, an increase in the maximum possible bit rate from 20 to 60 kbps. On the other hand, to use the advantages of the new 8PSK modulation technique, it was necessary to adapt the data format on the RLC/MAC layer, especially regarding the size of the transport blocks and the time transmission interval of the transport blocks. Different transport

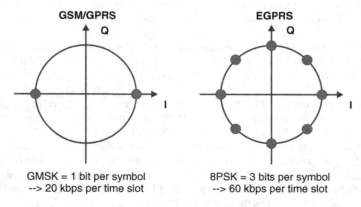

Figure 1.4   GSM/GPRS versus EGPRS modulation

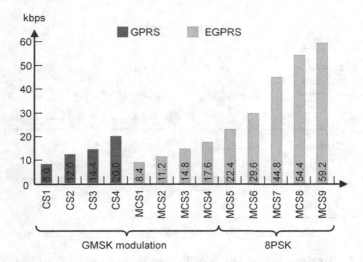

Figure 1.5 Modulation/coding scheme and maximum bit rate in GSM/GPRS versus EGPRS

block formats require a different CS. Thus, the so-called Modulation and Coding Scheme (MCS) and CS for GPRS and EGPRS as shown in Figure 1.4 have been defined. These MCSs stand for defined radio transmission capabilities on the UE and BTS (Base Transceiver Station) side. It is important to mention this, because in a similar way capability definition with UE physical layer categories instead of MCS was introduced for HSPA and will be found in LTE again.

In comparison to GSM/GPRS, the EGPRS technology also offered a more efficient retransmission of erroneous data blocks, mostly with a lower MCS than the one used previously. The retransmitted data also does not need to be sent in separate data blocks, but can be appended piece by piece to present regular data frames. This highly sophisticated error correction method, which is unique for EGPRS, is called Incremental Redundancy or Automatic Repeat Request (ARQ) II and is another reason why higher data transmission rates can be reached using EGPRS.

As a matter of fact, as shown in Figure 1.6, the risk of interference and transmission errors becomes much higher when the distance between a base station and a UE is large. Consequently, the MCS that

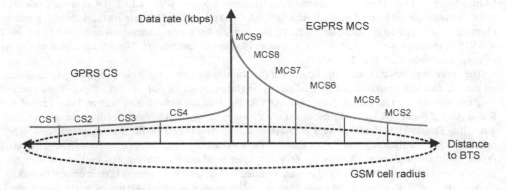

Figure 1.6 Cell footprint of maximum bit rate as function of MCS in (E)GPRS

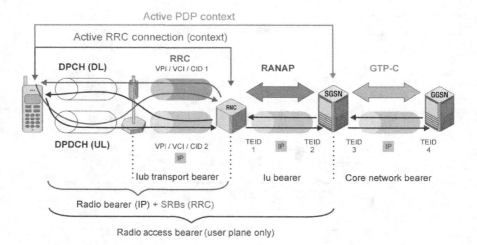

Figure 1.7   IP payload transmission using Release 99 bearers with UE in CELL_DCH state

allows the highest maximum bit rate cannot be used in the overall cell coverage area, but only in a smaller area close to the base station's antenna. Also, for this specific behavior, an adequate expression will be found in LTE radio access.

Recently, two key parameters have driven the evolution of packet services further toward LTE: higher data rates and shorter latency. EGPRS (or EDGE) focused mostly on higher bit rates, but did not include any latency requirements or algorithms to guarantee a defined Quality of Service (QoS) in early standard-ization releases. Meanwhile, in parallel to the development of UMTS standards, important enhancements to EDGE have been defined, which allow pre-emption of radio resources for packet services and control of QoS. Due to its easy integration in existing GSM networks, EDGE is widely deployed today in cellular networks and is expected to coexist with LTE on the long haul.

Nevertheless, the first standard that promised complete control of QoS was UMTS Release 99. In contrast to the TBFs of (E)GPRS, the user is assigned dedicated radio resources for PS data that are permanently available through a radio connection. These resources are called bearers.

In Release 99, when a PDP (Packet Data Protocol) context is activated, the UE is ordered by the RNC (Radio Network Controller) to enter the Radio Resource Control (RRC) CELL_DCH state. Dedicated resources are assigned by the Serving Radio Network Controller (SRNC): these are the dedicated physical channels established on the radio interface. Those channels are used for transmission of both IP payload and RRC signaling – see Figure 1.7. RRC signaling includes the exchange of Non-Access Stratum (NAS) messages between the UE and SGSN.

The spreading factor of the radio bearer (as the combination of several physical transport resources on the Air and Iub interfaces is called) depends on the expected UL/DL IP throughput. The expected data transfer rate can be found in the RANAP (Radio Access Network Application Part) part of the Radio Access Bearer (RAB) assignment request message that is used to establish the Iu bearer, a GPRS Tunneling Protocol (GTP) tunnel for transmission of an IP payload on the IuPS interface between SRNC and SGSN. While the spreading factor controls the bandwidth of the radio connection, a sophisticated power control algorithm guarantees the necessary quality of the radio transmission. For instance, this power control ensures that the number of retransmitted frames does not exceed a certain critical threshold.

Activation of PDP context results also in the establishment of another GTP tunnel on the Gn interface between SGSN and GGSN. In contrast to IuPS, where tunnel management is a task of RANAP, on the Gn interface – as in (E)GPRS – the GPRS Tunneling Protocol-Control (GTP-C) is responsible for context (or tunnel) activation, modification, and deletion.

However, in Release 99, the maximum possible bit rate is still limited to 384 kbps for a single connection and, more dramatically, the number of users per cell that can be served by this highest possible bit rate is very limited (only four simultaneous 384 kbps connections per cell are possible on the DL due to the shortness of DL spreading codes).

To increase the maximum possible bit rate per cell as well as for the individual user, HSPA was defined in Releases 5 and 6 of 3GPP.

In High-Speed Downlink Packet Access (HSDPA), the High-Speed Downlink Shared Channel (HSD-SCH), which bundles several High-Speed Physical Downlink Shared Channels (HS-PDSCHs), is used by several UEs simultaneously – that is why it is called a shared channel.

A single UE using HSDPA works in the RRC CELL_DCH state. For DL payload transport, the HSD-SCH is used, that is, mapped onto the HS-PDSCH. The UL IP payload is still transferred using a dedicated physical data channel (and appropriate Iub transport bearer); in addition, the RRC signaling is exchanged between the UE and RNC using the dedicated channels – see Figure 1.8.

All these channels have to be set up and (re)configured during the call. In all these cases, both parties of the radio connection, cell and UE, have to be informed about the required changes. While communication between NodeB (cell) and CRNC (Controlling Radio Network Controller) uses NBAP (Node B Application Part), the connection between the UE and SRNC (physically the same RNC unit, but different protocol entity) uses the RRC protocol.

The big advantage of using a shared channel is higher efficiency in the usage of available radio resources. There is no limitation due to the availability of codes and the individual data rate assigned to a UE can be adjusted quickly to the real needs. The only limitation is the availability of processing resources (represented by channel card elements) and buffer memory in the base station. In 3G networks,

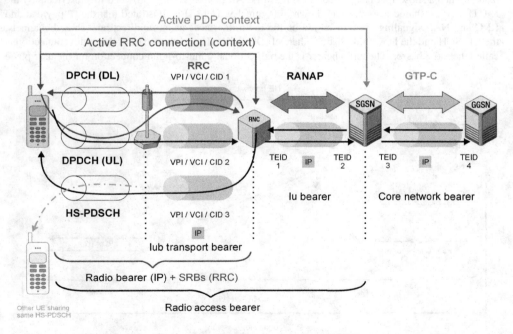

Figure 1.8 IP data transfer using HSDPA

the benefits of an Uplink Shared Channel (UL-SCH) have not yet been introduced due to the need for UL power control, that is, a basic constraint of Wideband CDMA (WCDMA) networks. Hence, the UL channel used for High-Speed Uplink Packet Access (HSUPA) is an Enhanced Dedicated Channel (E-DCH). The UL transmission data volume that can be transmitted by the UE on the UL is controlled by the network using the so-called grants to prevent buffer overflow in the base station and RNC. The same "grant" mechanism will be found in LTE.

All in all, with HSPA in the UTRAN, the data rates on the UL and DL have been significantly increased, but packet latency is still a critical factor. It takes quite a long time until the RRC connection in the first step and the radio bearer in the second step are established. Then, due to limited buffer memory and channel card resources in NodeB, often quite progressive settings of user inactivity timers lead to transport channel-type switching and RRC state change procedures that can be summarized as intra-cell hard handovers. Hard handovers are characterized by the fact that the active radio connection including the radio bearer is interrupted for a few hundred milliseconds. Similar interruptions of the data transmission stream are observed during serving HSDPA cell change procedures (often triggered by a previous soft handover) due to flushing of buffered data in NodeB and rescheduling of data to be transmitted by the RNC. That such interruptions (occurring in dense city center areas with a periodicity of 10–20 seconds) are a major threat for delay-sensitive services is self-explanatory.

Hence, from the user plane QoS perspective, the two major targets of LTE are:

- a further increase in the available bandwidth and maximum data rate per cell as well as for the individual subscriber;
- reducing the delays and interruptions in user data transfer to a minimum.

These are the reasons why LTE has an always-on concept in which the radio bearer is set up immediately when a subscriber is attached to the network. All radio resources provided to subscribers by the E-UTRAN are shared resources, as shown in Figure 1.9. Here, it is illustrated that the IP payload and RRC and NAS signaling are transmitted on the radio interfaces using unidirectional shared channels, the UL-SCH and the Downlink Shared Channel (DL-SCH). The payload part of this radio connection is called the radio bearer. The radio bearer is the bidirectional point-to-point connection for the user plane

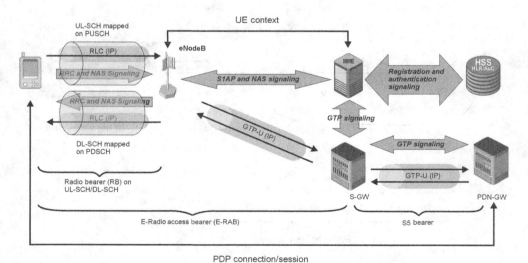

Figure 1.9   Packet data transfer in E-UTRAN/EPC

between the UE and eNodeB (eNB). The RAB is the user plane connection between the UE and the Serving Gateway (S-GW), and the S5 bearer is the user plane connection between the S-GW and public data network gateway (PDN-GW).

Note that a more detailed explanation of the LTE/EPC bearer concept is given in Section 1.6.

The end-to-end connection between the UE and PDN-GW, that is, the gateway to the IP world outside the operator's network, is called a PDN connection in the E-UTRAN standard documents and a session in the core network standards. Regardless, the main characteristic of this PDN connection is that the IP payload is transparently tunneled through the core and the radio access network.

To control the tunnels and radio resources, a set of control plane connections runs in parallel with the payload transport. On the radio interface, RRC and NAS signaling messages are transmitted using the same shared channels and the same RLC transport layer that is used to transport the IP payload.

RRC signaling terminates in the eNB (different from 3G UTRAN where RRC was transparently routed by NodeB to the RNC). The NAS signaling information is – as in 3G UTRAN – simply forwarded to the Mobility Management Entity (MME) and/or UE by the eNB.

For registration and authentication, the MME exchanges signaling messages with the central main subscriber databases of the network, the Home Subscriber Server (HSS).

To open, close, and modify the GTP/IP tunnel between the eNB and S-GW, the MME exchanges GTP signaling messages with the S-GW and the S-GW has the same kind of signaling connection with the PDN-GW to establish, release, and maintain the GTP/IP tunnel called the S5 bearer.

Between the MME and eNB, together with the E-RAB, a UE context is established to store connection-relevant parameters like the context information for ciphering and integrity protection. This UE context can be stored in multiple eNBs, all of them belonging to the list of registered tracking areas for a single subscriber. Using this tracking area list and UE contexts, the inter-eNB handover delay can be reduced to a minimum.

The two most basic LTE standard documents are 3GPP 23.401 "GPRS Enhancements for E-UTRAN Access" and 3GPP 36.300 "Overall Description Evolved Universal Terrestrial Radio Access (E-UTRA) and E-UTRAN." These two specs explain in a comprehensive way the major improvements in LTE that are pushed by an increasing demand for higher bandwidth and shorter latency of PS user plane services. The basic network functions and signaling procedures, as well as the network architecture, interfaces, and protocol stacks, are explained.

Although this book will not become simply a copy of what is already described in the standard documents, it is necessary to give a summary of the facts and parameters that are required to understand the signaling procedures and key performance indicators of the network and services. Additional explanations will be given to highlight facts that cannot be found in the specs.

1.2 LTE Radio Access Network Architecture

The E-UTRAN comes with a simple architecture that is illustrated in Figure 1.10. The base stations of the network are called eNodeB, and each eNB is connected to one or multiple MMEs. These MMEs in turn are connected to an S-GW that may also be co-located (comprising the same physical hardware) with the MME. The interface between the eNB and MME is the called the S1 interface. In case the MME and S-GW are not found in the same physical entity, the S1 control plane interface (S1-MME) will connect the eNB and MME while the S1 user plane interface (here S1-U) will connect the eNB with the S-GW.

In case one eNB is connected to multiple MMEs, these MMEs form a so-called MME pool and the appropriate network functionality is called S1 flex. The initial signaling procedure used to connect an eNB with an MME is the S1 setup procedure of the S1 Application Part (S1AP).

The X2 interface is used to connect eNBs with each other. The main purpose of this connectivity is intra-E-UTRAN handover. In the real world, it will not be possible for all eNBs of the network to be connected via X2 due to limited transport resources on the wired interfaces. It also must be expected

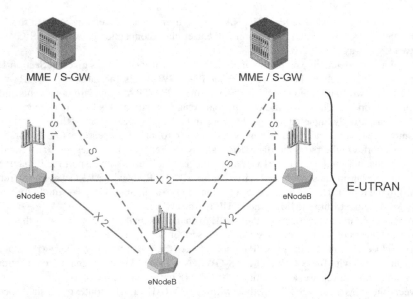

Figure 1.10 E-UTRAN network architecture (according to 3GPP 36.300). (*Source*: Reproduced with permission from © 2008 3GPP™.)

that, physically, the X2 links will lead from one eNB to the MME and then back to a second eNB. In other words, the hubs will be located at the physical location of the MME.

It is important to understand that only the base stations and their physical connections (wires or fibers) are defined by 3GPP as the E-UTRAN, while MME and S-GW are seen as elements of the EPC network.

1.3   Network Elements and Functions

The explanation given in the previous section indicates that, compared to base stations in GSM and UMTS UTRAN, the eNB will cover a set of new functions that are crucial to understand how the E-UTRAN is working.

In addition, the functionality of the MME and S-GW is different from that of their 2G/3G relatives, the RNC and the SGSN.

The following list of logical meta-functions performed within the overall network/system was defined by 3GPP:

- Network access control functions
- Packet routing and transfer functions
- Mobility management functions
- Security functions
- Radio resource management functions
- Network management functions.

These meta-functions are found in the different network elements with a more specific functionality definition.

1.3.1   The eNodeB (eNB)

The eNB is the network entity that is responsible for radio interface transmission and reception. This includes radio channel modulation/demodulation as well as channel coding/decoding and multiplexing/demultiplexing.

System information is broadcast in each cell on the radio interface DL to provide basic information to UEs as a prerequisite to access the network.

The LTE base station hosts all RRC functions such as broadcast of system information and RRC connection control including:

- Paging of subscribers.
- Establishment, modification, and release of RRC connection including the allocation of temporary UE identities (Radio Network Temporary Identifier, RNTI).
- Initial security activation, which means the initial configuration of the Access Stratum (AS) integrity protection for the control plane and AS ciphering for both control plane and user plane traffic.
- RRC connection mobility that includes all types of intra-LTE handover (intra-frequency and inter-frequency). In the case of handover, the source eNB will take care of the associated security handling and provide the necessary key and algorithm information to the handover target cell by sending specific RRC context information embedded in a transparent container to the handover target eNB.
- Establishment, modification, and release of DRBs (Dedicated Radio Bearers) carrying user data.
- Radio configuration control, especially the assignment and modification of ARQ and Hybrid Automatic Repeat Request (HARQ) parameters as well as Discontinuous Reception (DRX) configuration parameters.
- QoS control to ensure that, for example, user plane packets of different connections are scheduled with the required priority for DL transmission and that mobiles receive the scheduling grants for UL data transmission according to the QoS parameters of the radio bearers.
- Recovery functions that allow re-establishment of radio connections after physical channel failure or Radio Link Control Acknowledged Mode (RLC AM) retransmission errors.

The most crucial part for measuring the eNB performance is the UL/DL resource management and packet scheduling performed by the eNB. This is probably the most difficult function, which requires the eNB to cope with many different constraints like radio link quality, user priority, requested QoS, and UE capabilities. It is the task of the eNB to make use of the available resources in the most efficient way.

Furthermore, the RRC entity of the eNB covers all types of intra-LTE and inter-RAT measurements, in particular:

- Setup, modification, and release of measurements for intra-LTE intra-frequency, intra-LTE inter-frequency, inter-RAT mobility, transport channel quality, UE internal measurement reports to indicate, for example, current power consumption and GPS positioning reports sent by the handset.
- For compressed mode measurements, it is necessary to configure, activate, and deactivate the required measurement gaps.
- The evaluation of reported measurement results and start of necessary handover procedures are also eNB functions (while in 3G UMTS, all measurement evaluation and handover control functions have been embedded in the RNC). The many different parameters used in RRC measurement control functions like hysteresis values, time to trigger timer values, and event level threshold of RSRP and RSRQ (Received Signal Reference Power and Received Signal Reference Quality) are the focus of radio network optimization activities.

Other functions of the eNB comprise the transfer of dedicated NAS information and non-3GPP dedicated information, the transfer of UE radio access capability information, support for E-UTRAN sharing (multiple Public Land Mobile Network (PLMN) identities), and management of multi-cast/broadcast services.

The support of self-configuration and self-optimization is seen as one of the key features of the E-UTRAN. Among these functions we find, for example, intelligent learning functions for automatic updates of neighbor cell lists (handover candidates) as they are used for RRC measurement tasks and handover decisions.

The eNB is a critical part of the user plane connections. Here, the data is routed, multiplexed, ciphered/deciphered, segmented, and reassembled. It is correct to say that on the E-UTRAN transport layer level, the eNB acts as an IP router and switch. The eNB is also responsible for optional IP header compression. On the control plane level, the eNB selects the MME to which NAS signaling messages are routed.

1.3.2   Mobility Management Entity (MME)

The MME is responsible for the NAS connection with the UE. All NAS signaling messages are exchanged between the UE and MME to trigger further procedures in the core network if necessary.

A new function of the E-UTRAN is NAS signaling security. The purpose of this feature is to protect the signaling messages that could reveal the true subscriber's identity and location from unauthorized eavesdropping.

The MME is also responsible for paging subscribers in the EPS Connection Management (ECM) IDLE state (including control and execution of paging retransmission) and is concerned with tracking area list management. The list of tracking areas is the list of locations where the UE will be paged.

To route the user plane data streams, the MME will select the best fitting PDN-GW and S-GW. It will also connect the E-UTRAN with the 3G UTRAN using the S3 interface (MME to SGSN). When necessary, a relocation of gateways will be triggered and controlled by the MME.

As its name suggests, the MME will perform management of handovers by selecting a new (target) MME or SGSN for handovers to 2G or 3G 3GPP access networks. Also, it is the MME that hosts the connection to the HSS across the S6a interface and, hence, it is responsible for roaming management and authentication of subscribers.

Last but not least, the MME sets up, modifies, and releases default and dedicated bearers. This function is commonly known as the bearer management function.

According to standard documents, the MME will allow lawful interception of signaling traffic and transfer of warning messages (including selection of an appropriate eNB). The purpose of warning message transfer is to inform people living in a larger area about upcoming natural disasters like storms, bushfires, or tsunamis.

1.3.3   Serving Gateway (S-GW)

The S-GW is the gateway that terminates the interface to the E-UTRAN. A particular LTE subscriber will always be connected by a single S-GW.

In the case of inter-eNB handover, the S-GW acts as the mobility anchor of the connection and remains the same while the path for the transport of signaling and user plane will be switched onto the S1 interface. Once a handover is executed successfully and the associated UE has left the S-GW, the old S-GW will send one or more "end marker" packets to the source eNB, source SGSN, or source RNC of the handover to assist the reordering of user plane packets in these network elements.

Mobility anchoring of the S-GW is also defined for inter-3GPP mobility. Here, the S-GW acts as the terminating point of the S4 interface (see Chapter 2 for interface definitions and figures) and routes the traffic between the 2G/3G SGSN and the PDN-GW of the EPC. In other words, when it comes to inter-RAT handover involving the S4 interface, the S-GW acts as the Gateway GPRS Support Node (GGSN) (of the 2G/3G PS core network) that enables usage of the EPC transport and functions for UTRAN/GERAN PS services.

If the UE is in ECM-IDLE mode (see Section 1.12.9 for a description of different NAS states), the S-GW buffers user plane packets that will be sent to the UE after a successful paging response. The paging via the S1 and Uu interfaces is also triggered by the S-GW.

The S-GW is the network element that provides connectivity and software implementations for lawful interception.

On the IP transport layer, the S-GW acts as a packet router. User plane packets are forwarded transparently in the UL and DL direction and their underlying transport units are marked by S-GW with parameters like DiffServ Code Point, based on the QoS Class Indicator (QCI) of the associated EPS bearer.

Also embedded in the S-GW software are various charging functions for UL and DL charging per UE, PDN, and QCI. These functions are used to charge the operator's own subscribers as well as roaming users (inter-operator charging).

The S-GW can be connected to SGSNs in non-roaming and roaming scenarios. However, connectivity to a GGSN is not supported.

1.3.4   Packet Data Network Gateway (PDN-GW)

The PDN-GW provides access from the mobile operator's network to the PS networks that host the payload contents and operator's IP services. If a user has access to more than one packet data network, it is possible that this user is connected to more than just one PDN-GW. It is not possible for the same UE to simultaneously open connections to a PDN-GW and to a GGSN in the 3G PS domain, according to 3GPP standards.

The main function of the PDN-GW is to establish, maintain, and delete GTP tunnels to S-GW or SGSN in the case of inter-RAT mobility scenarios. The PDN-GW allocates the user's dynamic IP addresses and routes the user plane packets. In addition, it provides functions for lawful interception, policy/QoS control, and charging.

For policy control and charging, the PDN-GW can be connected to a Policy and Charging Rule Function (PCRF) via the Gx reference point. The PCRF provides guidance on how a particular service data flow should be treated in terms of priority, throughput, and other QoS parameters according to the user's subscription profile.

1.3.5   Interfaces and Reference Points[1]

As already explained, the E-UTRAN is an all-IP network. Figure 1.10 shows the network elements that are typically involved in the signaling procedures and routing of payload data from the UE to the PDN

---

[1] Reproduced with permission from © 3GPP™.

and vice versa. The figure also shows the reference points for inter-RAT handover (and inter-RAT packet routing) between E-UTRAN, UTRAN, and GERAN.

The pipeline symbols in the figure illustrate the different signaling connections and tunnels for IP payload transport established and maintained during the connection. The signaling on Gx and Rx used to negotiate specific QoS policies is ignored for reasons of better understandability. Besides, the existence of the PCRF is optional. Due to the fact that the MME and the S-GW may also be combined into a single physical entity, the S11 interface is also optional. The lab test scenarios existing at the time of writing (spring 2010) all have separated physical entities for the MME and S-GW.

The signaling connection across the LTE-Uu interface is the RRC signaling connection, represented by a set of Signaling Radio Bearers (SRBs). The user plane tunnel across LTE-Uu is the radio bearer (see also Section 1.6). The other user plane tunnels are named after the appropriate reference points: namely, S1 bearer and S5 bearer. After the PDN-GW, the connection is carried by the external bearer on SGi. S1AP signaling between the E-UTRAN and MME will be used to establish the tunnel on S1-U, and GTP-C signaling will be used to create the tunnel on S5. On SGi, we can already see plain IP traffic – pure payload, so to say.

The reference points shown in Figures 1.11–1.13 can be briefly described as follows:

- **S1-MME**: Reference point for the control plane protocol between the E-UTRAN and MME. This control plane protocol is the S1AP, which is quite similar to UTRAN RANAP. Indeed, in early drafts of LTE specs, this protocol was called "E-RANAP."
- **S1-U**: Reference point between the E-UTRAN and S-GW for the per bearer user plane tunneling and inter-eNB path switching during handover. The protocol used at this reference point is the GPRS Tunneling Protocol for the User Plane (GTP-U).
- **S3**: This is the reference point between the MME and SGSN. The SGSN may serve UTRAN, GERAN, or both. On S3, we can see plain control plane information for user and bearer information exchange for inter-3GPP access network mobility (inter-RAT handover) in the idle and/or active state. If the connection was set up originally in the E-UTRAN and is handed over to UTRAN/GERAN, the appropriate user plane streams are routed across the S4 reference point. What happens in the case of UTRAN/GERAN to E-UTRAN handover depends on whether S-GW also acts as the anchor for UTRAN/GERAN traffic. If this is true, the user plane tunnel can be switched smoothly between S4 and S1-U during the handover. The protocol used at the S3 reference point is the GTP-C.
- **S4**: The S4 reference point provides related control and mobility support between the GPRS core and the 3GPP anchor function of the S-GW using GTP-C. In addition, if a direct tunnel across S12 is not established, it provides user plane tunneling using GTP-U.
- **S5**: The S5 reference point provides user plane tunneling and tunnel management between the S-GW and PDN-GW. It is used in case of S-GW relocation due to UE mobility and if the S-GW needs to connect to a non-collocated PDN-GW for the required PDN connectivity. The protocol used at this reference point is GTP for both the control plane and user plane.
- **S6a**: The S6a reference point enables the transfer of subscription and authentication data for authorizing user access to the network. The reference point can be also described as the AAA interface between the MME and HSS. Compared to the legacy core network of 2G/3G standards, the functionality provided by S6a is similar to the one on the Gr interface, but due to the all-IP concept of EPC, the protocol used at this reference point is the DIAMETER protocol. In the IP world, DIAMETER is known as the successor of RADIUS, a protocol for granting access and authentication. However, the DIAMETER used on S6a does not have much in common with what is found in the IP world. The protocol header is based on IP standards, but the messages and parameters on the application layer are defined in a 3GPP-specific DIAMETER standard that has no meaning in the IP world.
- **Gx**: This point provides transfer of QoS policy and charging rules from the PCRF to the Policy and Charging Enforcement Function (PCEF) in the PDN-GW. This means that a set of rules for charging

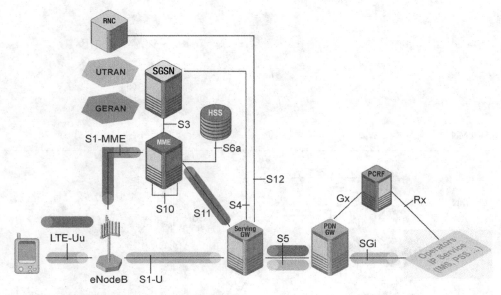

Connectivity for basic LTE call without mobility

Signaling / control plane
Payload / user plane

Figure 1.11   Connection via E-UTRAN non-roaming architecture

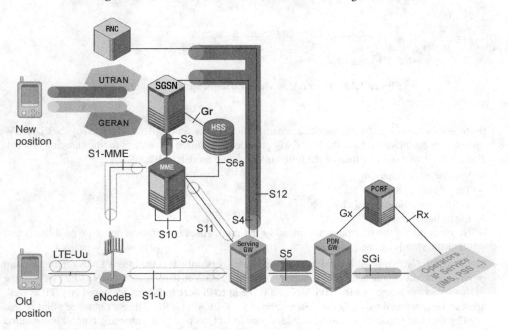

Figure 1.12   Connection after inter-RAT handover from E-UTRAN to UTRAN/GERAN

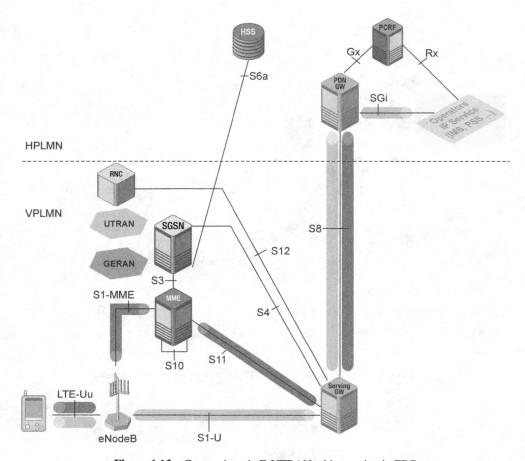

Figure 1.13   Connection via E-UTRAN with roaming in EPC

the transmission of a particular user data stream (called service flow) will be requested by the PDN-GW upon bearer establishment and the PCRF will provide the required parameters for the charging process. Especially, it will signal which of the following charging models will apply:

– Volume-based charging
– Time-based charging
– Volume- and time-based charging
– Event-based charging
– No charging (if the user pays at a monthly flat rate). Also, information about prepaid limits and other thresholds can be included

• **S8**: The S8 reference point is used by roaming subscribers only. It is the inter-PLMN reference point providing the user plane and control plane between the S-GW in the Visited PLMN (VPLMN) and the PDN-GW in the Home PLMN (HPLMN). S8 is the inter-PLMN variant of S5, based on GTP as well, and can be compared to the Gp interface defined for GERAN GPRS. The S8 reference point is also used by roaming subscribers only. It provides transfer of QoS policy and charging control information between the home PCRF and the visited PCRF in order to support the local breakout function. For example, imagine a prepaid limit that can only be known by the home PCRF and must be provided to the visited PCRF to allow roaming services for this user.

- **S10**: This is the reference point between MMEs for MME relocation and MME-to-MME information transfer. This reference point provides mobility functions for intra-E-UTRAN handover/relocation. In other words, signaling procedures on this interface are triggered by UE mobility. It should be noted that this kind of MME relocation in 3GPP 23.401 is called S1 handover. Hence, S10 is seen as special kind of S1 interface and the S1AP is used at this reference point.
- **S11**: This is the reference point between the MME and S-GW. The protocol used here is the GTP-C. The appropriate user plane is routed across S1-U.
- **S12**: The S12 reference point is located between the RNC in the 3G UTRAN and the S-GW for user plane tunneling when a "direct tunnel" is established. It is based on the Iu user plane and Gn user plane reference points using the GTP-U as defined between the SGSN and RNC or between the SGSN and GGSN in the 3G core network. Use of the S12 reference point is an operator configuration option. On S12, only GTP-U traffic can be monitored, as on S1-U.
- **S13**: This point enables a UE identity check procedure between the MME and EIR (Equipment Identity Register). Typically, there is no EIR installed in public networks due to the high administrative efforts, but this network element is found in some private networks. For instance, the GSM-based mobile network of the railway company Deutsche Bahn is equipped with an EIR. The purpose is to ensure that only staff of Deutsche Bahn can use the company's PLMN, but no private persons and staff of other European railway companies such as France's SNCF that also run trains through Germany.
- **SGi**: This is the reference point between the PDN-GW and the packet data network. This network may be an operator external public or private packet data network or an intra-operator packet data network, for example, for the provision of IP Multimedia Subsystem (IMS) services. To simplify the definition, it can be said that for many user plane connections, SGi is the interface to the public Internet. This reference point corresponds to Gi for 3GPP access. Typically, the complete TCP/IP suite can be monitored at this point.
- **Rx**: The Rx reference point resides between the Application Function (AF) and the PCRF defined in 3GPP 23.203. It is, for instance, mandatory if real-time communication services such as Voice over IP (VoIP) are to be charged differently than common PS data transfer.
- **SBc**: The SBc reference point lies between the Cell Broadcast Center (CBC) and MME for warning message delivery and control functions. This interface is used to broadcast warning messages to subscribers (not to send warning messages about network element status to the operation and maintenance center). A typical example of such warning messages could be the broadcast of bushfire or tsunami alarms.

The special anchor function of the S-GW can be illustrated when looking at a connection that was handed over from the E-UTRAN to UTRAN or GERAN as shown in Figure 1.12. In this case, the connections on S5 and SGi remain the same, but the payload is now routed through a tunnel across S4 or S12 while the signaling necessary to execute the inter-RAT mobility will be sent across S3. The old bearers and signaling connections on S1 and LTE-Uu will be deleted after successful handover of the connection.

Figure 1.13 illustrates the basic connection of a roaming subscriber. Signaling and payload take the same route as in Figure 1.12, but the HSS and PDN-GW and, thus, the connection to the public packet network are located in a foreign network. The IP tunneling from the S-GW to PDN-GW and vice versa is realized through the S8 interface, which has identical protocol structure and functions to S5. The only difference is that S8 must fulfill higher requirements in terms of inter-operability, because equipment from different manufacturers must be interconnected through this reference point.

1.4   Area and Subscriber Identities

1.4.1   Domains and Strati

For the EPC, a complete new NAS was designed including a new NAS protocol layer described in 3GPP 24.301.

In contrast to the core network of 3GPP Release 99 to Release 6 where a CS and a PS domain were defined as subdomains of the serving network domain, the EPC will not host any CS domain due to its all-IP character. However, it still distinguishes between AS and NAS signaling and functions as shown in Figure 1.14.

The AS comprises the radio chipset of the UE including the RRC protocol entity and all underlying transport layer entities. Here, all parameters that more or less frequently change during radio access can be found, including transport formats and radio-specific identities of serving cell and possible handover candidates (neighbor cells). The NAS covers all signaling exchanged between the USIM (UMTS Subscriber Identity Module) and the core network node, in case of LTE radio access: the MME. This is the home of all parameters that allow unambiguous identification of a subscriber or the handset hardware such as International Mobile Subscriber Identity (IMSI) and International Mobile Equipment Identity (IMEI). There are also temporary identities stored on the USIM card like Temporary Mobile Subscriber Identity (TMSI) and Globally Unique Temporary UE Identity (GUTI). From a protocol point of view, the NAS is the home of network access, initial subscriber registration, and mobility management procedures. Due to the all-IP concept of LTE/EPC, a new NAS protocol was defined, namely 3GPP 24.301, while similar functions for 2G/3G networks are defined in the standard 3GPP 24.008. The E-UTRAN NAS protocol 3GPP 24.301 does not contain any functions for CS call control and SMS. In the early planning stages of the E-UTRAN, it was assumed that all speech services via the E-UTRAN would use VoIP and the IMS architecture.

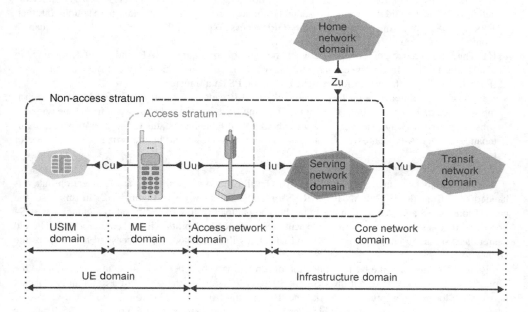

Figure 1.14   Domains and strati in E-UTRAN and EPC

As an alternative, the CS fallback option (implemented in the S1AP protocol) was designed, but obviously this did not satisfy the need for reliable and cost-efficient CS services in the E-UTRAN. Hence, an initiative formed of operators and Network Equipment Manufacturers (NEMs) started to work on the Voice over LTE via Generic Access standards (VoLGAs). VoLGA is beyond the scope of 3GPP. Its principle is to establish an IP connection between the UE and E-UTRAN and use the radio bearer for transparent forwarding of 3GPP 24.008 NAS signaling message and AMR (Adaptive Multirate) voice packets across the logical Z1 interface. Instead, in the S-GW, the RAB used for VoLGA is terminated in a special protocol converter and media gateway device, the VoLGA Access Network Controller (VANC), that is, the interconnecting point between the E-UTRAN/EPC and UTRAN/GERAN/Legacy Core Network.

1.4.2 IMSI

The IMSI allows unambiguous identification of a particular SIM or USIM card. The IMSI is composed of three parts (Figure 1.15):

- The Mobile Country Code (MCC), consisting of three digits. The MCC uniquely identifies the country of domicile of the mobile subscriber. MCC values are administrated and allocated by an international numbering plan.
- The Mobile Network Code (MNC), consisting of two or three digits for GSM/UMTS applications. The MNC identifies the home PLMN of the mobile subscriber. The length of the MNC (two or three digits) depends on the value of the MCC. A mixture of two- and three-digit MNC codes within a single MCC area is not recommended and is beyond the scope of this specification.
- The Mobile Subscriber Identification Number (MSIN), identifying the mobile subscriber within a PLMN. As a rule, the first two or three digits of the MSIN reveal the identity of the Home Location Register (HLR) or HSS that is used for Signaling Connection Control Part (SCCP) Global Title translation procedures when roaming subscribers register in foreign networks.

The National Mobile Subscriber Identity (NMSI) consists of the MNC and the MSIN.

A combination of MCC and MNC can be used to aggregate call-specific performance measurement data (such as cumulative counters) on IMSI groups. This will help to highlight problems of roaming subscribers such as network failures during registration procedures, as described later in this book. Table 1.1 shows some samples from an IMSI group mapping table with MCC/MNC combinations in

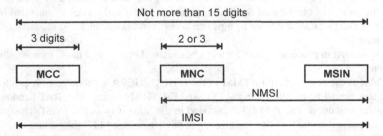

Figure 1.15 Structure of IMSI (according to 3GPP 23.303). (*Source*: Reproduced with permission from © 3GPP™.)

Table 1.1   IMSI group mapping table from Tektronix Communications NSA software

<IMSI IMSINumber= '26201' IMSIGroupName= 'T-MOBILE DEUTSCHLAND GMBH (GERMANY)' />
<IMSI IMSINumber= '26202' IMSIGroupName= 'VODAFONE D2 GMBH (GERMANY)' />
<IMSI IMSINumber= '26801' IMSIGroupName= 'VODAFONE TELECEL (PORTUGAL)' />
<IMSI IMSINumber= '27201' IMSIGroupName= 'VODAFONE IRELAND PLC (IRELAND)' />
<IMSI IMSINumber= '310560' IMSIGroupName= 'T-MOBILE USA, INC. (UNITED STATES)' />

"IMSINumber" fields and operator names in the "IMSIGroupName" field. Note the three-digit MNC used for the American operator.

It is possible that one-use equipment will work with more than just one (U)SIM. A good example is a mobile phone that has both business and private SIM cards as one device. Depending on the nature of the call (private or business), the owner of the handset can choose which (U)SIM should be used to make the call. Such a procedure might be required in case private phone calls need to be charged separately due to national income tax laws (as found, e.g., in Germany).

1.4.3   LMSI, TMSI, P-TMSI, M-TMSI, and S-TMSI

All temporary subscriber identities, Local Mobile Subscriber Identity (LMSI), TMSI, and P-TMSI, will not be seen in E-UTRAN signaling as long as there is no inter-RAT mobility between the EUTRAN and UTRAN/GERAN. Indeed, for LTE, a new NAS protocol was specified (3GPP 24.301) that introduces a new temporary subscriber identity for the E-UTRAN: the GUTI described in Section 1.4.4. However, to fulfill inter-RAT mobility requirements, TMSI, P-TMSI, and LMSI will still be found in E-UTRAN NAS messages, or at least it will be indicated if valid values of these parameters are stored on the USIM card.

The LMSI is a four-octet/byte number. It is a pointer to a database record for a particular IMSI in the Visitor Location Register (VLR) database. Although the VLR is no longer found in the EPC network architecture, there is a database with the same function hosted by the MME. The purpose of the LMSI was to speed up the search for particular database records. If this is still required, with the new powerful computer hardware used to build today's network elements, it is a design detail to be defined by NEMs. From definitions given in 3GPP 23.003, it can be guessed that the LMSI will not be used by the MME.

The TMSI is also encoded as a four-octet/byte hex number. It is allocated to a particular subscriber (more correctly, to a particular subscriber's (U)SIM card) during initial attach. The TMSI is used to mask the true subscriber's identity, which is the IMSI, in NAS signaling procedures. In the E-UTRAN, it is often used together with the GUTI. It can be coded using a full hexadecimal representation. Since the TMSI has only local significance (i.e., within a VLR and the area controlled by a VLR, or within an SGSN and the area controlled by an SGSN, or within an MME and the area controlled by an MME), it's structure and coding can be chosen by agreement between the operator and manufacturer in order to meet local needs.

The TMSI allocation procedure should always be executed in ciphered mode to prevent unauthorized eavesdropping.

The P-TMSI is the complement of TMSI in the UTRAN/GERAN PS domain. It is allocated by the SGSN and, hence, will be monitored in the EPC and E-UTRAN during inter-RAT handover/relocation preparation and execution. The P-TMSI is encoded in the same way as the TMSI. The difference is in defining value ranges. If the first two leading digits have the value "11," the parameter is identified as a P-TMSI. Thus, in the hexadecimal format, all TMSI values starting with C, D, E, or F as the first hex number are P-TMSIs.

The M-TMSI is a 32-digit binary number that is part of the GUTI and exclusively used in the E-UTRAN.

The S-TMSI consists of the Mobility Management Entity Code (MMEC) and M-TMSI. Indeed, it is just a shorter variant of the GUTI.

1.4.4   GUTI

The GUTI is assigned only by the MME during initial attach of a UE to the E-UTRAN.

The purpose of the GUTI is to provide an unambiguous identification of the UE that does not reveal the UE or the user's permanent identity in the E-UTRAN. It also allows identification of the MME and network to which the UE attaches. The GUTI can be used by the network to identify each UE unambiguously during signaling connections.

The GUTI has two main components: the Globally Unique Mobility Management Entity Identifier (GUMMEI) that uniquely identifies the MME that allocated the GUTI and the M-TMSI that uniquely identifies the UE within the MME that allocated the GUTI. The GUMMEI is constructed from the MCC, MNC, and Mobility Management Entity Identifier (MMEI).

The MMEI should be constructed from a Mobility Management Entity Group ID (MMEGI) and an MMEC. The GUTI should be constructed from the GUMMEI and the M-TMSI as shown in Figure 1.16.

For paging purposes, the mobile is paged with the S-TMSI. The S-TMSI is constructed from the MMEC and the M-TMSI. It is correct to say that the S-TMSI is a shorter format of GUTI that can be used because, after successful registration of a UE, the serving network and the serving MME group are known and stored in the core network databases.

The operator needs to ensure that the MMEC is unique within the MME pool area and, if overlapping pool areas are in use, unique within the area of overlapping MME pools.

The GUTI should be used to support subscriber identity confidentiality and, in the shortened S-TMSI form, to enable more efficient radio signaling procedures (e.g., paging and service request).

MCC and MNC should have the same field size as described for the IMSI.

The M-TMSI has a length of 32 bits, MMEGI is 16 bits in length, and MMEC is 8 bits in length.

It is important to understand that on the S1 interface, the IMSI is typically not seen, just like the GUTI. Exceptions are initial attach to the network when no old GUTI is stored on the USIM card or the true subscriber's identity is checked using NAS signaling, which regularly happens when roaming subscribers attach. Also, in the case of the paging procedure, the IMSI might be seen.

For monitoring and network performance measurement, the IMSI on S1 can only be revealed if the changing temporary identities are tracked with a quite sophisticated architecture. Full IMSI tracking can only be ensured by monitoring *all* S1 interfaces of an operator's E-UTRAN and ideally all S6a interfaces and storing the current GUTI/IMSI relations in a central point as stored in the HSS.

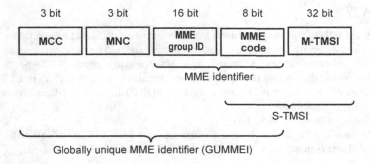

Figure 1.16   Format of GUTI and S-TMSI

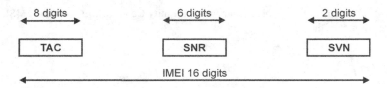

Figure 1.17   Structure of IMEISV (according to 3GPP 23.303). (*Source*: Reproduced with permission from © 3GPP™.)

Table 1.2   Example of handset name mapping table

<Handset IMEI='35942100' HandsetName='MOTOROLA V3 RAZR' />
<Handset IMEI='35942200' HandsetName='MOTOROLA V3 RAZR' />
<Handset IMEI='35942300' HandsetName='MOTOROLA V3 RAZR' />
<Handset IMEI='01161200' HandsetName='APPLE IPHONE 3G' />
<Handset IMEI='01161300' HandsetName='APPLE IPHONE 3G' />
<Handset IMEI='01161400' HandsetName='APPLE IPHONE 3G' />
<Handset IMEI='35179700' HandsetName='SAMSUNG SGH-E100' />
<Handset IMEI='35179800' HandsetName='SAMSUNG SGH-E100' />
<Handset IMEI = '35179900' HandsetName = 'SAMSUNG SGH-A800' />

1.4.5   IMEI

The IMEI or IMEISV is used to unambiguously identify the hardware and (with IMSISV) software version of a mobile phone. The IMEISV that is expected to be used for all 4G and UEs consists of 16 bits as shown in Figure 1.17.

The eight leading digits stand for the Type Approval Code (TAC). This TAC indicates the manufacturer of the equipment. The next six digits stand for the Serial Number (SNR), and finally, the two last digits represent the software version.

As shown in Table 1.2 (which gives a list of 3G handsets), the TAC (in the table named "Handset IMEI") is always unique for a particular equipment type, but due to large manufacturing series, several TACs are assigned to the same type if the number of manufactured units exceeds the threshold of 100 000 that can be numbered with the six-digit SNR.

1.4.6   RNTI

In 3G UMTS, the RNTIs are always used to identify information dedicated to a particular subscriber on the radio interface, especially if common or shared channels are used for data transmission. Now, in LTE, it is the rule that common channels and shared channels are used to transmit all UE-specific data, but also some network-specific data across the radio interface. For this reason, the RNTI in LTE is not always related to a particular subscriber, but sometimes also used to distinguish broadcast network information from data streams of subscribers.

The RNTI is signaled in the MAC layer.

When MAC uses the Physical Downlink Control Channel (PDCCH) to indicate radio resource allocation, the RNTI that is mapped on the PDCCH depends on the logical channel type:

- C-RNTI, Temporary Cell Radio Network Temporary Identifier (temp C-RNTI), and Semi-Persistent Scheduling (SPS) C-RNTI for Dedicated Control Channel (DCCH) and DTCH

Table 1.3 RNTI values (according to 3GPP 36.321)

Value (hexadecimal)		RNTI
FDD	TDD	
0000–0009	0000–003B	RA-RNTI
000A–FFF2	003C–FFF2	C-RNTI, semi-persistent scheduling C-RNTI, temporary C-RNTI, TPC-PUCCH-RNTI, and TPC-PUSCH-RNTI
FFF3–FFFC		Reserved for future use
FFFE		P-RNTI
FFFF		SI-RNTI

Source: Reproduced with permission from © 3GPP™.

- Paging Radio Network Temporary Identity (P-RNTI) for Paging Control Channel (PCCH)
- Random Access Radio Network Temporary Identifier (RA-RNTI) for Random Access Response (RAR) on DL-SCH
- Temporary C-RNTI for Common Control Channel (CCCH) during the random access procedure
- System Information Radio Network Temporary Identifier (SI-RNTI) for Broadcast Control Channel (BCCH)

All RNTIs are encoded using the same 16-bit format (2 octets = 2 bytes).[2]
The following values (given in Table 1.3) are defined for the different types of RNTI.

1.4.6.1 P-RNTI

The P-RNTI is the 4G complement of the paging indicator known from 3G UMTS. It does not refer to a particular UE, but to a group of UEs.

The P-RNTI is derived from the IMSI of the subscriber to be paged and constructed by the eNB. For this reason, the IMSI is transmitted in an S1AP paging message from the MME to eNB, although in other S1 signaling only, the GUTI is used to mask the true identity of the subscriber.

1.4.6.2 RA-RNTI

The RA-RNTI is assigned by the eNB to a particular UE after this UE has sent a random access preamble on the Physical Random Access Channel (PRACH). If this random access preamble is received by the eNB and network access granted, the base station sends an acquisition indication back to the mobile and this acquisition indication message contains the RA-RNTI. In turn, the UE will use the RA-RNTI to send a RRC connection request message on the radio interface UL and the parameter will help to distinguish messages sent by different UEs on the Random Access Channel (RACH).

1.4.6.3 C-RNTI

The C-RNTI is a 16-bit numeric value. Its format and encoding are specified in 3GPP 36.321 (MAC). The C-RNTI is part of the MAC Logical Channel Group ID field (LCG ID). It defines unambiguously

---

[2] The terms octet and byte have the same meaning, but the origin is different. While "octet" was used in the telecommunication standards of CCITT and ITU to describe a field of 8 bits, in computer science and hence in the TCP/IP standardization, the term "byte" was introduced.

which data sent in a DL direction within a particular LTE cell belongs to a particular subscriber. For instance, all RRC messages belonging to a single connection between a UE and the network are marked with the same C-RNTI value by the MAC entity that provided transport services to the RRC and NAS. Thus, C-RNTI is an important parameter for call tracing.

The C-RNTI comes in three different forms: temp C-RNTI, semi-persistent scheduling C-RNTI, and permanent C-RNTI.

The temp C-RNTI is allocated to the UE during random access procedure (with a RRC connection setup message) and may turn into a permanent C-RNTI depending on the result of a subsequently performed contention resolution procedure or in the case of contention-free random access.

The semi-persistent scheduling C-RNTI is used if the subscriber is running services with a predictable unchanging QoS profile. A typical example is VoIP for which the required bit rate will not change during the entire connection. In such a case, the dynamic (re)scheduling of radio resources, which is mandatory in the case of bursty payload traffic to ensure optimal usage of resource blocks, is not required. The SPS C-RNTI is used to indicate an area of resource blocks that will be used by the same UE for a longer time frame without any expected change.

1.4.6.4 SI-RNTI

The SI-RNTI is sent on the PDCCH. It does not stand for a particular UE identity. Instead, it signals to all mobiles in a cell where the broadcast System Information Blocks (SIBs) are found on the Physical Downlink Shared Channel (PDSCH). This is necessary since the PDSCH is used to transport both broadcast system information for all UEs and signaling/payload for particular mobiles. In other words, the SI-RNTI indicates which DL resource blocks are used to carry SIBs that in 3G UMTS have been sent on the broadcast (transport) channel mapped onto the Primary Common Control Physical Channel (P-CCPCH). In LTE, there is no CCPCH, only DL-SCH.

1.4.7 Location Area, Routing Area, Service Area, Tracking Area, and Cell Global Identity

The Location Area (LA) and Routing Area (RA), known from 2G and 3G RAN, will be used in the E-UTRAN only if the UE was involved in inter-RAT mobility procedures.

The LA is a set of cells (defined by the mobile operator) throughout which a mobile that is camping on UTRAN or GERAN will be paged. The LA is identified by the Location Area Identity (LAI) within a PLMN. The LAI consists of the MCC, MNC, and Location Area Code (LAC) – see Figure 1.18.

The RA is defined as a sub-area of a LA with specific means for PS services. Each UE informs the SGSN about the current RA. RAs can consist of one or more cells. Each RA is identified by a Routing Area Identity (RAI). The RAI is used for paging and registration purposes and consists of the LAC and

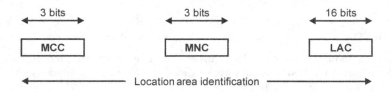

Figure 1.18 Structure of location area identification (according to 3GPP 23.303). (*Source*: Reproduced with permission from © 3GPP™.)

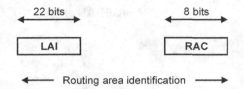

Figure 1.19 Structure of routing area identification (according to 3GPP 23.303). (*Source*: Reproduced with permission from © 3GPP™.)

Routing Area Code (RAC). The RAC (length 1 octet, fixed) identifies an RA within an LA and is part of the RAI.

The RAI is composed of the following elements, shown in Figure 1.19:

`<MCC><MNC><LAC><RAC>`

The Tracking Area Identity (TAI) is the identity used to identify tracking areas. The TAI is constructed from the MCC, MNC, and TAC (Tracking Area Code).

A Tracking Area (TA) includes one or several E-UTRAN cells. Details of TA design are subject to individual radio network planning. In typical plans, a tracking area consists of 1 to 100 eNodeB. How large a tracking area can be also depends on the cluster structure, for example, if it is an indoor cell or covers rather a rural area than a city center hot spot. . In theory, a single eNB or even a single cell may also belong to multiple TAs, for example, in the case of network sharing. In any case, the details of area configuration are defined in radio network planning using proprietary definitions and rules. The scheme shown in Figure 1.20 is based on the assumption that one TA covers all cells of one eNB.

A UTRAN/GERAN LA corresponds to an MME group as defined in 3GPP standards – see Section 1.4.8. Surely, each MME may also work as an MME group on its own.

The network allocates a list of one or more TAs to the UE. In certain operating modes, the UE may move freely in all TAs of the list without updating the MME. The E-UTRAN Cell Global Identity (CGI) consists of the PLMN-ID (MCC + MNC) plus the E-UTRAN Cell Identity (CI), a 28-bit string. The leftmost bits of the CI correspond to the eNB ID. Using the E-UTRAN CGI, any cell in any E-UTRAN in the world can be unambiguously identified. To have a unique format for all cells in all RATs also for GERAN and UTRAN cells, a global CI is defined; starting with (this is true for all the releases 8, 9, 10, 11, 12, ...) Release 8, it is requested that these identities are broadcast in each cell and included in RRC measurement reports.

Indeed, on the radio interface, the cells of the different RATs are identified by the following parameters:

- **GERAN**: Absolute Radio Frequency Channel Number (ARFCN) + Base Station Color Code (BCC).
- **UTRAN**: UMTS Absolute Radio Frequency Channel Number (uARFCN) + primary scrambling code.
- **E-UTRAN**: eARFCN + physical Cell ID (c-ID) (in fact, also a scrambling code).

The ARFCN values are used to identify the frequency of the cell.

1.4.8 Mapping between Temporary and Area Identities for EUTRAN and UTRAN/GERAN-Based Systems[3]

For the construction of the RA update request message in GERAN/UTRAN or TA update request message in E-UTRAN, the following identities should be included.

---

[3] Reproduced with permission from © 3GPP™.

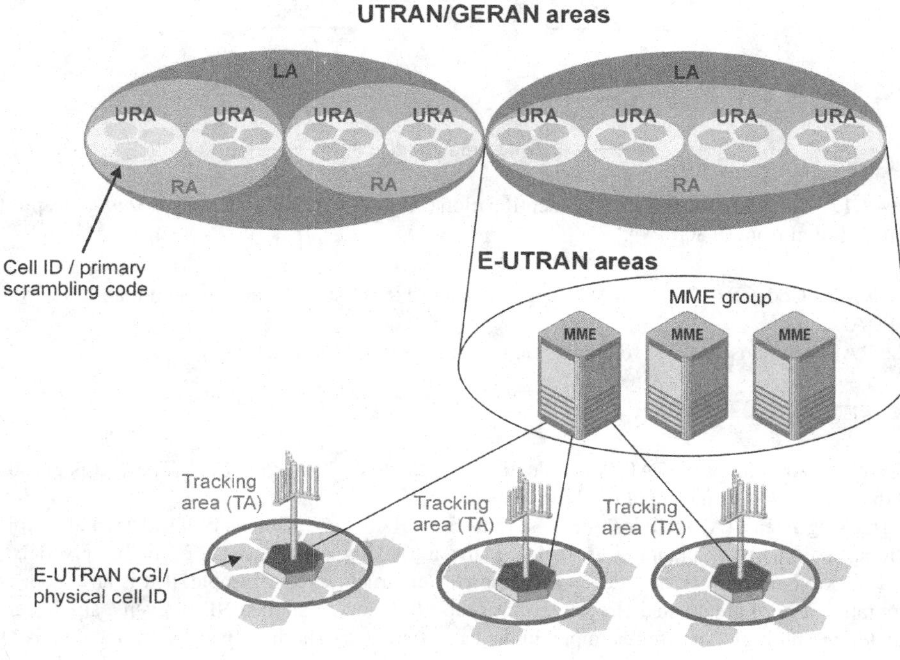

UTRAN/GERAN areas

Figure 1.20   Areas in UTRAN/GERAN and E-UTRAN

In GERAN and UTRAN, the RAI is constructed from the MCC, MNC, LAC, and RAC. In addition, the routing area update request message contains the P-TMSI that includes the mapped Network Resource Identifier (NRI) that is used in Iu flex networks. (An example of NRI usage is given in Kreher and Ruedebusch, 2007.)

P-TMSI should be of 32-bit length where the two topmost bits are reserved and always set to 11. These are needed since the GERAN representation of P-TMSI, of the form TLLI (Temporary Logical Link Identity), imposes this restriction. Hence, for a UE that may hand over to GERAN/UTRAN (based on subscription and UE capabilities), the corresponding bits in the M-TMSI are set to 11.

The NRI field is of variable length and should be mapped into the P-TMSI starting at bit 23 and down to bit 14. The most significant bit of the NRI is located at bit 23 of the P-TMSI regardless of the configured length of the NRI.

In the case of a combined MME-SGSN node, the NRI of the SGSN part and the MMEC of the MME part refer to the same combined node. The RAN configuration allows NAS messages on GERAN/UTRAN and E-UTRAN to be routed to the same combined node. The same or different values of NRI and MMEC may be used for a combined node.

The mapping of the GUTI should be done to a combination of the RAI of GERAN/UTRAN and the P-TMSI as follows:

E-UTRAN <MCC>	maps to GERAN/UTRAN <MCC>
E-UTRAN <MNC>	maps to GERAN/UTRAN <MNC>
E-UTRAN <MME Group ID>	maps to GERAN/UTRAN <LAC>
E-UTRAN <MME Code>	maps to GERAN/UTRAN <RAC>

It is also copied into the eight most significant bits of the NRI field within the P-TMSI.

E-UTRAN <S-TMSI> maps as follows:

- 22 bits of the E-UTRAN <M-TMSI> starting at bit 30 and down to bit 9 are mapped into the remaining 22 bits of the GERAN/UTRAN <P-TMSI>;
- the remaining 8 bits of the E-UTRAN <M-TMSI> are copied into 8 bits of the <P-TMSI signature> field.

For the UTRAN, the 10-bit-long NRI bits are masked out from the P-TMSI and also supplied to the RAN node as an Intra-Domain NAS Node Selector (IDNNS).

The mapping of P-TMSI (TLLI) and RAI in GERAN/UTRAN to GUTI in E-UTRAN should be performed as follows:

GERAN/UTRAN <MCC>	maps to E-UTRAN <MCC>
GERAN/UTRAN <MNC>	maps to E-UTRAN <MNC>
GERAN/UTRAN <LAC>	maps to E-UTRAN <MME Group ID>
GERAN/UTRAN <RAC>	maps to 8 bits of the M-TMSI

The eight most significant bits of GERAN/UTRAN <NRI> map to the MMEC.

GERAN/UTRAN <P-TMSI or TLLI> excluding the eight most significant bits at the NRI position maps to the remaining bits of the M-TMSI.

The values of <LAC> and <MME group id> should be disjoint, so that they can be differentiated. It is recommended that the most significant bit of the <LAC> be set to zero and the most significant bit of <MME group id> set to one.

1.4.9 GSM Base Station Identification[4]

To identify the target cell of an inter-RAT handover from the E-UTRAN to GERAN and possible handover candidates in RRC measurement reports, it is necessary to know how these cells are identified on the radio interface and in core network signaling procedures.

1.4.9.1 CI and CGI

The BSS and cell within the BSS are identified within an LA or RA by adding a CI to the LA or RA identification, as shown in Figure 1.21. The CI is of fixed length with two octets and can be coded using a full hexadecimal representation.

The CGI is the concatenation of the LA identification and the CI. The CI should be unique within a LA.

1.4.9.2 Base Station Identity Code (BSIC)

The BSIC is a local color code that allows an MS to distinguish between different neighboring base stations. The BSIC is a 6-bit code, which is structured as shown in Figure 1.22.

---

[4] Reproduced with permission from © 3GPP™.

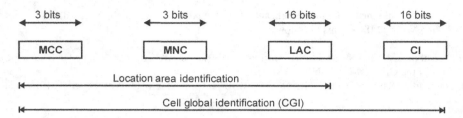

Figure 1.21 Structure of cell global identification (according to 3GPP 23.303). (*Source*: Reproduced with permission from © 3GPP™.)

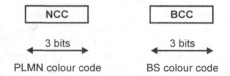

Figure 1.22 Structure of BSIC (according to 3GPP 23.303). (*Source*: Reproduced with permission from © 3GPP™.)

In the definition of the Network Color Code (NCC), care should be taken to ensure that the same NCC is not used in adjacent PLMNs, which may use the same BCCH carrier frequencies in neighboring areas.

Imagine a border area between different countries such as Germany, Poland, and the Czech Republic where, in this area, the radio signals of different international mobile carriers overlap. Now, the NCC is used to distinguish between cells of the German, Polish, and Czech network operators that may operate on the same frequency (BCCH carrier) while the BCC is used to identify cells of different German carriers (e.g., Vodafone and T-Mobile) from each other.

1.4.10 UTRA Base Station Identification

The 3G Universal Terrestrial Radio Access (UTRA) cells are identified on the radio interface unambiguously by a combination of uARFCN and primary scrambling code.

In signaling procedures between network elements, the NBAP c-ID is unambiguous within a single Radio Network Subsystem (RNS), which means an area controlled by a single RNC. The same NBAP c-ID might be reused by network architects in different RNC areas.

In signaling communication between RNCs (e.g., to prepare handover/relocation of subscribers) and toward core network elements, the Service Area Identity (SAI) is commonly used.

1.4.10.1 uARFCN and Primary/Secondary Scrambling Codes

In RRC measurement reports and RRC handover messages, the primary scrambling code together with the uARFCN will be used to identify target cells for handover to the UTRAN and possible handover to UTRAN candidates.

The primary scrambling codes are integer values in the range 0–511. Radio network planners must ensure that cells with the same primary scrambling code and same uARFCN never overlap.

1.4.10.2   SAI

The SAI is constructed as follows:

`<MCC><MNC><LAC><SAC>`

It is encountered during initial NAS signaling transport from the RNC to core network elements and might be later updated according to changes in the best cell of an active set using the RANAP update location procedure. The SAI is also used a source and/or target ID for 3G–3G and 3G–2G relocation/handover procedures.

A Service Area Code (SAC) is often used to identify a single UTRAN cell. However, the authors have also seen configurations where one SAC is used to identify a particular antenna sector (area covered by one antenna) that is covered by two different UMTS frequencies. In this scenario, the SAC addressed two cells that used the same primary scrambling code but different uARFCN and different NBAP c-ID.

1.4.10.3   Shared Network Area Identifier

The Shared Network Area Identifier (SNA-ID) is a new identity introduced in Release 8 standards and used to identify an area consisting of one or more LAs. Such an area can be used to grant access rights to parts of a shared network to a UE in connected mode.

The SNA-ID consists of the PLMN-Id followed by the Shared Network Area Code (SNAC)

`<MCC><MCN><SNAC>`

where the SNAC is defined by the operator.

1.4.11   Numbering, Addressing, and Identification in the Session Initiation Protocol

The Session Initiation Protocol (SIP) addresses the so-called Uniform Resource Identifiers (URIs). The generic format is defined as follows:

`sip:user:password@host:port:uri-parameters?headers.`

This format can be compared to the structure of an e-mail address (e.g., ralf@tektronix.com). Optionally, some specific SIP parameters may be included in the SIP address. Among such parameters, it can be defined which transport protocol should be used to exchange SIP messages, for instance. Also, a priority can be given to messages, authentication procedures can be invoked, errors can be reported, and routing instructions can be given by using the header fields. Some header fields only make sense in requests or responses. These are called request header fields and response header fields, respectively. A complete list of header fields and their relation to different SIP methods can be found in RFC (Request for Comments) 3261.

At the end, a SIP address monitored in a live network may look like this:

`sip:ralf@tektronix.com;transport=tcp?priority=urgent.`

Besides the recipient's name and host, this address contains a URI parameter to request transport of this SIP information using the reliable TCP, and the header information indicates that this SIP information will be treated with urgent priority.

1.4.12   Access Point Name

In the GPRS backbone, an Access Point Name (APN) is a reference to a GGSN. To support inter-PLMN roaming, the internal GPRS DNS (Domain Name System) functionality is used to translate the APN into the IP address of the GGSN.

In the EPC network, the APN is found in GTP-C signaling when packet contexts are established, but it is no longer found in LTE NAS signaling. This means in turn that for 2.5G and 3G phones, the APN is an important parameter to be stored on the (U)SIM card, but for 4G phones, the APN does not need to be configured by the end user. This will also resolve the problem where many PDP context setup failures seen currently in the GERAN and UTRAN are due to an unknown or missing APN.

The APN is composed of two parts as follows:

- The APN network identifier; this defines to which external network the GGSN is connected and optionally a requested service by the MS. This part of the APN is mandatory.
- The APN operator identifier; this defines in which PLMN GPRS backbone the GGSN is located. This
  part of the APN is optional.

The APN operator identifier is placed after the APN network identifier. An APN consisting of both the network identifier and operator identifier corresponds to the DNS name of a GGSN; the APN has, after encoding as defined next, a maximum length of 100 octets.

The encoding of the APN follows the name syntax defined in RFC 2181 [18], RFC 1035 [19], and RFC 1123 [20]. The APN consists of one or more labels. Each label is coded as a one-octet length field followed by that number of octets coded as 8-bit ASCII characters. Following RFC 1035 [19], the labels should consist only of the alphabetic characters (A–Z and a–z), digits (0–9), and the hyphen (-). Following RFC 1123 [20], the labels should begin and end with either an alphabetic character or a digit. The case of alphabetic characters is not significant. The APN is not terminated by a length byte of zero.

Typical APNs are

- mms.tim.net
- wap.eplus.net
- wap.beeline.ru
- wap.debitel.de
- web.vodafone.de
- internet.t-mobile.

The APN of a 3G subscriber is stored in the USIM. In LTE NAS signaling, the APN is no longer used, because it is chosen by the intelligent packet routing functions of MME. However, the APN Average Maximum Bit Rate (APN-AMBR) may be signaled to the UE by the MME to control the amount and bandwidth of UL traffic.

1.5   User Equipment

As in UMTS, the LTE mobile station is called User Equipment (UE). It is constructed using a modular architecture that consists of three main components (see Figure 1.23):

- **Mobile Termination**: The MT represents termination of the radio interface. In this entity, the RRC
  signaling is terminated and RRC messages are sent/received.
- **Terminal Adapter**: The terminal adapter represents the termination of the application-specific service protocols, for example, SIP signaling for VoIP. The terminal adapter might be constructed as an

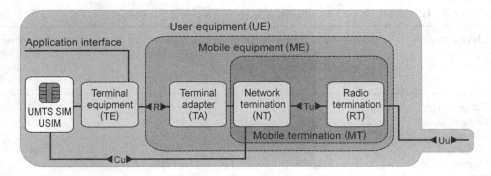

Figure 1.23   Modular architecture of a UE

external interface, for example, USB to connect a laptop PC using LTE technology with a mobile network.

• **Terminal Equipment**: The TE represents termination of the service. Depending on the UE's application capabilities, it may act as the TE or not. For instance, the Apple iPhone with its browser functionalities has full TE capability while a simple USB stick for mobile data transmission has no TE capability at all. In the case of the USB stick, the connected laptop PC is the TE.

1.5.1   UE Categories

The UE categories stand for an abstract grouping of common UE radio access capabilities and are defined in 3GPP 36.306.

In particular, the handset-type groups vary in maximum possible throughput (the maximum number of DL-SCH transport blocks bits received within a Time Transmission Interval (TTI)). Assuming a TTI of 1 ms for category 1, the maximum possible throughput is 10296 bits/1 ms, which is approximately 10 Mbps of physical layer DL throughput (including the RLC/MAC header information – so the payload throughput will be slightly less).

Category 5 mobiles are the only handsets that support 64 Quadrature Amplitude Modulation (QAM) on the UL as highlighted in Tables 1.4 and 1.5. The maximum possible bit rate ranges from 5 Mbps (Cat. 1) to 75 Mbps (Cat. 5).

Table 1.4   UE categories and DL capabilities (according to 3GPP 36.306)

UE category	Maximum number of DL-SCH transport block bits received within a TTI	Maximum number of bits of a DL-SCH transport block received within a TTI	Approximate maximum bit rate DL (Mbps)
Category 1	10296	10296	10
Category 2	51024	51024	50
Category 3	102048	75376	75
Category 4	150752	75376	75
Category 5	302752	151376	150

Source: Reproduced with permission from © 3GPP™.

Table 1.5 UE categories and UL capabilities (according to 3GPP 36.306)

UE category	Maximum number of bits of an UL-SCH transport block transmitted within a TTI	Support for 64QAM in UL	Approximate maximum bit rate UL (Mbps)
Category 1	5160	No	5
Category 2	25 456	No	25
Category 3	51 024	No	50
Category 4	51 024	No	50
Category 5	75 376	Yes	75

Source: Reproduced with permission from © 3GPP™.

1.6 QoS Architecture

The EPS bearer service layered architecture is depicted in Figure 1.24. Besides the different names of bearers and reference points, this architecture does not look very different from the bearer service architecture defined in Release 99. However, there is a major difference that is not obvious at first sight.

In 3G UMTS, the request of a subscriber for a defined QoS of an end-to-end service starts the QoS negotiation procedure. It depends on the subscriber's subscribed QoS stored in the HLR and the available network resources which QoS is granted to a particular connection at the end. The QoS negotiation and control process starts on the NAS layer with the first SM message sent by the UE.

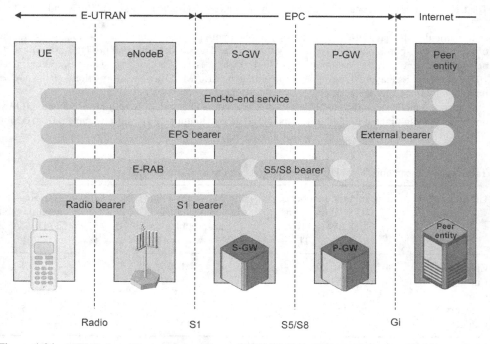

Figure 1.24 LTE QoS architecture (according to 3GPP 23.401). (*Source*: Reproduced with permission from © 3GPP™.)

In LTE – different from 2.5 and 3G PS connections – a default bearer with a default QoS is already established when the UE attaches to the network. The QoS attributes of this default bearer are determined by the subscribed QoS parameters stored in the HSS. This is still as seen in 2.5/3G networks.

However, if now the first user plane packet is sent by the UE, it is routed toward the PDN where the PCRF analyzes the requested end-to-end service. Depending on this service, the PCRF may now trigger a modification of QoS parameters in all the involved bearers. There is no option for the subscriber to request a particular QoS; only the network is in charge of QoS control. There is also no way for the UE to request something known as a secondary context in 3G (see Section 3.26 in Kreher and Ruedebusch, 2007). In LTE, all QoS management is tied to the application, not to SM signaling.

It is important to understand that one UE in LTE can have multiple end-to-end services active and each of these services will have its own individual bearer. It is not intended by LTE standards that, for example, non-real-time services like web browsing and e-mail will be mapped onto the same bearer (e.g., the same S1-U GTP tunnel) as we have seen in 3G UMTS. For this reason also 256 individual E-RABs for a single UE can be addressed by E-UTRAN protocols while in UMTS only 15 different RAB-IDs had been defined by the standard organizations.

In the 3GPP specs, there is also a Traffic Flow Template (TFT) mentioned for the UL as well as for the DL part of the connection. These TFTs are bound to the EPS bearers. In general, a TFT can be described as a set of filters for a particular end-to-end service. Each TFT consists of a destination IP address and a set of source/destination port numbers. On the DL, the IP address is the address assigned to the UE; on the UL, it is the address of a server on the PDN. If we assume, for example, an HTTP 1.1 end-to-end service, the DL TFT of this service consists of the UE's IP address, the TCP source port number is 80, and the TCP destination port number is 80. On the UL, the port numbers are the same, but the IP address is the address of the server that hosts the web site.

To standardize the QoS handling, a set of nine QCIs have been defined by 3GPP. There are four classes with a Guaranteed Bit Rate (GBR) and five classes with a Non-Guaranteed Bit Rate (Non-GBR).

Besides the bit rate, the parameter priority, packet delay budget, and packet error loss rate are critical factors as given in Table 1.6.

Table 1.6 Standardized QCI, QoS parameter thresholds, and example services (according to 3GPP 23.203)

QCI	Resource type	Priority	Packet delay budget (ms)	Packet error loss rate	Example services
1	GBR	2	100	10^{-2}	Conversational voice
2		4	150	10^{-3}	Conversational video (live streaming)
3		3	50	10^{-3}	Real-time gaming
4		5	300	10^{-6}	Non-conversational video (buffered streaming)
5	Non-GBR	1	100	10^{-6}	IMS signaling
6		6	300	10^{-6}	Video (buffered streaming) TCP based (e.g., www, e-mail, chat, ftp, p2p file sharing, progressive video, etc.)
7		7	100	10^{-3}	Voice, video (live streaming) interactive gaming
8		8	300	10^{-6}	Video (buffered streaming) TCP based (e.g., www, e-mail, chat, ftp, p2p file sharing, progressive video)
9		9			

Source: Reproduced with permission from © 3GPP™.

1.7   LTE Security[5]

What are the new security functions in the E-UTRAN? This question can briefly be answered as follows.

The first feature we see is a completely new ciphering mechanism and integrity protection for NAS signaling messages that was never seen in any 2G or 3G radio access network. On the radio interface, this new NAS security leads to situations with double ciphering. On top of the protocol stack, the NAS messages exchanged between the UE and MME are encrypted and the underlying RRC that acts as the transport layer for NAS is secured by ciphering mechanisms as well, so that the ciphered NAS message is ciphered together with its RRC transport message a second time.

The second new security feature is the option to secure the complete IP-based transport of the control plane and user plane on the S1 reference point using Secure IP (IPsec). There is no way to decipher IPsec by just monitoring the data that is exchanged between two endpoints of an IPsec connection. To decipher IPsec requires the monitoring software to be informed about which IPsec ciphering parameters (which can be changed frequently) are currently used in each of the involved endpoints of the IP connection. In a typical case, these endpoints are the eNB and the MME or S-GW. To allow deciphering, there must be a dedicated Application Programming Interface (API) installed that allows the monitoring software to access IPsec-relevant parameters for deciphering. To design such an API requires close cooperation between the NEMs of eNB and MME/S-GW and the manufacturers of the monitoring software. The conclusion related to this fact is that free-of-charge monitoring software like Wireshark will not be able to decipher IPsec. However, to obtain statistics of S1 control plane and user plane performance, it is crucial to have metrics for E-UTRAN QoS and QoE (Quality of Experience). Consequently, IPsec deciphering will become one of the key differentiators for E-UTRAN monitoring software.

Besides these new security features, all the security elements from previous standards such as mutual authentication and masking of subscriber identity by using temporary identities can be found in the E-UTRAN. There is only a minor change here: the TMSI will be replaced by the new GUTI parameter.

To understand how the overall LTE security concept works, it is crucial to understand the hierarchy of LTE security keys first. This LTE security key hierarchy, shown in Figure 1.25, includes the following keys: K_{eNB}, $K_{NAS_{int}}$, $K_{NAS_{enc}}$, $K_{UP_{enc}}$, $K_{RRC_{int}}$, and $K_{RRC_{enc}}$:

- K_{eNB} is a key derived by the UE and MME from K_{ASME} or by the UE and target eNB from K_{eNB}* during eNB handover. K_{eNB} should only be used for the derivation of keys for RRC traffic and the derivation of keys for UP (User Plane) traffic, or to derive a transition key K_{eNB}* during an eNB handover.

Keys for NAS traffic:

- $K_{NAS_{int}}$ is a key that should only be used for the protection of NAS traffic with a particular integrity algorithm. This key is derived by the UE and MME from K_{ASME}, as well as an identifier for the integrity algorithm.
- $K_{NAS_{enc}}$ is a key that should only be used for the protection of NAS traffic with a particular encryption algorithm. This key is derived by the UE and MME from KASME, as well as an identifier for the encryption algorithm.

Keys for UP traffic:

- $K_{UP_{enc}}$ is a key that should only be used for the protection of UP traffic with a particular encryption algorithm. This key is derived by the UE and eNB from K_{eNB}, as well as an identifier for the encryption algorithm.

---

[5] Reproduced with permission from © 3GPP™.

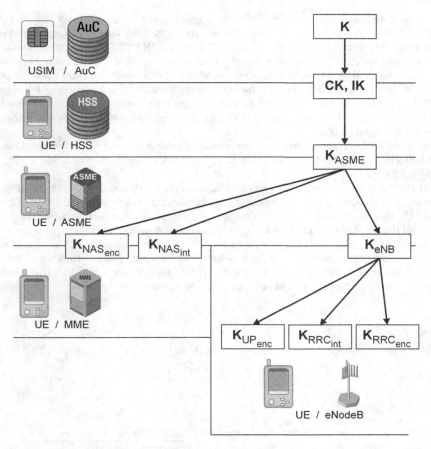

Figure 1.25   LTE security key hierarchy (according to 3GPP 33.401). (*Source*: Reproduced with permission from © 3GPP™.)

Keys for RRC traffic:

- $K_{RRC_{int}}$ is a key that should only be used for the protection of RRC traffic with a particular integrity algorithm. $K_{RRC_{int}}$ is derived by the UE and eNB from K_{eNB}, as well as an identifier for the integrity algorithm.
- $K_{RRC_{enc}}$ is a key that should only be used for the protection of RRC traffic with a particular encryption algorithm. $K_{RRC_{enc}}$ is derived by the UE and eNB from K_{eNB} as well as an identifier for the encryption algorithm.

Now, whenever a call is established, the security functions will work as shown in Figures 1.26–1.28. The start trigger of the security functions is when an initial NAS signaling message sent by the UE that contains UE security capability information arrives at the MME. The security capability list informs the MME, for instance, about which ciphering and integrity protection algorithms are supported by this UE.

After the MME has received the initial NAS message and it has not been in contact with this subscriber before, or if all previously received security tokens sent by the HSS have been used, the MME must contact the HSS to receive new tokens. Thus, the MME sends a DIAMETER authentication information request message to the HSS that contains the subscriber's identity. The HSS holds the secret network key "K" that is also stored on the USIM card of each subscriber. "K" is unique to every network operator.

From "K" and the subscriber's identity, the HSS derives three of the four parameters found inside the DIAMETER authentication information response message: the security key K_{ASME}, the Authentication Token (AUTN), and the Expected Response (XRES) parameter. The random number parameter RAND is truly just a random number.

After the MME has received these four parameters, it produces three more derivatives from K_{ASME}. These derivatives are the NAS encryption key $K_{NAS_{enc}}$, the NAS integrity protection key $K_{NAS_{int}}$, and the security key for the eNB K_{eNB}.

What follows is the authentication procedure between the MME and the UE. The MME sends the unciphered NAS authentication request message that includes the random number RAND and the AUTN. Now, the UE must use its secret key "K" from the USIM card to calculate another number based on "K," AUTN, and RAND. The number is the UE's authentication response number RES.

RES is sent back to the MME by using the authentication response message, and in the last step of the authentication procedure, the MME compares the value of RES to the value of XRES, which is the XRES value computed previously by the HSS. If RES and XRES have the same value, the UE has successfully authenticated itself to the network and the NAS signaling connection can proceed.

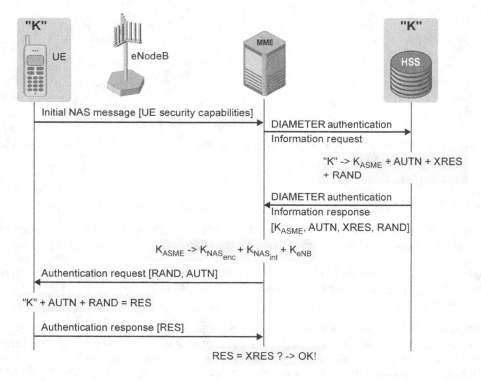

Figure 1.26   Subscriber authentication

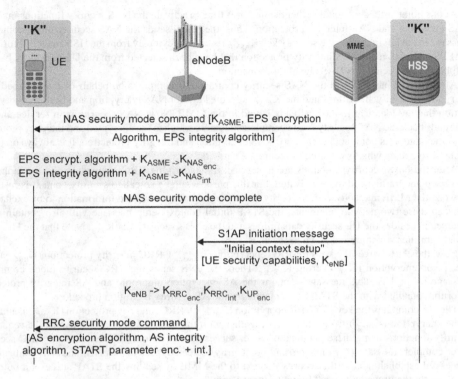

Figure 1.27   NAS security initiation and RRC security initiation

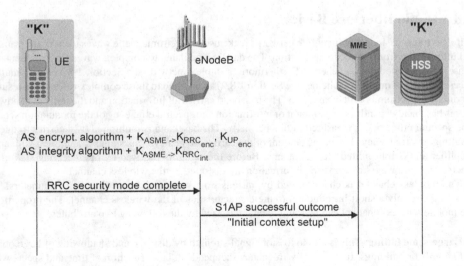

Figure 1.28   RRC security completion

At this point, after successful authentication, it is time to activate the NAS security functions: namely, NAS ciphering and NAS integrity protection. Thus, the MME sends the NAS security mode command message to the UE including the security key K_{ASME} received previously from the HSS and the algorithms for EPS encryption and EPS integrity protection that have been selected from the UE capability list and will be used to secure this NAS signaling connection.

After the UE has received the NAS security command, it computes on behalf of the assigned EPS encryption/integrity algorithms and the K_{ASME} key, the keys for NAS encryption and NAS integrity protection that are identical to those already stored in the MME. Now, NAS security is in service and the UE sends back the NAS security mode complete message, which is the encrypted and integrity protected NAS message. It is not mandatory to use NAS encryption and integrity protection. It is always up to the operator to decide what is required to secure the network.

After the NAS security functions are in service, the underlying RRC connection and the ciphering for user plane traffic need to be activated. For this purpose, first a so-called security context is installed between the MME and eNB. Since security is not the only context-related information to be exchanged between these two network elements, the S1AP initial context setup message will also contain other parameters besides the UE security capabilities and the eNB's security key K_{eNB}. Note that the UE security capabilities so far are unknown to the eNB.

Now, the eNB derives the keys for RRC encryption ($K_{RRC_{enc}}$), RRC integrity protection ($K_{RRC_{int}}$), and user plane encryption ($K_{UP_{enc}}$) from K_{ASME}. Then, the eNB sends the RRC security mode command message to the UE. This message contains the AS encryption algorithm and AS integrity protection algorithm bundled with the START parameters for the AS security activation procedure.

The UE computes the keys for RRC encryption ($K_{RRC_{enc}}$), RRC integrity protection ($K_{RRC_{int}}$), and user plane encryption ($K_{UP_{enc}}$) from the K_{ASME} together with the received keys and activates the requested security functions using these parameters. After successful activation, the UE sends the RRC security mode complete message (i.e., ciphered and/or integrity protected) back to the eNB. The eNB confirms the successful establishment of the security context to the MME by sending the S1AP successful outcome message for the procedure code "Initial Context Setup."

1.8 Radio Interface Basics

Wireless transmissions within mobile networks make use of electromagnetic waves to carry the transmitted information from source to destination. The data is modulated to complex waves (in a mathematical sense) using a modulation scheme. Furthermore, a duplex method (see Section 1.8.1) and a multiple access scheme or multiplex scheme (see Section 1.8.2) are applied to those complex waves. The resulting modulated spectrum is called baseband. The baseband carries all information and the utilized bandwidth of this baseband depends on the amount of information and spectral efficiency of the modulation scheme. The spectral efficiency is measured in bits per hertz. The baseband is multiplied by a carrier frequency, resulting in a frequency shift in the amount of the carrier frequency. This signal is amplified with an RF amplifier and is transmitted via an antenna. Before the signal is received by the receiving station, the electromagnetic waves carrying the information are distorted by the wireless channel.

The wireless channel is characterized by various time-variant and time-invariant parameters. This section gives only a short introduction to the characteristics of the wireless channel. The properties of the mobile wireless channel can be roughly characterized by the following two attributes:

- **Large-scale fading**: This is due to loss of signal strength by distance and shadowing of large objects like hills or buildings. It is typically frequency independent, but a function of time and space, which fluctuates by means of cell areas.

● **Small-scale fading**: This is due to the constructive and destructive interference of the multiple signal paths between the transmitting node and the receiving one, resulting in signal strength changes on a spatial scale of the wavelength. Therefore, signal strength variation greatly increases with faster moving stations and is frequency selective.

Free space attenuation as a function of distance d and wavelength λ is shown in the following equation. This basic attenuation (part of large-scale fading) denotes the signal decrease between source and destination without taking any shadowing, multipath fading, or scattering into account:

$$L = 10 \log \left(\frac{4\pi d}{\lambda} \right)^2 [dB]$$

Especially, the small-scale fading introduces distortion in the received signal to such an extent that it needs to be eliminated, or at least reduced by entities called channel estimator and channel equalizer described later in this section.

Figure 1.29 shows a typical received wireless channel quality as a function of frequency and time. The channel quality (received signal strength of certain frequencies) changes on a large scale (large-scale fading) and is superposed by the small-scale fading of a moving node. The physical layer of mobile wireless transmission systems has to deal with these characteristics of mobile channels to ensure data transmission to a specific subscriber velocity. Because small-scale fading spatially changes by means of the wavelength, typically by several centimeters in mobile networks, the user velocity introduces fast

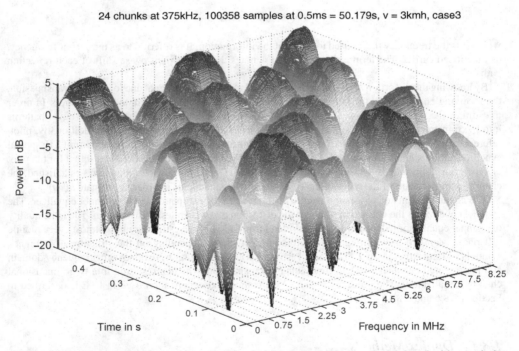

Figure 1.29   Time-variant frequency-selective wireless channel. (*Source*: Reproduced with permission from Nomor.)

fading to the received signal. As an example, a user velocity of 100 km/h (27.8 m/s) can result in signal fading changes of 250 times a second. As a result, the received signal is additionally amplitude modulated (a fast amplitude change) caused by the fast fading of a moving user. The amplitude modulated due to the time-variant wireless channel causes additionally frequency dispersion. Entities like fast power control and fast frequency-selective scheduling (see Sections 1.8.7 and 1.8.9) are introduced into mobile systems in order to counteract this.

A multipath channel is time dispersive, which means that a single transmitted signal is received more than once with different strong echoes (reflections). The electromagnetic waves are reflected by obstacles like buildings, hills, and mountains. The direct beam between the transmit antenna and receiver is called the LOS (Line Of Sight). The LOS is usually the strongest pattern within the brought field of received reflections. $h(\tau,t)$ is the time-variant (t and τ) impulse response of the wireless channel with i paths (reflections) and an attenuation $a_i(t)$ of each path i. The impulse response denotes the characteristic behavior of a system (in a mathematical sense), like the wireless channel when an impulse (delta peak) is given as input. In theory, this impulse is so steep that all possible frequencies are included; thus, it shows the behavior of all transmitted patterns. τ is the delay of the signal between the source and receiver. This delay is called the propagation delay. τ_i is the additional delay of the reflection path i. Thus,

$$h(\tau, t) = \sum_{\forall i} a_i(t)\delta(\tau - \tau_i(t))$$

The transmitted baseband signal is distorted by the wireless channel because different reflections of the signal are interfering at the receiver. Thus, the time-continuous received complex baseband signal $y_b(t)$ of a transmitted signal $x_b(t)$ with additive white Gaussian noise $n(t)$ is

$$y_b(t) = \sum_{\forall i} a_i(t)\exp[-j2\pi f_c \tau_i(t)]x_b(t - \tau_i(t)) + n(t)$$

where f_c is the frequency that is used to transmit the information. It is referred to as the carrier frequency, as mentioned earlier. The term $\exp[-j2\pi f_c\tau_i(t)]$ denotes the time-variant phase shift of each reflection path i.

By knowing the distortion $h(\tau,t)$, which was applied to the received signal due to wireless transmission, the receiving entity is able to reverse this distortion and retrieve the transmitted information. In order to estimate the wireless impulse response by the receiver, the transmitter entity inserts known patterns into the transmit signal. Those signals are referred to as pilot or reference signals. Additionally, pilot, reference, or synchronization signals can be applied for time and frame synchronization (in LTE, special synchronization signals are used, see Section 1.8.15). The receiver scans the received signal for the pilot or synchronization signals by using correlation functions. Once frame synchronization is established, the channel estimator unit of the receiver analyzes the known signal part in order to estimate $h(\tau,t)$.

The channel estimator passes the results of the process of estimation to the channel equalizer. The channel equalizer is the entity that removes the distortion due to wireless transmission. Thus, the quality of channel equalization depends on the provided information of the estimator unit. Almost every mobile cell phone standard uses different estimators and equalizers depending on the structure of reference symbols and slot structure. In LTE, channel estimation and equalization are done in the frequency domain and interpolated between adjacent time domain transmission symbols, resulting in a two-dimensional channel equalization. The process of channel estimation and equalization used in LTE is described in Sections 1.8.6 and 1.8.15.

1.8.1   Duplex Methods

A valuable communication feature is bidirectional communication between peers. Feedback is the major difference to unicast or broadcast systems, such as TV or radio transmission. Bidirectional information transfer can occur either simultaneously or consecutively. Both in data networks and especially for human communication, simultaneous bidirectional communication is very basic and essential. Systems with the nature of simultaneous bidirectional information transfer are full-duplex systems, in contrast to half-duplex systems, which allow only one-by-one bidirectional communication.

A full-duplex application visible to the user does not need to imply a full-duplex transmission scheme on lower layers. In mobile networks, two basic methods of multiplexing and handling the duplex streams are used. Here, duplex means the sense of UL and DL transmission from the handset to the base station and vice versa.

Frequency Division Duplex (FDD) divides the available frequency spectrum in a frequency range dedicated to UL transmission and a separate range for DL transmission only. A guard band is used between the frequency bands allocated for the UL and DL direction in order to prevent UL and DL interference. Figure 1.30 depicts a divided frequency spectrum as is typically used with FDD.

Time Division Duplex (TDD) makes use of the same frequency resources for both transmission directions. This method divides the time domain into slots allocated for UL and DL transmission as shown in Figure 1.31. A guard interval is implemented to prevent UL and DL interference, especially for larger cells with a longer propagation delay spread (larger cells have a longer duration behavior until all echo reflections of the transmitted signal are received) and cases of non-ideal UL synchronization.

TDD defines various slot configurations of switching between UL and DL transmission (see Table 1.10). Different configurations are possible depending on the mixture between UL and DL traffic; usually, the UL data volume is much smaller compared to DL data consumption. A change in transmission resources for the UL and DL direction is just one method of soft system configuration with TDD, instead of FDD, which uses fixed frequency resources.

TDD systems have the advantage of a synchronous channel between the UL and DL direction because the wireless channel characteristic is frequency dependent and TDD uses the same frequency band for UL and DL transmission. This is especially interesting in systems that use frequency-selective scheduling (see Section 1.8.7) and smart antennas (MIMO (Multiple Input, Multiple Output) transmission with multiple antennas) as they are defined in LTE. The base station needs to know the reception conditions

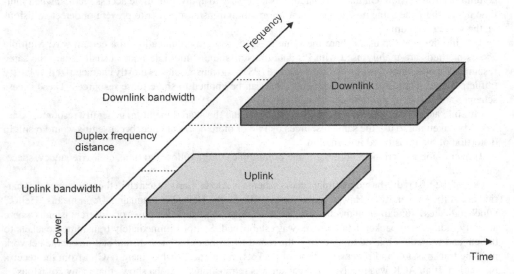

Figure 1.30   FDD system

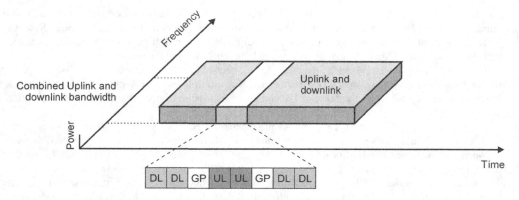

Figure 1.31 TDD system with example duplex slot structure

of the UE in order to make an efficient frequency resource scheduling decision and for selecting the best antenna configuration for multiple antenna layer precoding.

1.8.2   Multiple Access Methods

Multiple access has to be performed in systems where a medium is shared for transmission and reception by multiple users or entities. Those entities or users, sometimes also called nodes, are accessing the very same medium to transmit their information, other than, for example, in CS communication schemes. For example, within classical CS communications like Plain Old Telephone Systems (POTSs), each communication node gets a dedicated resource (a telephone landline) to dedicatedly access for the complete communication session. On the other hand, it is necessary to apply a multiple access method when multiple nodes share the same medium for their information transfer, in order to prevent or detect collisions on the shared medium.

This implies that the users share the same resources for communication in a certain way. Multiple access methods are mainly used with PS data transmission, as multiple nodes usually share the same resources for efficiency reasons. Packet Switched (PS) data transmission is mostly characterized by bursty traffic patterns. There are different schemes that can be applied to share those resources. These access schemes are known as multiple access methods.

Multiple access methods introduce rules for accessing the shared medium, generally resources. Care has to be taken not to use the same resources by two or more nodes at once, because this would result in distortion of the transferred information.

Transmission/reception resources are one or multiples of the following: time, code, frequency, space, and so on.

One of the first radio-based multiple access scheme is Aloha. Early research in these schemes was carried out at the University of Hawaii in the early 1970s. In the Hawaiian language, *Aloha* means "Hello," which indicates a fundamental mechanism: the university ran several campuses on different islands where an early radio-based packet data network was established. Stations immediately transmitted packets to be sent and waited for a fixed time (double the round-trip time of the most distant stations in the network plus the transmission and processing time of packets) for an acknowledgment (ACK) from the receiving station. If an ACK was received, the packet was retransmitted. Aloha shows that many collisions of packets occur when applying this scheme.

Most modern access methods use mechanisms of avoiding, detecting, or preventing collisions within the shared medium, in order to reach a certain system efficiency. A basic method to avoid collisions is sensing the medium before starting a transmission, to avoid interrupting or interfering with an ongoing transmission of other communication peers.

This scheme is known as Carrier Sense Multiple Access (CSMA). Sensing the medium before transmission adds Collision Avoidance functionality (i.e., CSMA/CA). In addition to sensing the medium before transmission, one has to take care of detecting collisions when two terminals have started a transmission at the same time. This scheme, for example, is used with Ethernet local area networks, CSMA/CD. If a collision is detected from both transmit entities, a collision resolution mechanism must be applied. Both peers select a random time in a defined range in order to restart their transmission after this randomly selected period of time. The process starts by sensing the medium again, as the other collision peer (or a new transmission of a third node) could already have (re)started its transmission as it has selected a shorter back-off period. The efficiency can be increased by introducing discrete back-off slots.

A special effect of wireless networks without infrastructure has to be taken care of. As mentioned earlier, a station listens to the channel before sending data to avoid a collision with an ongoing transmission from two other stations at that time. But this behavior is not fully sufficient to avoid collisions at all stations within the transmission range. If a node within the transmission range receives a data frame from another station, which is not in the range of a node also trying to allocate the channel, this node will interfere with reception of the other node without warning. This unrecognized collision scenario is called the hidden terminal effect, because the sending node is "hidden" or out of range.

Figure 1.32 shows a typical scenario for the hidden terminal effect. The circles around the stations demonstrate the transmission ranges of the nodes. Node A has a link to node B but does not know about

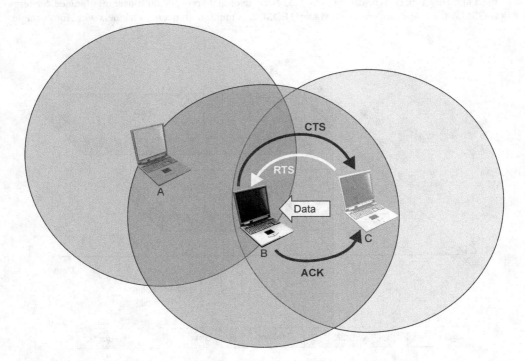

Figure 1.32   Illustration of the hidden terminal effect

the existence of node C, which is the hidden terminal from the point of view of node A. Node B has a link to both node A and node C. In this scenario, node C attempts to transmit data to node B and indicates this with a Request to Send (RTS) packet with the destination address of node B. The designated destination node B confirms this request with a Clear to Send (CTS) packet to node C; this CTS packet is also received by node A. Thereby, node A detects that there is another station while transmitting, until the reception node B sends an ACK packet to complete this transmission. Within that time, node A will not initiate any transmission, not to node B nor to any other possible node, in order not to corrupt reception at node B.

Note that this effect only occurs in mobile networks without fixed infrastructure, for example, mobile ad-hoc networks. Thus, the hidden terminal effect does not affect LTE.

One way of sharing the same resources between communication entities is to introduce a master entity, which takes care of the usage of the shared medium. This scheme is widely used within mobile cell phone networks, as the base station controls and grants the access of resources within its cell. Within a Time Division Multiple Access (TDMA) mobile network, one frequency resource is divided into time slots that are used by different users. Mobile networks of the second generation (GSM) share eight time slots within a certain frequency band. Figure 1.33 illustrates the DL frame of a TDMA system, which uses FDD. Thus, UL and DL utilize different frequency bands and both use TDMA. A Guard Period (GP) used between time slots serves to reduce the risk of multi-user interference. The training sequence in the middle of the time slot is used to estimate the wireless channel conditions. This information is extrapolated time-wise to the adjacent data sections.

Non-overlapping frequency bands are assigned to different UEs within Frequency Division Multiple Access (FDMA). A single user allocates one frequency resource, which is used for the complete active time. Guard bands need to be designed for the system in order to prevent multi-user interference. Systems (as GSM) often use a mixture of TDMA and FDMA, as multiple frequency channels with, for example,

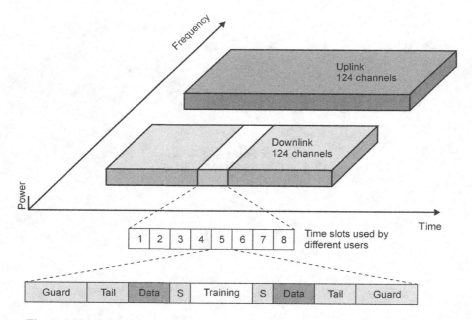

Figure 1.33   Schematic example of a Time Division Multiple Access (TDMA) system

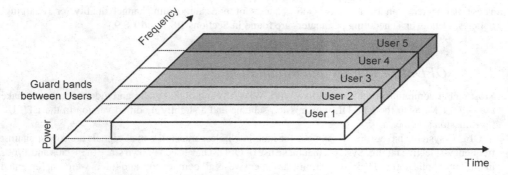

Figure 1.34 Schematic of Frequency Division Multiple Access (FDMA)

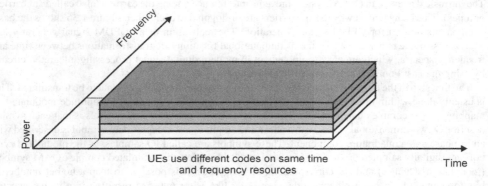

Figure 1.35 Code Division Multiple Access (CDMA)

eight time slots serve as cell resources. Figure 1.34 depicts an FDMA system with five users assigned to five different frequency bands.

UMTS users share orthogonal codes, which are used to spread transmission data in order to be transmitted on the same frequency resources within one cell. Picture this as a metaphor: each user uses the same time, space, and frequency resources, but communicates in a different language. This multiple access method is known as CDMA and illustrated in Figure 1.35.

The shared resources in LTE are very small frequency bands and small transmission time slots. Thus, the method combines FDMA and TDMA behavior, but in a very agile way. Frequency and time resources are often reassigned for diversity or efficiency reasons during ongoing user transmissions. This method is known as Orthogonal Frequency Division Multiple Access (OFDMA). OFDMA is described in Section 1.8.4 in detail, and OFDM signal generation and characteristics are described in Section 1.8.3.

Within mobile cell phone network systems, time slot resources, frequency resources, and code resources are controlled by the base station. This controlled multiple access method is known as scheduling. The scheduling method applied in LTE is described for the DL direction in Section 1.8.7 and for the UL in Section 1.8.9. This scheduling case allows active transmission collision protection, as well as other parameters, to be taken into account when granting transmissions between users.

These parameters can be fairness, QoS requests of nodes, medium/channel quality, or (charging) policies. Details on scheduling parameters are found in Sections 1.8.7 and 1.8.9.

1.8.3   OFDM Principles and Modulation

Most recent communication systems like WiFi, WiMAX, and digital audio and video broadcasts make use of OFDM, as in the LTE DL transmission scheme and a slightly modified version in the LTE UL transmission scheme.

OFDM systems have some advantages for mobile wireless transmission as signals are robust against frequency-selective fading. Systems that make use of OFDM have been known since the 1950s and 1960s in military applications. Their realization was expensive as all components and filters were implemented as analog circuits. Nowadays, a wide range of applications profit from the benefits of OFDM systems since digital signal processing has become inexpensive and available in consumer products.

Information is modulated on very small adjacent carriers within the allocated bandwidth (baseband). The intrinsic design of an OFDM system prevents interference among the carriers (also called subcarriers or tones). This is the reason why the subcarriers are orthogonal to each other. Figure 1.36 shows the basic components needed for OFDM signal generation. The realization of an OFDM signal generator and analyzer is simple to achieve as the main computational functions are transformations between time and frequency spectra, which are easy to implement in modern digital signal processing integrated circuits by using the Fast Fourier Transform (FFT) algorithm.

The first step in the transmit chain is the serial-to-parallel conversion of the data to be transmitted. This is usually done within the transmit buffer. This binary data is now quadrature amplitude modulated by mapping bits to complex data symbols. The characteristic of complex data symbols is that each symbol describes a two-dimensional vector with a phase and amplitude. A complex data symbol is described with an in-phase and a quadrature component. These symbols are called IQ samples, as the modulated symbols are digitally sampled. Figure 1.54 in Section 1.8.11 depicts the modulated complex QAM symbols (modulation schemes) and the corresponding bit mappings. It is possible to map a higher number of bits to symbols by using a higher modulation order like 16 or 64QAM resulting in a higher spectral efficiency, which means transmitting more bits per hertz of the utilized bandwidth. A higher spectral efficiency allows greater user and cell data throughput. The number of bits that are carried by the different modulation schemes can be seen in Table 1.7. Those numbers are OFDM independent and are equal to other transmission schemes.

Mobile cell phone standards, which do not use OFDM, like GSM, CDMA2000, or UMTS, modulate the data to complex symbols in the time domain. This means that the resulting sinusoid over time after

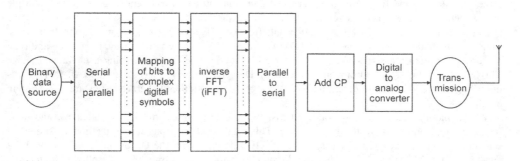

Figure 1.36   Block diagram of OFDM signal generation

Table 1.7 Bits to be carried by the modulation schemes used with LTE

Modulation scheme	Number of bits that can be carried by one complex symbol
BPSK	1
QPSK	2
16QAM	4
64QAM	6

modulation is the time domain signal of the baseband to be transmitted on the RF carrier frequency. OFDM systems interpret the modulated symbols as modulated frequency tones, which are to be transformed to a signal over time in order to be transmitted. Thus, the modulated symbols are mapped to orthogonal subcarriers (tones) of the baseband spectrum. The transformation to the time domain is done with an n-point inverse Fast Fourier Transform (iFFT). The Fourier transformation adds the orthogonal spectrum of each subcarrier to the resulting baseband spectrum. The spectrum of each subtone is a $si(x)$ function $(\sin(x)/x)$; thus, the resulting spectrum is an addition of $si(x)$ functions as depicted in Figure 1.37. The inherent behavior of the Fourier transformation lets each $si(x)$ maximum match zero transitions of all other $si(x)$ functions, resulting in non-interfering subtones since the data was modulated to individual subcarriers (peaks of the $si(x)$ functions). This characteristic is known as orthogonal behavior, which

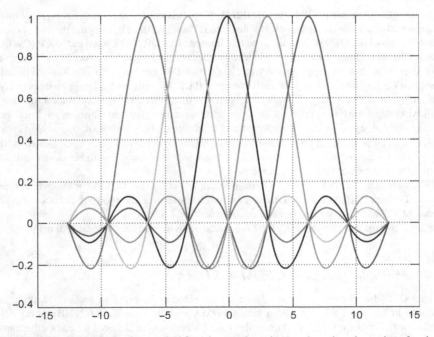

Figure 1.37 OFDM signal of orthogonal Si functions (subcarriers); subcarriers do not interfere because at each subcarrier, the signals from other subcarriers are zero

means the data is perfectly demodulatable and no guard band between subcarriers is needed, in contrast to FDM, where inter-carrier interference needs to be taken care of, for example, with guard bands.

The designated system bandwidth (baseband bandwidth) is divided into m subcarriers, which are sampled with an n-point Fourier transformation, where $n > m$, indicating oversampling. System bandwidth is defined by a number of resource blocks (see Section 1.8.5) from 6 to 110, each resource block grouping 12 subcarriers. For approximately 20 MHz system bandwidth (100 resource blocks), 1200 subcarriers are defined and a common FFT size is 2048 samples. LTE defines the sampling frequency as $f_s = 1/T_s = 30.72$ MHz, which leads to an LTE OFDM symbol length of 66.67 μs for normal CP (Cyclic Prefix). The CP is 5.2 μs or 160 samples in the first symbol and 4.7 μs or 144 samples in the other symbols for normal CP. The CP lasts 16.7 μs or 512 samples for extended CP. Details of LTE slot definitions are given in Section 1.8.6.

OFDM systems show robust characteristics against frequency-selective fading caused by the wireless channel, because fading holes are bigger compared to the subcarrier bandwidth, leading to a flat fading of individual subcarriers, which is equalized by interpolating between defined reference symbols (reference subcarriers).

LTE defines a set of reference symbols in order to distinguish between various entities: cell-specific, UE-specific, antenna-port-specific, and MBMS-specific (Multimedia Broadcast/Multicast Service) reference symbols. MBMS data and reference symbols are always transmitted on antenna port 4 if MBMS data transmission is enabled.

In other words, the time domain is just the "transmit domain" for OFDM systems. The resulting time domain signal after transforming the modulated frequency signal representing the data is, so to say, "just noise" and cannot be interpreted without transformation back to the frequency domain. All channel estimation, equalization, and interpretation of the data are done in the frequency domain within OFDM systems.

A timely guard interval or GP between OFDM symbols is needed to prevent intersymbol interference due to channel delay spread (arrival of all reflections). This is realized by copying the end of each OFDM symbol in front of the OFDM samples to be transmitted. This GP, also known as the CP (Cyclic Prefix), decreases the alias effects caused by a windowing effect of the Fourier spectrum as the Fourier transformation expects an infinite repeated spectrum, but the OFDM symbol has a time-limited duration. LTE defines two CP lengths, a normal CP and an extended CP, for cells with a larger channel delay spread. Additionally, the CP is used for frame synchronization using an auto/cross-correlation function.

OFDM systems have the drawback of a high dynamic range after transforming the frequency signal to a time domain signal, which is amplified and transmitted. This high Peak-to-Average Power Ratio (PAPR) (squared peak signal amplitude to average signal power level) leads to cost-intense RF amplifiers and shorter battery life. This is especially a disadvantage for mobile handset devices; thus, another transmission scheme needs to be found for the UL.

LTE uses Single Carrier Frequency Division Multiple Access (SC-FDMA) or Discrete Fourier Transform (DFT) spread OFDM as the UL transmission scheme to overcome some drawbacks of pure OFDM systems. SC-FDMA is described in more detail in Section 1.8.8.

Figure 1.38 presents an overview of the functional steps needed for physical channel processing.

1.8.4   Multiple Access in OFDM–OFDMA

LTE uses OFDM as the transmission scheme, as described in the previous section. Multiple access, as introduced in Section 1.8.2, is realized with OFDMA with the base station (eNB) taking care of the resources within its cell; this procedure is also called scheduling as the eNB schedules the transmission of user data in the DL and UL direction on the transmission medium used by all users within this cell. The transmission is done on the basis of a shared channel. The control information for granting an UL

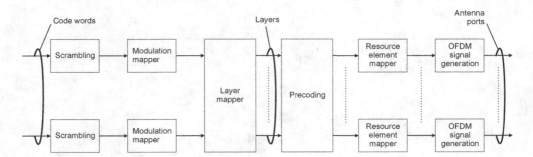

Figure 1.38 Overview of physical channel processing. (*Source*: Reproduced with permission from © 3GPP™.)

transmission on the UL-SCH, or informing a UE about data that is transmitted for it on the DL-SCH, is done within the DL control channel with designated control information. Physical channels defined with LTE are described in more detail in Section 1.8.12. Figure 1.39 shows the principle of an OFDM shared channel-based multiple user communication.

Resources to be scheduled with OFDMA systems are units of frequency resources and time units describing a time slot for which the scheduled frequency units are valid.

This scheduling procedure not only adds overhead, but also enables the system to be more efficient by introducing a frequency-selective scheduling algorithm with feedback from the UEs regarding current reception quality, rather than use only diversity gains by spreading transmitted data.

Figure 1.40 compares a shared channel of an OFDMA system using localized and distributed scheduling of user data. In the localized mode, granted areas belonging to one user allocate adjacent frequency resources within one block. The distributed mode is used when frequency diversity is to be used by spreading the user data of the shared channel to distributed non-adjacent frequency resources. Simulations show that systems with frequency-selective scheduling with a fast channel quality feedback report from users can achieve higher cell throughput compared to systems just spreading data over the spectrum in order to achieve frequency diversity.

LTE defines both localized and distributed scheduling in the DL direction but only localized scheduling in the UL direction in order to keep the PAPR small in the SC-FDMA symbols of each user. Details of scheduling the LTE DL can be found in Sections 1.8.7 and 1.8.9, which describe the LTE UL scheduling.

1.8.5   Resource Blocks

This section describes how LTE defines and divides the bandwidth from physical subcarriers of the OFDM symbol in the logical abstract sense used for scheduling shared channel data.

The smallest division of the LTE spectrum carrying data is a subcarrier, as described in more detail in the previous sections. OFDM systems modulate all data in the frequency domain on the subcarrier of the OFDM spectrum. A modulated subcarrier is defined as a Resource Element (RE) and is the smallest logical unit of the LTE spectrum. One subcarrier or RE has a bandwidth of 15 kHz in normal and extended CP mode, but a special 7.5 kHz subcarrier spacing mode is defined with extended CP transmission. All physical LTE channels use REs to modulate the data. Each RE of the shared channel is modulated using a variable modulation scheme from Quadrature Phase Shift Keying (QPSK) to 64QAM as assigned by the scheduling process of the eNB. REs carry a variable number of data bits, due to the variable number of bits mapped to an RE because of the modulation order, but also because of applied CSs. CSs are used to make

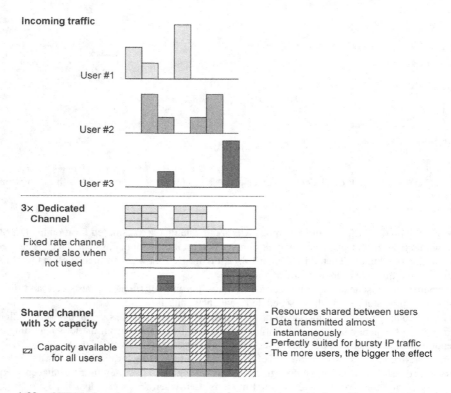

Figure 1.39 OFDM shared channel-based multiple user communication. (*Source*: Reproduced with permission from Nomor.)

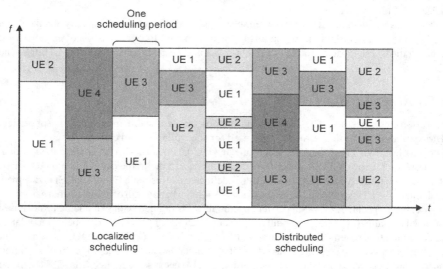

Figure 1.40 Localized versus distributed shared channel scheduling

Table 1.8 Configuration of cyclic prefix

Configuration		N_{sc}^{RB}	N_{symb}^{DL}
Normal cyclic prefix	ff = 15 kHz	12	7
Extended cyclic prefix	ff = 15 kHz	24	6
	ff = 7.5 kHz		3

Source: Reproduced with permission from © 3GPP™.

the data transmission more robust against transmission errors. They add redundancy to the transmitted information, which increases the probability of the receiver to retrieve the information error-free. LTE defines an Adaptive Modulation and Coding (AMC) process, which is described in Section 1.8.11 in more detail.

A physical Resource Block (RB) defines the smallest unit used by the scheduling algorithm. Therefore, the minimal scheduled user transmission on the shared channels is one RB. An RB consists of 12 adjacent REs on the frequency axis. Consequently, it has a bandwidth of 180 kHz, since one RE is 15 kHz wide in normal and extended CP mode (additionally, a mode with 7.5 kHz is defined for extended CP). The possible configurations of CPs are given in Table 1.8.

From a time perspective, an RB spans one scheduling period, which is defined as one subframe. One subframe has a duration of 1 ms. A subframe is divided into two slots of 0.5 ms. Within a subframe, 14 OFDM symbols are transmitted in the case of normal CP length and 12 OFDM symbols in the case of extended CP length; hence, an RB covers an area of, respectively, 12×14 and 12×12 REs. The time-wise LTE slot and frame structure is described in Section 1.8.6.

A third dimension is introduced by using multiple antenna ports with MIMO. The MIMO transport layers depend on and correlate with the number of used transmit antenna ports. Each antenna port layer adds additional RB elements to the antenna port dimension. Figure 1.41 shows the two-dimensional (frequency and time) area of an RB.

RBs have a primary role in the scheduling process, but are also used for describing the LTE overall cell bandwidth. The cell bandwidth is announced in the data transmitted in the Physical Broadcast Channel (PBCH) in number of resource blocks. Table 1.9 maps the number of RBs to the LTE spectrum bandwidth in megahertz.

The scheduling procedure defines virtual resource blocks. Virtual resource blocks and physical resource blocks are of equal size. The scheduler always uses virtual resource blocks for defining user allocations. There are two different types of virtual resource blocks:

- Localized virtual resource blocks
- Distributed virtual resource blocks

Localized virtual RBs are equal to physical RBs. Therefore, localized virtual RBs address physical RBs directly. Distributed RB mapping enables the usage of frequency diversity without scheduling distributed RBs directly. Distributed virtual RBs split a physical RB at the slot boundary into two halves. The first half of the scheduled distributed virtual RB directly equals the physical RBs. The second slot is hopped to another second slot of another UE, which is virtually scheduled in the distributed way. There is one hopping gap between the scheduled RBs at system bandwidths with smaller 50 RBs and two gaps at systems with a larger number of RBs. Virtual RBs are used with resource allocation type 2 described in Section 1.8.7.

Resource Element Groups (REGs) are defined to map physical channels to OFDM symbols; this is done especially in the first OFDM symbols used by the PDCCH. The maximum length of the PDCCH is

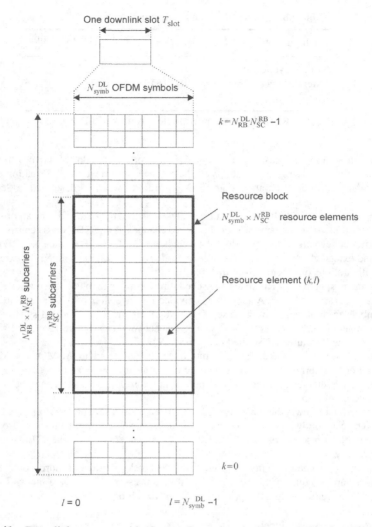

Figure 1.41   Downlink resource grid. (*Source*: Reproduced with permission from © 3GPP™.)

Table 1.9   Commonly used number of resource blocks

Bandwidth (MHz)	Resource blocks
1.4	6
3	13
5	25
10	50
15	75
20	100

three OFDM symbols; thus, other PHY channels are mapped into the resources of the PDCCH. A REG is defined in such a way that it spreads the information over a bigger frequency range to gain frequency diversity. Figure 1.59 in Section 1.8.12 illustrates REGs in the PDCCH.

1.8.6 Downlink Slot Structure

This section introduces the LTE DL structure with respect to time. In the time domain, the DL is divided into slots and frames. A radio frame is the largest unit and lasts 10 ms. As both duplex modes, FDD and TDD, use similar timings, this makes a UE starting to synchronize to an LTE cell unaware of the duplex method used. Thus, TDD and FDD both introduce radio frames of 10 ms timing.

Both duplex methods define different types of radio frames:

• Frame type 1 is the frame type used with FDD.
• Frame type 2 is the frame type used with TDD.

One radio frame is split into 10 subframes of 1 ms duration. A subframe is also the most important unit of scheduling and for physical control channel durations, the PDCCH is found in the first OFDM symbols of each subframe. Furthermore, subframes are divided into two slots of 0.5 ms. The smallest unit is an OFDM symbol. Depending on the CP length (normal or extended) used (see Section 1.8.3), the length will be seven or six OFDM symbols transmitted within one slot, resulting in 14 or 12 OFDM symbols within one subframe, respectively. Figure 1.42 shows one radio frame of type 1 (FDD) with 20 slots and 10 subframes. The length of one radio frame is defined as $T_f = 307\,200 \times T_s$, where T_s is the sampling period. Therefore, the base sampling frequency of 30.72 MHz ($T_s = 1/30.72\,\text{MHz}$) is used.

Figure 1.43 shows a complete radio frame of type 1 as used with FDD. It illustrates all 10 subframes with areas of the PDSCH. The first OFDM symbols are allocated by the PDCCH. In the area of the PDCCH, additional physical channels for control information are embedded. These channels are the Physical HARQ Indicator Channel (PHICH) and the Physical Control Format Indicator Channel (PCFICH).

The PCFICH indicates the number of OFDM symbols allocated for the complete PDCCH in its current subframe. Thus, the PDCCH is transmitted on a variable number of OFDM symbols depending on how much control or scheduling information has to be transmitted, and the resulting number of OFDM symbols, which are used for the PDSCH, is variable as well. The PDCCH allocates between 1 and 3 OFDM symbols and the PDSCH between 13 and 11 for normal CP and between 11 and 9 OFDM symbols for extended CP, respectively.

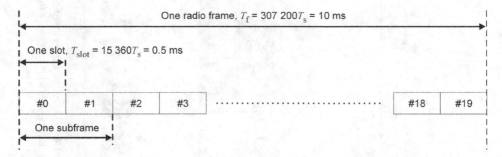

Figure 1.42 Frame structure type 1 used with FDD TS36.211. (*Source*: Reproduced with permission from © 3GPP™.)

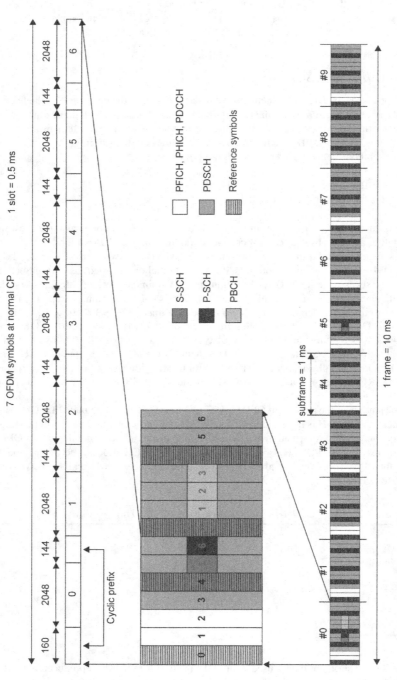

Figure 1.43 Downlink FDD radio frame (normal cyclic prefix) with PDCCH, PDSCH, PBCH, reference signals, and synchronization signals. (*Source*: Reproduced with permission from Nomor.)

In the middle (around the DC subcarrier) of the bandwidth are six RBs used for some common signals and channels in some subframes. These locations are used for initial cell search and cell synchronization. This central location enables a bandwidth-independent cell and frame synchronization as well as initial cell access. Two-step hierarchical synchronization signals are defined and located in the first and sixth subframes. The Primary Synchronization Signal (PSS) is transmitted on the seventh OFDM symbol and the Secondary Synchronization Signal (SSS) is transmitted on the sixth OFDM symbol of subframes described earlier. PSS and SSS can be seen in Figure 1.43, and Figure 1.62 shows the bandwidth of the synchronization signals. Only 62 subcarriers of the 72 provided by six allocated RBs are used for synchronization signals. Initial UE radio access is described in Section 1.8.15 as well as the information carried by those signals.

Resources for the PBCH are allocated apart from the SSS. The PBCH is transmitted in each first subframe of all radio frames. Other than the synchronization signals, the PBCH uses all 72 subcarriers of the six central RBs (see Figures 1.43 and 1.62).

Reference signals are needed by the channel estimation process in order to correct the wireless channel distortion in the signal at the receiver. Four sets of reference signals are specified for each transmit antenna as LTE defines multi-antenna transmissions (DL MIMO). The receiver needs to know the propagation

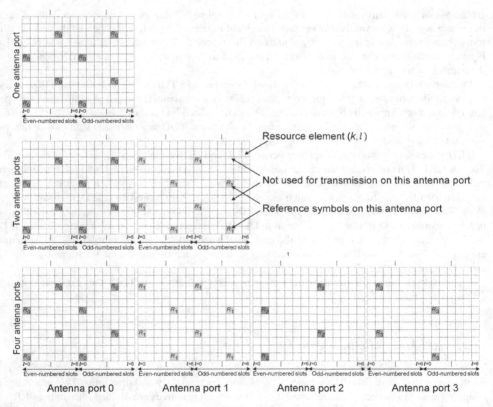

Figure 1.44 Mapping of downlink reference signals (normal cyclic prefix) (TS36.211). (*Source*: Reproduced with permission from © 3GPP™.)

Table 1.10   Uplink–downlink configurations for TDD

Uplink–downlink configuration	Downlink–uplink switch-point periodicity (ms)	Subframe number									
		0	1	2	3	4	5	6	7	8	9
0	5	D	S	U	U	U	D	S	U	U	U
1	5	D	S	U	U	D	D	S	U	U	D
2	5	D	S	U	D	D	D	S	U	D	D
3	10	D	S	U	U	U	D	D	D	D	D
4	10	D	S	U	U	D	D	D	D	D	D
5	10	D	S	U	D	D	D	D	D	D	D
6	5	D	S	U	U	U	D	S	U	U	D

S = Switching point.
Source: Reproduced with permission from © 3GPP™.

conditions of each transmit antenna for using the complete MIMO gain. Therefore, a defined signal (reference signal) is transmitted from each individual antenna completely independently and all other transmit antennas do not transmit any signals on those specific frequencies and time resources (REs). Figure 1.44 illustrates the frequency and time RE used for transmitting reference signals depending on the number of antennas configured.

Frame type 2 is used with TDD. The general slot structure of a TDD frame is similar to an FDD frame, since a mobile is not aware of the duplex method before synchronizing to the cell. A radio frame of type 2 has a duration of 10 ms as its FDD equivalent. The TDD radio frame is also divided into 10 transmission time intervals of 1 ms duration called subframes. Furthermore, TDD subframes are split as well into two slots of 0.5 ms period.

TDD systems switch on one frequency between DL and UL transmission. Therefore, as described in Section 1.8.1, TDD needs defined switch points and guard intervals between UL and DL transmission.

LTE defines two basic switch point interval durations of 5 and 10 ms. A switch point is a designated subframe divided into three zones, which are already known from basic UMTS TDD: Downlink Pilot Time Slot (DwPTS), GP, and Uplink Pilot Time Slot (UpPTS). Other subframes are used for either UL or DL transmission. Operators can decide from UL–DL configurations which subframes are used as DL and UL depending on the UL and DL traffic mixture of a network. Seven different UL and DL structures are defined. The possible UL–DL configurations with the switch points and the subframe used for UL and DL transmission are given in Table 1.10.

Figure 1.45 shows a radio frame with frame structure type 2 used with TDD. This is an example with a 5 ms switch-point periodicity.

1.8.7   OFDM Scheduling on LTE DL

The eNB advises each UE when and on which resources to transmit its data or informs a UE where it should listen to receive data. The resources are defined by frequency and time units. This procedure of assigning system resource is called scheduling. The system resources are divided into units of RBs, as described in Section 1.8.5. Only integer numbers of RBs can be assigned to one user. Localized and distributed RB allocations are possible. Localized allocations assign adjacent RBs to one UE, and distributed

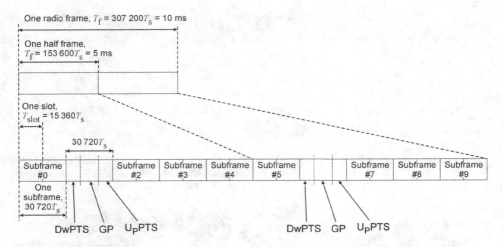

Figure 1.45 Frame structure type 2 used with TDD (for 5 ms switch-point periodicity) (TS36.211). (*Source*: Reproduced with permission from © 3GPP™.)

allocations distribute the scheduled RBs over the spectrum with gaps, for example, in order to achieve frequency diversity.

A new scheduling assignment is transmitted for each subframe; thus, the scheduling period on the time axis is 1 ms. The DL scheduling information is transmitted in the PDCCH. The assignments on the frequency scale vary between one RB (minimal scheduled transmission) and the maximum number of available RBs in respect of the system bandwidth.

Generally, the LTE scheduling algorithm is not defined by the standard; it is a matter for the eNB vendors. This enables the base station vendors to differentiate between each other and use different optimization goals. Various parameters can be used as input for the scheduling decisions: channel quality of different users (measured or reported by the UE with the Channel Quality Indicator, CQI), QoS, congestion/resource situation, fairness, charging policies, and so on. Most schedulers aim to maximize the cell throughput under consideration of fairness metrics between cell edge users and users with very good channel conditions. Figure 1.46 shows a screenshot of a typical scheduling and cell resource allocation analysis tool. It gives insight into the scheduling process of the eNB and is able to evaluate the scheduler performance and the utilization of cell resources.

The CQI is reported in the UL direction and can be derived periodically or upon request by the eNB. It gives reception quality feedback to the scheduling algorithm in the eNB in order to schedule data on those frequency regions with the best possible reception characteristics. CQI is described further in Section 1.8.11.

The scheduling information is encoded as Downlink Control Information (DCI). The DCI is then mapped to REGs of the PDCCH. The length of the PDCCH can vary between one and three OFDM symbols depending on the load to be transmitted on the PDCCH. The number of used OFDM symbols is indicated in the PCFICH.

The DCI does not just transmit RB assignments and its assignment type, but also other information needed for the transmission or reception of data. This information is, for example, the MCS, HARQ feedback information, or power control commands for UL transmission of the Physical Uplink Control Channel (PUCCH) or Physical Uplink Shared Channel (PUSCH) (see below).

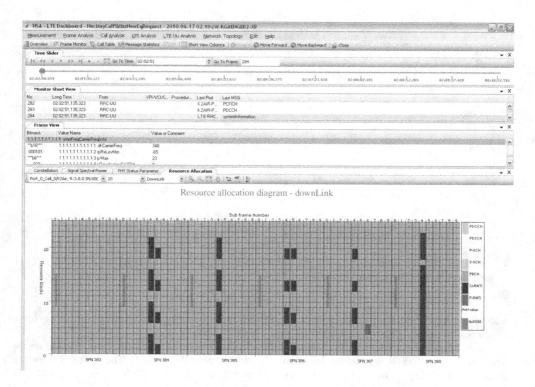

Figure 1.46   Scheduling and cell resource allocation analysis

The following DCI formats are defined and used for scheduling and UL Transmit Power Control (TPC) commands:

- **DCI format 0**: UL scheduling grant, see Section 1.8.9.
- **DCI format 1**: Single transport block (code word) scheduling assignment, for example, used for assigning resources to system information, paging, or random access. Further information: MCS, HARQ (New Data Indicator (NDI), redundancy version, HARQ process number), and TPC for PUCCH.
- **DCI format 1A**: Compact single transport block scheduling assignment. Further information: MCS, HARQ (NDI, redundancy version, HARQ process number), and TPC for PUCCH.
- **DCI format 1B**: A special DCI format for transmission mode 6 (MIMO closed-loop rank 1 pre-coding). Further information: precoding vector, MCS, HARQ (NDI, redundancy version, HARQ process number), and TPC for PUCCH.
- **DCI format 1C**: Even more compact scheduling format than DCI 1A. For example, it is used for assigning resources to SIBs, paging, or RARs. This DCI is always transmitted using frequency diversity via distributed virtual resource block assignments using resource allocation type 2, as shown in the following. This is done because channel feedback cannot be derived for such common information, as it is received by multiple users. The modulation is fixed to QPSK. Further information: MCS, HARQ (NDI, redundancy version, HARQ process number), and TPC for PUCCH.

- **DCI format 1D**: A special DCI format for transmission mode 5 (multi-user MIMO). Further information: MCS, HARQ (NDI, redundancy version, HARQ process number), power offset indicator if two UEs share power resources, and TPC for PUCCH.
- **DCI format 2**: Scheduling for transmission mode 4 (closed-loop MIMO), multiple antenna port transmission operation, addressing multiple transport blocks (code words) to be transmitted on different antenna ports (layers). Further information: MCS for each transport block, HARQ (NDI, redundancy version, HARQ process number), number of transmission layers, precoding, and TPC for PUCCH.
- **DCI format 2A**: Used with transmission mode 3 (open-loop MIMO using Cyclic Delay Diversity (CDD)), multiple antenna port transmission operation, addressing multiple transport blocks (code words) to be transmitted on different antenna ports (layers). Further information: MCS for each transport block, HARQ (NDI, redundancy version, HARQ process number), number of transmission layers, precoding, and TPC for PUCCH.
- **DCI format 3**: A 2-bit UL TPC command applying for PUSCH and PUCCH. Multiple users are addressed.
- **DCI format 3A**: A 1-bit UL TPC command applying for PUSCH and PUCCH. Multiple users are addressed.

The resource allocation assignments of the aforementioned DCI formats can use different resource allocation types. Table 1.11 maps the DCI formats to the allowed resource allocation types:

- **Resource allocation type 0**: With resource allocation type 0, a bit map is transmitted describing Resource Block Groups (RBGs). An RBG is a number of consecutive physical resource blocks (RBs). The number depends on the system bandwidth and has a range between one and four physical RBs. Table 1.12 maps the size of an RBG to the system bandwidth. The allocated RBGs do not have to be adjacent.

Table 1.11 Resource allocation types and the applying DCI formats TS36.213

Resource allocation type	Applying DCI formats
Type 0	1, 2, 2A, and 2B
Type 1	1, 2, 2A, and 2B
Type 2	1A, 1B, 1C, and 1D

Source: Reproduced with permission from © 3GPP™.

Table 1.12 Type 0 resource allocation RBG size versus DL system bandwidth

System bandwidth N_{RB}^{DL}	RBG size (P)
≤10	1
11–26	2
27–63	3
64–110	4

- **Resource allocation type 1**: The bit map transmitted with resource allocation type 1 makes use also of RBGs but can address single physical RBs by introducing additional flags. The number of RBGs is smaller than the ones used with resource allocation type 0, thus not reaching the complete bandwidth. The bit map addresses not whole RBGs but a subset within each RBG, which is pointed to by the bit map. A selection flag indicates the position within the RBG regions and a shift flag shows the position of the numbered RBGs within the system bandwidth as the number of RBGs does not address the complete system bandwidth: shift flag = 0 indicates that the RBGs start at the beginning of the system bandwidth leaving an unaddressable region at the end of the system bandwidth; shift flag = 1 indicates that the RBGs are shifted to the end of the system bandwidth leaving the unaddressable region at the beginning of the system bandwidth.
- **Resource allocation type 2**: This resource allocation type uses virtual RBs as scheduling units as described in Section 1.8.5. Two types of virtual RB scheduling assignments are used:
  - A localized type, where the allocated virtual RBs equal a number of consecutive physical RBs addressed with a starting RB and a number of adjacently assigned RBs. This information is encoded into an 11-bit Resource Indication Value (RIV).
  - A distributed type, where the addressed virtual RBs are distributed over the frequency with one or two gaps (depending on the system bandwidth) hopping at slot boundaries as described in Section 1.8.5. Virtually distributed RB assignments are always used with DCI format 1C. There is a 1-bit flag indicating whether virtual distributed or virtual localized RB assignment is used in the case of DCI formats 1A, 1B, and 1D.

Instead of addressing a UE with a PDCCH scheduling assignment (DCI) directly by adding a UE ID (e.g., a RNTI) to the DCI, the 16-bit CRC of the PDCCH message is scrambled with the RNTI, introducing common and UE-specific search spaces. This CRC scrambling saves additional resources in DCIs, but increases slightly the chance of decoding a DCI for a different UE, which is not intended to be addressed.

Figure 1.47 depicts an example PDCCH message of DCI format type 1 with all the transmitted information.

Some special applications require the transmission of small data chunks in equidistant periods of time. An example is a VoIP application. In order to minimize the signaling overhead in such cases, a mode Semi Persistent Scheduling (SRS) is introduced. SPS parameters are configured by the RRC layer, enabling the transmission of data on defined RBs in frequency and time without further scheduling on the PDCCH.

Battery energy saving is always an important topic with mobile handset systems. A potential scheduling assignment could be sent in each PDCCH, which occurs every millisecond. Therefore, each attached UE would need to monitor the PDCCH each millisecond for scheduling information. The DRX mode enables the UE just to listen to defined subframes for scheduling assignments and turn off its receiver in between, in order to save battery consumption. Short and long DRX cycle periods are defined. The DRX parameters are set by MAC and RRC. Figure 1.48 depicts a DRX cycle.

1.8.8   SC-FDMA Principles and Modulation

The OFDM transmission scheme shows robustness against multipath fading and is especially useful for mobile communication systems due to various reasons. An example is the complexity of the receiver where fairly simple and channel estimation/equalization is done in the frequency domain. Why is another transmission scheme selected for the LTE UL? A major disadvantage of OFDM systems is that the time domain signal, which is amplified and transmitted, shows a large dynamic range after modulating symbols on subcarriers and transformation to a time signal. This leads to a high Peak-to-Average Power Ratio (PARP) of the signal, which should be avoided for battery-powered handsets, which underlies a limit on

```
**B24***   K2AIR-PHY  (PDCCH)                             Tektronix K2Air LTE PHY Data Message Header
**B24***   1 PDCCH Message (Single UE Mode)
***B8***   1.1 Common Message Header
00000000   Protocol Version                               0
----1101   Transport Channel Type                         PDCCH/PUCCH Single UE mode
0100----   Physical Channel Type                          PDCCH
**b10***   System Frame Number                            327
-----0--   Direction                                      Downlink
----1---   Radio Mode                                     FDD
---0----   Internal use                                   0
-00-----   Status                                         Original data
0-------   Reserved                                       0
***b9***   Physical Cell ID                               0
----011-   UE ID/RNTI Type                                C-RNTI
0111----   Subframe Number                                7
***B2***   UE ID/RNTI Value                               'a277'H
**B16***   1.2 List of DCI for this UE
**B16***   1.2.1 DCI
----0001   DCI Format                                     Format 1 - DL schedule
---1----   Last Format indicator                          last DCI Format
--1-----   New Data Indicator DL 1                        new data
01------   Transmit Power Control                         +0 dB
---00000   Modulation Scheme Index DL 1                   0
-00-----   Redundancy Version DL 1                        0
0-------   Reserved                                       0
----0000   HARQ process number                            0
-010----   Modulation Order DL 1                          QPSK
0-------   Resource Allocation Type                       Type 0
00000000   Reserved                                       0
**B11***   1.2.1.1 List of allocated Downlink Resource Blocks
-0000000   Resource Block ID                              0
-0000001   Resource Block ID                              1
-0000010   Resource Block ID                              2
-0000011   Resource Block ID                              3
-0000100   Resource Block ID                              4
-0000101   Resource Block ID                              5
-0010100   Resource Block ID                              20
-0010101   Resource Block ID                              21
-0010110   Resource Block ID                              22
-0010111   Resource Block ID                              23
-0011000   Resource Block ID                              24
00000000   Padding                                        '00'H
```

Figure 1.47   Example PDCCH message of DCI format 1 (downlink scheduling assignment)

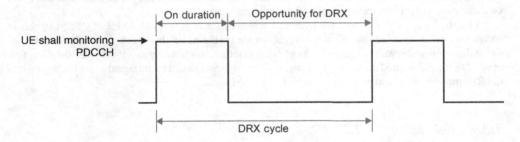

Figure 1.48   DRX cycle. (*Source*: Reproduced with permission from © 3GPP™.)

the budget to justify the business case. The linear transmission of such a signal needs a highly complex RF amplifier for the handsets in order not to run into the nonlinear region of the transmitter. Additionally, the power consumption is larger compared to a transmitter running a smaller linear dynamic range. The PAPR even increases with a wider OFDM bandwidth for a larger number of modulated subcarriers. This results in another transmission scheme for the UL: Single-Carrier FDMA (SC-FDMA).

Actually, SC-FDMA is very similar to OFDMA but shows a better PAPR, leading to a longer battery lifetime and a cost-effective RF amplifier design. Figure 1.49 illustrates the components of an SC-FDMA

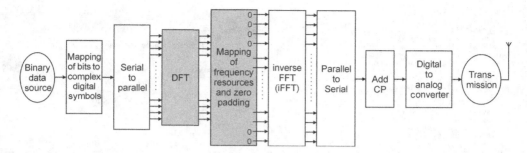

Figure 1.49   Block diagram of SC-FDMA transmitter with localized mapping to frequency resources

transmitter system with its block entities. Highlighted in the figure are the main differences from a regular OFDM system. The main difference is the DFT in the transmitter and the inverse DFT (iDFT) in the receiver, respectively. Thus, SC-FDMA is sometimes called DFT-spread-OFDM. Due to this difference, we can picture the information to be transmitted as modulated (bits mapped to two-dimensional QAM symbols with an I and Q component) to a time domain signal instead of modulating subcarriers of a frequency domain signal. The output of the DFT can be interpreted as a spectrum of the previously modulated data symbols. This spectrum has the characteristic of consecutive modulated subcarriers; it has, therefore, no scattered spectral distribution. Thus, it has the inherent behavior of localized RE usage as described in Section 1.8.5. This localized spectrum is now mapped to the consecutive frequency REs, which are specified in the UL scheduling grant, as only localized frequency resource assignments are allowed with UL transmission, which refers to the intrinsic signal characteristic of DFT-spread-OFDM. The rest of the spectrum for the full system bandwidth is filled with zeros. This zero patched spectrum is fed to an iFFT unit transforming a full system-wide spectrum "back" to the time domain for transmission. Figure 1.49 shows this mapping of the DFT symbols to the full iFFT width by adding zeros to frequency positions at the block in the center. The zero patched frequency areas are not used by this user and could be assigned to other users transmitting in the same time slot.

As for the future, a current research item is to overcome the lack of a high PAPR of OFDM signals in such a way that a digital reverse distorted signal is added to the signal that is to be transmitted. The non-linear distortion is known for a given RF amplifier. Therefore, it is possible to pre-calculate the distortion applied to the transmitted signal. This distortion is inverted and joined to the signal to such a degree that the RF amplifier distortion eliminates the inverted signal again.

1.8.9   Scheduling on LTE UL

The LTE UL scheduling is very similar to the DL scheduling described in Section 1.8.7, although the UL scheduler is a distinct entity. This section describes the difference from the DL scheduling only.

UL scheduling grants are indicated to the UE by transmitting all relevant UL scheduling information within the PDCCH. This is done by using a dedicated DCI type, DCI 0. Each UE has to monitor the PDCCH in every subframe for DCI types 0 scrambled with their RNTI (for a message example, see Figure 1.50). This does not apply for the case when power saving mode DRX is enabled, which switches off the UE's receiver periodically. UL resources are allocated without a designated PDCCH UL grant in the case of SPS (see Section 1.8.7) or for nonadaptive HARQ retransmissions. A nonadaptive HARQ retransmission is triggered by the transmission of a Negative Acknowledgment (NACK) by the UE.

```
BITMASK              ID Name                           Comment or Value
**B65***  LTE-RLC/MAC  (MAC-PDU (UL))                  3GPP LTE-RLC/MAC Rel.8 (MAC TS 36.321 V8.5.0, 2009-03, RLC  TS 36.322 V8.5.0, 2009-03)
**B65***  1 MAC PDU (Uplink)
***B4***  1.1 MAC Header Part
00111101  1.1.1 MAC PDU Subheader
00------  Reserved                                     0
--1-----  Extension field                             Another MAC PDU subheader follows
---11101  LCID-UP                                      Short BSR
***B2***  1.1.2 MAC PDU Subheader
00------  Reserved                                     0
--1-----  Extension field                             Another MAC PDU subheader follows
---00010  LCID-UP                                      Logical Channel 2
0-------  Format field                                7-Bits Lengthfield
-0111000  SDU Length                                  56
00011111  1.1.3 MAC PDU Subheader
00------  Reserved                                     0
--0-----  Extension field                             Last MAC PDU subheader
---11111  LCID-UP                                      Padding
**B61***  1.2 MAC Payload Part
00000111  1.2.1 MAC Control Short BSR
00------  Logical Channel Group ID                       22 < BS <=    26
--000111  Buffer Size
          Logical Channel ID                          2
**B56***  1.2.2 RLC Data PDU (acknowledged mode)
1-------  D/C                                          Data
-0------  RF                                           AMD PDU
--1-----  P                                            Status report is requested
---00---  FI                                           First byte is first of SDU, Last byte is last...
-----1--  E                                            A set of E & LI field follows on next octet
**b10***  SN                                           0
**b12***  1.2.2.1 E LI Set
0-------  E                                            Data field follows on next octet
**b11***  LI                                           26
----0000  Padding                                      0
**B26***  1.2.2.2 RLC Data with LI
**B26***  Data with LI                                 00 48 02 41 00 41 e6 86 86 06 06 06 06 06...
**B26***  1.2.2.3 RLC Data
**B26***  Data                                         01 48 02 41 00 41 e6 86 86 06 06 06 06 06...
***B4***  1.2.3 MAC Padding
***B4***  macPad                                       '00000000'H
```

Figure 1.50   Example PDCCH message of DCI format 0 (uplink scheduling grant)

In the UL only localized scheduling is allowed, which means that an integer number of consecutive RBs is allocated to one UE. Furthermore, there is only one scheduling process per UE; thus, there is no dedicated scheduling process per radio bearer.

The UE feeds the UL scheduler with CQI, Buffer Status Reports (BSRs), ACKs/NACKs, and Scheduling Requests (SRs). BSRs indicate the fill level status of the current transmit buffer. This buffer status is reported in bins, quantizing the fill level in bytes. Figure 1.51 shows an example of a MAC BSR message.

```
BITMASK              ID Name                           Comment or Value
**B65***  LTE-RLC/MAC  (MAC-PDU (UL))                  3GPP LTE-RLC/MAC Rel.8 (MAC TS 36.321 V8.5.0, 2009-03, RLC  TS 36.322 V8.5.0, 2009-03)
**B65***  1 MAC PDU (Uplink)
***B4***  1.1 MAC Header Part
00111101  1.1.1 MAC PDU Subheader
00------  Reserved                                     0
--1-----  Extension field                             Another MAC PDU subheader follows
---11101  LCID-UP                                      Short BSR
***B2***  1.1.2 MAC PDU Subheader
00------  Reserved                                     0
--1-----  Extension field                             Another MAC PDU subheader follows
---00010  LCID-UP                                      Logical Channel 2
0-------  Format field                                7-Bits Lengthfield
-0111000  SDU Length                                  56
00011111  1.1.3 MAC PDU Subheader
00------  Reserved                                     0
--0-----  Extension field                             Last MAC PDU subheader
---11111  LCID-UP                                      Padding
**B61***  1.2 MAC Payload Part
00000111  1.2.1 MAC Control Short BSR
00------  Logical Channel Group ID                       22 < BS <=    26
--000111  Buffer Size
          Logical Channel ID                          2
**B56***  1.2.2 RLC Data PDU (acknowledged mode)
1-------  D/C                                          Data
-0------  RF                                           AMD PDU
--1-----  P                                            Status report is requested
---00---  FI                                           First byte is first of SDU, Last byte is last...
-----1--  E                                            A set of E & LI field follows on next octet
**b10***  SN                                           0
**b12***  1.2.2.1 E LI Set
0-------  E                                            Data field follows on next octet
**b11***  LI                                           26
----0000  Padding                                      0
**B2***   1.2.2.2 RLC Data with LI
**B26***  Data with LI                                 00 48 02 41 00 41 e6 86 86 06 06 06 06 06...
**B26***  1.2.2.3 RLC Data
**B26***  Data                                         01 48 02 41 00 41 e6 86 86 06 06 06 06 06...
***B4***  1.2.3 MAC Padding
***B4***  macPad                                       '00000000'H
```

Figure 1.51   Example MAC Buffer Status Report (BSR) message

LTE trial studies show that there are various cases of scheduled empty UL grants. This happens when the eNB assigns UL resources to a UE and the designated UE does not use those UL resources for an UL transmission. Operators should minimize such empty scheduled UL resource in order to optimize the usage of UL radio resources.

1.8.10   Uplink Slot Structure

The UL slot structure is similar to the DL slot structure. Differences are mainly due to simplifications of reference symbols, robustness, and physical UL channel multiplexing. A radio frame lasts 10 ms and is divided into 10 subframes with two slots of 0.5 ms duration as a DL radio frame is split into subframes and slots. Figure 1.52 shows a UL radio frame with its subframe and slot structure. A UL RB has 12 subcarriers on the frequency axis and in the time domain, it has seven SC-FDMA symbols per slot when normal CP is used and six SC-FDMA symbols when extended CP is used (see Figure 1.53).

UL synchronization signals are not used because all UL signals from UEs transmitting within the cell are time aligned at the eNB with a UL timing control procedure. The eNB signals a Timing Advance (TA) command to each UE to track the UL alignment. Timing varies due to the different special distribution of UEs within a cell region. Signals are delayed because of the propagation delay. Firstly, during the random access procedure, a total timing offset to the UE is transmitted. After this, there is a control loop just tracking and signaling differential timing offsets in steps of 0.52 µs ($16 \times T_s$).

The PRACH always uses six consecutive RBs on the frequency axis and is timewise one subframe wide. Which six RBs are used is variable and set in SIB type 2. The PRACH configuration index is signaled in SIB type 2 as well, which defines the subframes that carry the PRACH's six RBs. This subframe configuration applies for either even or any radio frame. There are 64 possible PRACH configuration index permutations.

Resources at the upper and lower edges of the system bandwidth are used to carry the PUCCH. The frequency resources in between the PUCCH bands are designated to PUSCH transmission.

UL reference signals are used for UL channel estimation. DL reference signals are spread over frequency and time as single REs, which leads to a two-dimensional spheric channel estimation. UL reference signals are similar to TDMA pilots in the middle of each time slot. LTE defines reference signals in the middle of each slot (on the fourth OFDM symbol assuming normal CP duration) and spanning via the complete allocated frequency range of each UE.

The eNB can instruct UEs to transmit special reference signals over the complete system bandwidth, or parts of it, independently of UL data transmission on either PUSCH or PUCCH. These reference signals are called Sounding Reference Signals (SRSs). SRSs are used to estimate UL channel quality of a wider frequency range in order to optimize frequency-selective UL scheduling.

Figure 1.52 depicts, in addition to the radio frame, a zoomed UL subframe with a PUCCH and PUSCH example configuration with UL demodulation reference symbols (DMRSs) and SRS symbols.

1.8.11   Link Adaptation in LTE

Mobile wireless reception conditions vary greatly over frequency and time as described in Section 1.8. In order to cope with these circumstances and guarantee best possible QoS, a procedure is implemented known as Adaptive Modulation and Coding (AMC). AMC controls and changes transmission parameters to achieve a defined Transport Block Error Rate (BLER) of below 10%, in order to keep retransmissions in a suitable range. This is done by adapting the modulation scheme, in LTE on the shared channels between QPSK and 64QAM, and the Forward Error Correction (FEC) coding rate.

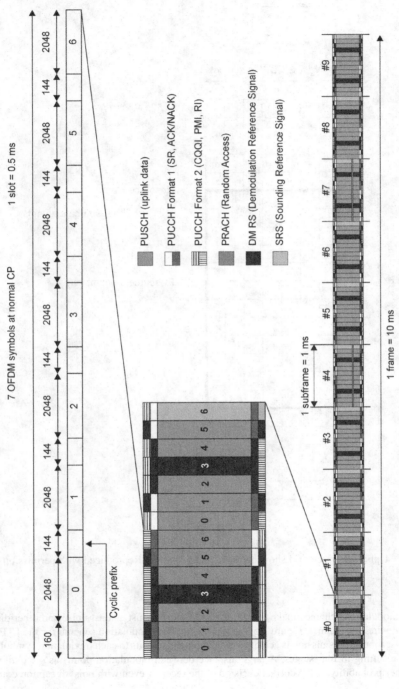

Figure 1.52 Uplink radio frame and subframe with two slots including PUSCH, PUCCH, PRACH, DMRS, and SRS. (*Source:* Reproduced with permission from Nomor.)

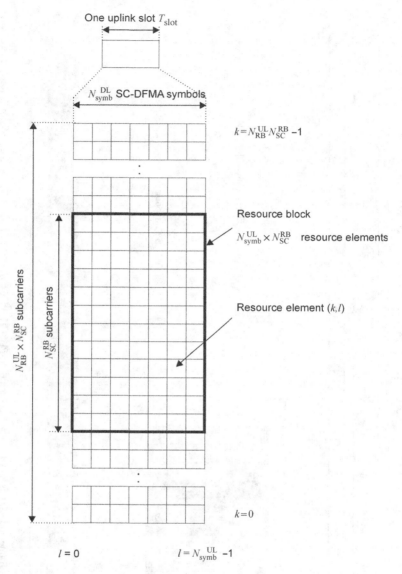

One uplink slot T_{slot}

$N_{\text{symb}}^{\text{DL}}$ SC-DFMA symbols

$k = N_{\text{RB}}^{\text{UL}} N_{\text{SC}}^{\text{RB}} - 1$

Resource block

$N_{\text{symb}}^{\text{UL}} \times N_{\text{SC}}^{\text{RB}}$ resource elements

Resource element (k,l)

$N_{\text{RB}}^{\text{UL}} \times N_{\text{SC}}^{\text{RB}}$ subcarriers

$N_{\text{SC}}^{\text{RB}}$ subcarriers

$k = 0$

$l = 0$ $l = N_{\text{symb}}^{\text{UL}} - 1$

Figure 1.53 Uplink resource grid showing on UL RB. (*Source*: Reproduced with permission from © 3GPP™.)

Different modulation schemes make the bit detection more robust against noise and other distortion caused by the wireless channel. Figure 1.54 shows the applied modulation scheme for the LTE shared channels. Most robust transmission is achieved by mapping just 2 bits to each modulation symbol as seen with QPSK, resulting in four stages. A large distance between modulation points as seen with QPSK allows a higher probability of the correct decision at the receiver even with noisy reception conditions.

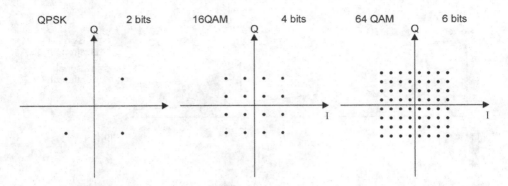

Figure 1.54 Different QAM schemes used with LTE and the number of bits mapped to each scheme

Both 16QAM and 64QAM map 4 and 6 bits, respectively, to one modulation symbol used with better wireless channel conditions to achieve a higher data throughput. It is to find the best compromise between the modulation scheme and code rate for a given channel quality. LTE defines a list of MCS combinations and just signals an MCS index.

The data modulated with the different modulation schemes to subcarriers needs to be protected against transmission errors. LTE defines a turbo de-/encoder with trellis termination of a native code rate of one-third. The turbo coder adds redundancy bits to the data, which makes it possible to correct some bit errors. The code rate is a fraction of source data rate to resulting protected data rate; thus, a code rate of one-third encodes 1 bit into 3 bits. Other code rates are needed in order to optimize the trade-off between protection and efficiency. This is done by puncturing the native coded bit stream to a higher (less protection) code rate by deterministically leaving out coded bits, or by deterministically repeating coded bits if a smaller code rate is desired (more protection).

Additionally, one parameter being controlled is the UL transmit power. UL power control is implemented to deal with the near–far effect. Figure 1.55 shows a UL scenario with a near–far effect compared to a DL scenario without power differences between user signals as they are equally attenuated because the mix of the signal is transmitted from one position (eNB). This occurs when a user is close to the base station (near) and another user is far away from the base station, introducing a higher power path loss, which leads to a lower receive power of the signal of the cell edge user. All UL receive signals should have equal power in order to have the same analog-to-digital converter saturation of each signal to reduce the quantization noise of users with low received signals, reducing inter-subcarrier interference between the users. This happens with imperfect UL synchronization within real-life scenarios.

UL TPC commands are sent via designated DCI formats 3 and 3A. DCI 3 and 3A are differential power control commands for PUCCH and PUSCH transmission in steps of decibels. DCI 3 is a 2-bit assignment as opposed to DCI 3A, which is a single-bit command. These dedicated DCIs with TPC commands are only used when there is no data to be transmitted to the UE; otherwise, the TPC command is transmitted embedded in other control information on the PDCCH for this UE. An initial 3-bit TPC command is embedded in the RAR message, see Section 1.8.16. The different TPC commands are listed in Tables 1.13 and 1.14.

UEs report their received channel quality to the eNB by transmitting a CQI value (see also Section 1.8.7). The CQI value represents either a wideband receive quality as a scalar or a more detailed report about frequency sections (sub-bands) as a vector. CQI reports are transmitted periodically or aperiodically configured by higher layers. Sub-band CQI reports indicate the receive quality of each sub-band relative to the wideband average with four steps: worse, equal, better, and much better. The UE

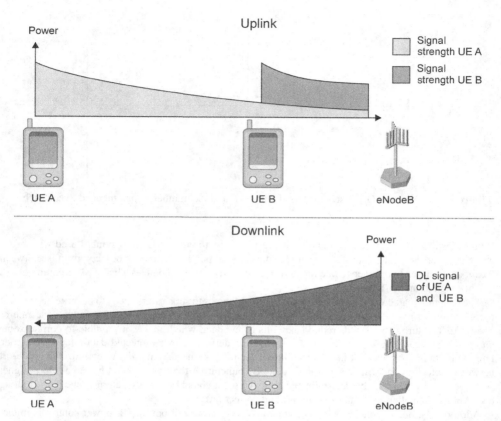

Figure 1.55   Near–far effect occurring in uplink direction, compared to equal signal strength reception in downlink

Table 1.13   Mapping of TPC command field in DCI format 1A/1B/1D/1/2A/2/3 to δ_{PUCCH} values

TPC command field in DCI format 1A/1B/1D/1/2A/2/3	δ_{PUCCH} (dB)
0	−1
1	0
2	1
3	3

Source: Reproduced with permission from © 3GPP™.

Table 1.14 Mapping of TPC command field in DCI format 3A to δ_{PUCCH} values

TPC command field in DCI format 3A	$\delta_{PUCCH}(dB)$
0	−1
1	1

Source: Reproduced with permission from © 3GPP™.

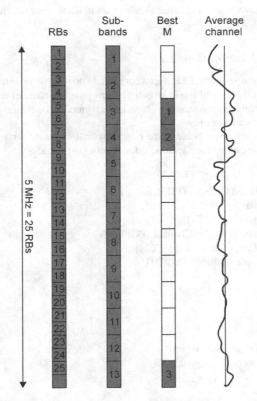

Figure 1.56 CQI illustration with sub-bands and best M reporting. (*Source*: Reproduced with permission from Nomor.)

reports the best M sub-bands compared to the average channel quality with the best M method as depicted in Figure 1.56.

The UE should derive an MCS scheme from the aforementioned measurement information, which is indexing 1 MCS out of 16 to suit the target BLER of below 10%. This enables differentiation between high- and low-cost handsets, which use more or less expensive RF hardware and/or a more sophisticated IQ signal processing engine.

Recapitulating, LTE link adaptation uses various stacked link adaptation techniques, securing transmission, or, in order to make it more effective, using different control loop delays regarding the process's
dimension. The following items summarize the LTE link adaptation functions with their responsiveness:

- **Adaptive frequency-selective scheduling**: Assign frequency resource to UEs on a 1 ms basis,
  which provides each UE with the individual best reception quality.
- **AMC**: Obtain the most efficient modulation and FEC code rate in order to balance retransmissions versus maximization of throughput.
- **HARQ**: Multiple retransmission process using prior transmission to increase the correct decoding probability.
- **TPC**: Provides UL power control in order to minimize multiple user interference.

1.8.12   Physical Channels in LTE

As indicated in the previous sections, LTE uses a completely new transmission scheme and physical layer
design compared to WCDMA, resulting in an entirely new physical channel layout. Main user data, both
control plane and user plane, is transmitted via shared channel architecture. Physical control channels
control physical layer procedures only.

Physical DL channels and signals are listed next. Physical channels carry user-relevant information
and physical signals do not carry user-specific information, just the SSS.

Physical DL channels:

- Physical Downlink Shared Channel (PDSCH)
- Physical Broadcast Channel (PBCH)
- Physical Multicast Channel (PMCH)
- Physical Control Format Indicator Channel (PCFICH)
- Physical Downlink Control Channel (PDCCH)
- Physical HARQ Indicator Channel (PHICH)

DL signals:

- Reference signals
- Synchronization signals:
  - Primary Synchronization Signal (PSS)
  - Secondary Synchronization Signal (SSS)

1.8.12.1   PDSCH

As LTE relies fundamentally on shared channel architecture, the PDSCH carries all data in the DL
direction to the UEs within a cell. RBs with their REs are modulated with user data (MAC PDUs,
Packet Data Units) as described in the previous sections. The PDSCH bears both control plane (in-band
signaling) and user plane data, which is multiplexed on logical channels on the DL-SCH as described
in Section 1.8.14. Data is transmitted using adaptive modulation schemes of QPSK, 16QAM, or
64QAM.

UE-specific RRC messages are transmitted with the PDSCH together with user plane packets and also with paging and system broadcast information (SIBs).

The different Transmission Modes (TMs), for example, multiple antenna transmission or single antenna transmission, apply for the PDSCH only. Different TMs can be used dynamically between UEs signaled within the rrcConnectionSetup message.

Defined TM:

- **TM 1**: Single antenna transmission.
- **TM 2**: Transmit diversity.
- **TM 3**: Open-loop spatial multiplexing (MIMO) using Cyclic Delay Diversity (CDD). The signal is transmitted on another antenna with a cyclic delay, in order to increase the multipath reception, which leads to a MIMO gain.
- **TM 4**: Closed-loop spatial multiplexing (MIMO).
- **TM 5**: Multi-user MIMO (or "virtual" MIMO).
- **TM 6**: Closed-loop rank 1 precoding.
- **TM 7**: Beamforming (Space Division Multiple Access (SDMA)) with UE-specific reference symbols.

Each transport block transmitted on the PDSCH is transmitted with a 24-bit checksum for error detection.

1.8.12.2   PBCH

The PBCH carries the Master Information Block (MIB), which is transmitted logically in an interval of 40 ms. FEC is protecting the MIB in such a way that four equal-sized blocks are derived. It is possible to decode each block individually and retrieve all the information in the MIB. Each block is transmitted in the PBCH resources in every radio frame (see Figure 1.57); thus, a MIB is decodable every 10 ms.

Information being transmitted is mainly the system bandwidth in RBs. An example of a PBCH for 5 MHz bandwidth with 25 RBs is shown in Figure 1.58.

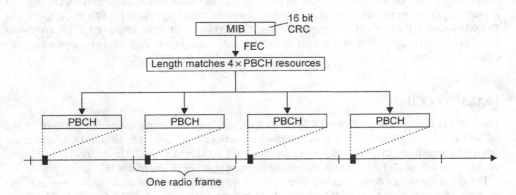

Figure 1.57   Master information block encoding and mapping to four DL radio frames

```
***B3***  LTE-RLC/MAC  (MAC-TM-PDU (DL))                    3GPP LTE-RLC/MAC Rel.8 (MAC TS 36.321 U8.5.0, 2009
***B3***  1 MAC PDU (Transparent Content Downlink)
***B3***  MAC Transparent Data                              49 80 00
***B3***  LTE-RRC_BCCH_BCH  (bCCH-BCH-Message)              RRC (BCCH to BCH)  3GPP TS 36.331 U8.5.0 (2009-03)
***B3***  bCCH-BCH-Message
***B3***  1 bCCH-BCH-Message
***B3***  1.1 message
010-----  1.1.1 dl-Bandwidth                                n25
---010--  1.1.2 phich-Config
---0----  1.1.2.1 phich-Duration                            normal
----10--  1.1.2.2 phich-Resource                            one
***b8***  1.1.3 systemFrameNumber                           '60'H
**b10***  1.1.4 spare                                       '0000000000'B
```

Figure 1.58   Example of a MIB (PBCH) broadcast channel message

As illustrated in Figures 1.60 and 1.61, the PBCH is located in each first subframe of every radio frame. It allocates the six center RBs (72 REs) regardless of the system bandwidth of the first four OFDM symbols in the second slot.

1.8.12.3   PMCH

Multicast transmission (e.g., MBMS) is a feature of later LTE releases, although it is necessary to define all physical channels already at this stage, in order to define a backward-compatible physical layer. The PMCH is transmitted on the fourth antenna port only with designated reference symbols. This channel is basically similar to the PDSCH, but designed for a multicell reception.

1.8.12.4   PCFICH

The PCFICH is transmitted in the first OFDM symbol of each subframe and indicates the number of OFDM symbols used for the PDCCH. Control format indicator values run from 1 to 4, but only a maximum of three OFDM symbols is defined by the first release of LTE, reserving value 4 for future use. Because it is needed to correctly decode the PCFICH, the information carried is very well protected. Values 1–4 are encoded with a 32-bit pattern and modulated with QPSK to 16 subcarrier (REs), as 2 bits can be mapped to a QPSK symbol. Those 16 REs are mapped to four REGs. The REGs are not adjacent in order to gain frequency diversity. Figure 1.59 shows an example mapping of the PDCFICH to the first OFDM symbol of the PDCCH. The position of the PCFICH information is shifted and scrambled by the physical c-ID, in order to protect it against inter-cell interference. Figure 1.59 shows examples for three different c-ID ranges.

1.8.12.5   PDCCH

The PDCCH is needed to transport DCIs. DCIs carry information on scheduling assignments for the DL and scheduling grants for the UL as described in Sections 1.8.7 and 1.8.9. Various DCI formats are defined for scheduling tasks and DCI 3 and 3A (Section 1.8.11) for UL TPC. This channel is mapped to the first one to three OFDM symbols of each subframe as signaled by the PCFICH.

Figure 1.60 depicts a detailed example mapping of the PDCCH in which each square represents one RE (one subcarrier within an OFDM symbol). Each column represents one OFDM symbol with respect to time (horizontal) and frequency (vertical). White areas (REs) are used for the user data transmission (PDSCH). Black REs are fixed and allocated for DL reference signals (see Figure 1.44 for detailed illustrations of antenna mappings in the case of multiple antenna ports).

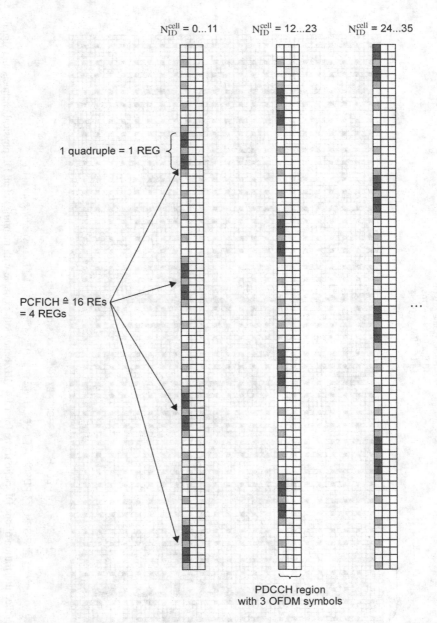

Figure 1.59 Mapping examples of PCFICH into resource element groups (always within the first OFDM symbol of a subframe) within the PDCCH area. (*Source*: Reproduced with permission from Nomor.)

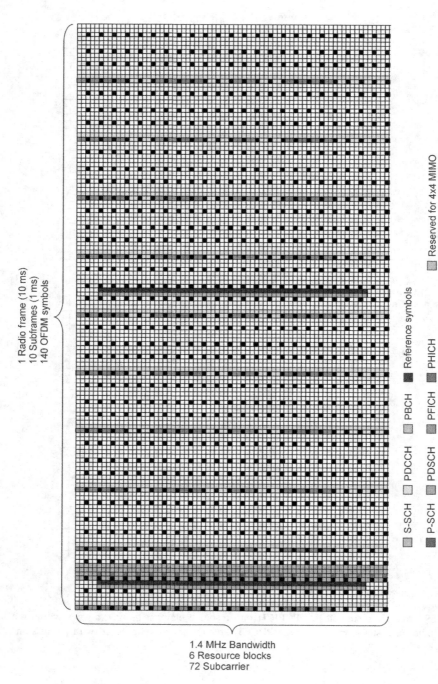

1 Radio frame (10 ms)
10 Subframes (1 ms)
140 OFDM symbols

1.4 MHz Bandwidth
6 Resource blocks
72 Subcarrier

S-SCH PDCCH PBCH Reference symbols

P-SCH PDSCH PFICH PHICH Reserved for 4x4 MIMO

Figure 1.60 Detailed illustration of one DL radio frame for a 1.4 MHz bandwidth example with mappings of all DL physical channels and signals. (*Source*: Reproduced with permission from Nomor.)

Table 1.15 Supported PDCCH formats with different FEC protection

PDCCH format	Number of CCEs	Number of resource element groups	Number of PDCCH bits
0	1	9	72
1	2	18	144
2	4	36	288
3	8	72	576

Source: Reproduced with permission from © 3GPP™.

The PDCCH is transmitted in the DwPTS subframe with TDD.

In Sections 1.8.7 and 1.8.9, example PDCCH messages are described for DL scheduling and UL scheduling assignments, respectively.

UE-specific DCI is addressed to UEs implicitly by scrambling the CRC to save signaling resources. The DCI can be located in various positions; thus, UEs have to blind-decode and "search" for their designated information. Twelve common search spaces and 32 UE-specific search spaces are defined, resulting in a worst case of 44 decoding attempts.

Different PDCCH formats are specified, which use different FEC protection code rates in order to achieve higher protection, for example, for cell edge users with poor reception conditions (see Table 1.15).

1.8.12.6 PHICH

The PHICH carries HARQ feedback information in the DL to the UEs. In other words, the ACK or NACK of a previous UL transmission is signaled to the sending UE via the PHICH. This feedback information (1 bit) is repeated three times and each triple is orthogonally Walsh spread to four complex symbols. This orthogonal sequence carrying the HARQ feedback information is then robustly modulated using Binary Phase Shift Keying (BPSK) on REGs within the PDCCH. REGs used by the PHICH are shown in Figure 1.60.

1.8.12.7 Reference Signals

Reference signals are used for estimating the wireless channel leading to distortion of the received signals as described in Section 1.8. The LTE DL uses a two-dimensional spherical channel estimation procedure on frequency and time. Reference symbols are interleaved in frequency and time as can be seen in Figure 1.44, which shows the DL reference signal grid for each transmit antenna port. Furthermore, reference signals are depicted in Figure 1.60.

1.8.12.8 PSS

The PSS is a basic signal for synchronizing an LTE cell. It is primarily used for cell search and time slot synchronization. Zadoff–Chu sequences are used as PSS signals, which are sequences with an amplitude of one and varying phase shifts. Three types of PSS signals are defined, which indicate the physical c-ID group. Figure 1.62 shows an example PSS and SSS signal as a screenshot of an IQ sample analysis

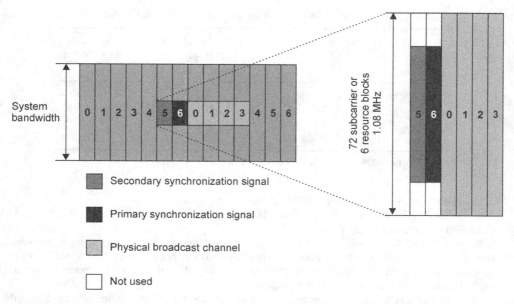

Figure 1.61 Illustration of the location of the physical broadcast channel and the primary and secondary synchronization signals within the first subframe of each radio frame. (*Source*: Reproduced with permission from Nomor.)

dashboard. Note that the gray plane shows the IQ layer. In the third dimension, the occurrence of a specific IQ combination (specific IQ vector) is depicted.

1.8.12.9 SSS

The SSS already carries the first information modulated with BPSK. After synchronizing with the SSS, the mobile has not only retrieved slot and radio frame timing, but also detected the duplex method (TDD or FDD), CP length, and the 9-bit physical layer c-ID. Figures 1.61 and 1.62 show the location of the synchronization signals within a radio frame and a subframe, respectively (Figure 1.61 depicts the first subframe of a radio frame with its 14 OFDM symbols in the case of FDD and normal CP). Furthermore, note that both synchronization signals use the centered six RBs of the spectrum, but only use 62 of the 72 REs in order to gain more robustness with this spectral guard gap.

Table 1.16 lists the applied modulation schemes for each DL channel. Physical UL channels:

- Physical Uplink Shared Channel (PUSCH)
- Physical Uplink Control Channel (PUCCH)
- Physical Random Access Channel (PRACH)

    UL signals:

- Reference signal:
  - Demodulation Reference Signal (DMRS)
  - Sounding Reference Signal (SRS)

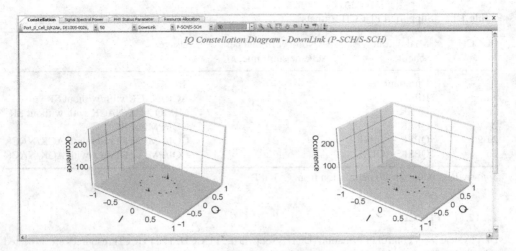

Figure 1.62 Vector IQ analysis of a primary and secondary synchronization signal on two transmit antennas

Table 1.16 Physical downlink channels and their modulation schemes

Physical downlink channel	Modulation schemes
PDSCH	QPSK, 16QAM, 64QAM
PDCCH	QPSK
PMCH	QPSK, 16QAM, 64QAM
PHICH	BPSK
PCFICH	QPSK
PBCH	QPSK

1.8.12.10 PUSCH

The PUSCH is the physical channel carrying the transport channel UL-SCH. Therefore, UL data transmission is shared channel based, similar to the DL-SCH approach. The eNB assigns UL-SCH resources through UL grants transmitted via the PDCCH, see Section 1.8.9.

All UL higher layer control plane (in-band signaling) and user plane information is sent via the PUSCH using the modulation schemes of QPSK, 16QAM, or 64QAM. UL 64QAM is UE and system (signaled cell-wide in an SIB) optional. PUSCH transport blocks are transmitted with a 24-bit CRC checksum.

Physical layer control information is transmitted via the PUSCH as well, in the case of parallel PUSCH and UL control information transmission.

There is no MIMO defined in the UL direction; therefore, the aforementioned transmission modes do not apply for PUSCH. The eNB has the option of selecting a transmit antenna port of the UE for the best reception.

Table 1.17 Supported PUCCH formats

PUCCH	Modulation scheme	Number of bits per scheme subframe, M_{bit}	Usage
1	Unmodulated	N/A	SR
1a	BPSK	1	ACK/NACK with/without SR
1b	QPSK	2	MIMO ACK/NACK with/without SR
2	QPSK	20	CQI/PMI or RI
2a	QPSK + BPSK	21	CQI/PMI or RI mux with ACK/NACK
2b	QPSK + QPSK	22	CQI/PMI or RI mux with ACK/NACK

Source: Reproduced with permission from © 3GPP™.

1.8.12.11 PUCCH

UL physical layer control information (UCI) is sent via the PUCCH. The PUCCH resources are allocated at both system bandwidth edges in order to decrease interference and gain frequency diversity. UL control information is:

- **HARQ feedback**: Inner loop retransmission/reception ACK/NACK feedback.
- **Precoding Matrix Indicator (PMI)**: Used with multiple antenna transmission.
- **SR**: Used to request UL resources.
- **CQI**: DL reception quality feedback.
- **Rank Indicator (RI)**: Used with multiple antenna transmission.

Figure 1.52 depicts an example PUCCH configuration in the middle of the figure. The most important information is transmitted adjacent to the demodulation reference symbols as the channel estimation here is most accurate.

Table 1.17 lists the defined PUCCH UCI formats with their allowed modulation schemes, number of bits per subframe, and the usage of each UL control format.

1.8.12.12 PRACH

The frequency and time location of the six consecutive RBs used for PRACH are cell-wide signaled using the SIB type 2. Zadoff–Chu sequences, forming the preamble, are transmitted after a CP as random access inquiry. Figure 1.63 shows the random access preamble format. PRACH transmissions are not UL synchronized as no UL synchronization is established yet. There are 64 preambles used within one cell, which are generated from a RACH preamble root sequence indicated in SIB type 1.

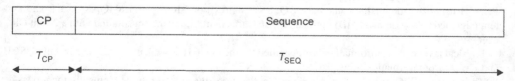

Figure 1.63 Random access preamble format (TS36.211). (*Source*: Reproduced with permission from © 3GPP™.)

Table 1.18 Physical uplink channels with their modulation schemes

Physical uplink channel	Modulation schemes
PUSCH	QPSK, 16QAM, 64QAM
PUCCH	QPSK
PRACH	Zadoff–Chu sequences

There are five random access formats defined, specifying different CP and preamble length. Four formats are used with frame type 1, which is used for FDD and the fifth format is used with frame type 2 for TDD. Diverse formats are used for different cell sizes, load, and interference conditions.

Table 1.18 lists the physical UL channels and maps the modulation schemes that are allowed for use.

1.8.12.13 DMRS

The Demodulation Reference Signals (DMRSs) are defined symbols for UL channel estimation of data transmitted on the PUSCH and PUCCH. The DMRSs are located in the middle of each slot, thus on the fourth OFDM symbol. UL reference signals do not carry any information.

1.8.12.14 SRS

The Sounding Reference Signal (SRS) is a signal transmitted upon request by the eNB. It is used for UE-specific UL channel quality also on frequencies that were not used for transmission previously. The eNB requests the UEs to transmit an SRS symbol on parts of the bandwidth or on the complete system bandwidth. SRS transmission takes place always on the last OFDM symbol of the subframe, if it is requested.

1.8.13 Transport Channels in LTE

The physical layer provides a transport service for MAC PDUs. This service is accessed by transport channels. Most transport channels are directly mapped to physical channels as described in Section 1.8.14. Thus, in other words, transport channels are the gateway to physical channels and a selection for MAC PDUs where they are to be transmitted.

The following DL transport channel types are defined:

- **Broadcast Channel (BCH):**
  - Uses a static transport format and has the requirement that all UEs within the cell have to receive its information error-free. The reception of the BCH is mandatory for accessing any service of a cell.
- **DL-SCH:**
  - Carries all semi-static broadcast information (SIB) and all UE-specific traffic channels.
  - DL-SCH is secured with HARQ algorithms.
  - Efficiency is realized with AMC link adaptation.
  - Various TMs are defined to meet different environment scenarios to increase efficiency in respect of current conditions.
  - DRX is available in order to increase handset operating time.
  - Makes use of spatial algorithms like beamforming or MIMO.

- **Paging Channel (PCH)**:
  - – Needs to be received in complete cell coverage area.
  - – Supports DRX in order to increase battery operating cycle.
  - – Dynamically allocated via own physical identifier (P-RNTI).
- **Multicast Channel (MCH)**:
  - – Broadcast to entire cell coverage area.
  - – MBMS transmission with use of multiple cells.

A designated DL control channel is not defined as the PDCCH is used for physical channel control only. The higher layer control plane is transmitted via the DL-SCH.

Defined UL transport channel types are shown in the following items:

- **UL-SCH**:
  - – UL-SCH is secured with HARQ algorithms.
  - – Fully dynamic and semi-static resource allocation schemes.
  - – Can make use of multi-user MIMO (UL "virtual" MIMO).
  - – Uses dynamic link adaptation like AMC.
- **RACH**:
  - – Accessible without UL synchronization.
  - – Collision-based and collision-free operating modes.
  - – Various modes depending on cell size and interference.

As in the DL direction, no UL control transport channel is defined as the higher layer control plane is transmitted on the UL-SCH. The PUCCH is a control channel used by the physical layer only.

1.8.14   Channel Mapping and Multiplexing

Besides physical and transport channels, LTE also defines logical channels. Logical channels are multiplexed to transport channels within the MAC layer. Logical channels map different content connections to transport channels, like CCCHs to multiple UEs or DCCHs to a specific UE or dedicated transport channels carrying higher layer application data.

Logical channels are addressed with a logical channel ID. The logical ID is a field within the MAC header PDU. Logical channels are multiplexed by using logical channel IDs to transport channels specifying where the information should be transmitted. Finally, transport channels are transferred with physical channels as a service provided by the physical channel.

Figures 1.64 and 1.65 show the above-described channel architecture from the basic physical channels via transport channels to logical channels bearing higher layer messages for DL and UL, respectively.

Two basic sets of logical channels are defined:

- **Control channels**: CCCH and DCCHs
- **Traffic channels**: CCCH and DTCHs.

The nature of common channels is such that no specific UE is addressed, but the information is either general for all cell-wide subscribers or a message from a UE that has not yet established a dedicated control/traffic channel. A typical example of a CCCH is the broadcast of SIBs.

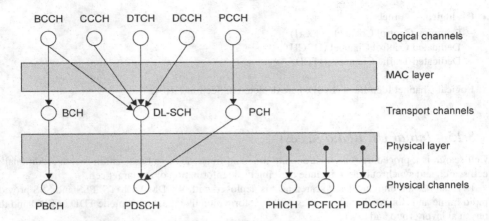

Figure 1.64   Downlink channel mapping and multiplexing from logical channels via transport channels to physical channels

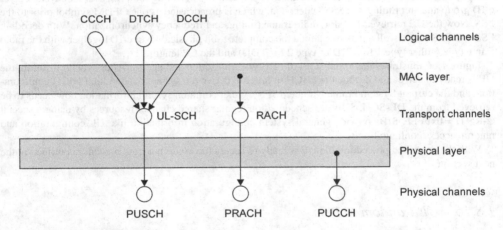

Figure 1.65   Uplink channel mapping and multiplexing from logical channels via transport channels to physical channels

Traffic channels carry user plane protocols like the Packet Data Convergence Protocol (PDCP) and application IP packets, while control channels carry control plane protocols such as RRC and NAS.

- **DL logical channels**:
    - Broadcast Control Channel (BCCH)
    - Paging Control Channel (PCCH)
    - Common Control Channel (CCCH)
    - Dedicated Control Channel (DCCH)
    - Dedicated Traffic Channel (DTCH)

- **UL logical channels**:
  - Common Control Channel (CCCH)
  - Dedicated Control Channel (DCCH)
  - Dedicated Traffic Channel (DTCH)

Logical Channel Identifiers (LCIDs) are depicted in Section 1.10.4.

1.8.15   Initial UE Radio Access

Cell search is a procedure for synchronizing time and frequency to a base station sector. Additionally, cell search and synchronization include deriving basic information of the target cell.

LTE defines a hierarchical cell search as it is deployed with WCDMA UMTS. PSS and SSS provide radio frame and slot synchronization, as well as information like the duplex mode TDD or FDD and the physical layer group and c-ID.

UEs synchronizing to a new LTE cell start searching for a PSS, which is a Zadoff–Chu sequence. Three sequences are defined indicating the physical cell group ID. Three physical layer c-ID groups with 168 physical layer c-IDs each are defined. After successfully detecting the PSS with its physical layer c-ID group and slot timing, the SSS is decoded, which is broadcasted as one OFDM symbol prior to the PSS. Now the UE retrieved DL slot, radio frame timing, and frequency synchronization. With decoding PSS and SSS successfully, it also obtained the complete 9-bit physical layer c-ID together with the radio frame type (either type 1 for FDD or type 2 for TDD) and the CP length.

Figure 1.66 illustrates the initial steps of cell synchronization and access. After synchronization, the UE is ready to detect and decode the PBCH in order to derive the system bandwidth, PHICH configuration, and the current System Frame Number (SFN). Other common system information now needs to be retrieved from the DL-SCH. SIBs are scheduled on regular shared channel resources by using a special C-RNTI = 0xFFFF. SIBs provide general system configuration information like UL configuration and random access configuration.

With the described procedure, the UE is ready to access the random access procedure outlined in the next section.

1.8.16   UE Random Access

A UE random access is performed not just with the UE initial access to the network, but for various other reasons. Especially in LTE, the random access procedure is a central UE procedure as the physical transport channel architecture is shared channel based and only two RRC states are defined: RRC_CONNECTED and RRC_IDLE. FDD and TDD define a common random access procedure. The PRACH is designed in such a way that the same routine and format is used for all cell sizes (long propagation delay and long channel echo spread) and just some configuration parameters can be changed in order to cope with various cell conditions, as described in Section 1.8.12.

The following items outline the events in which a random access procedure should be made:

- At UE initial access.
- For RRC connection re-establishment, for example, when radio link failures occur.
- When handovers are triggered, usually a contention-free RACH procedure is configured.
- In case of no UL synchronization and arrival of DL data.
- In case of no UL synchronization and UL data is to be transmitted.
- In case UL data is to be transmitted but no PUCCH resources are available to indicate an SR.

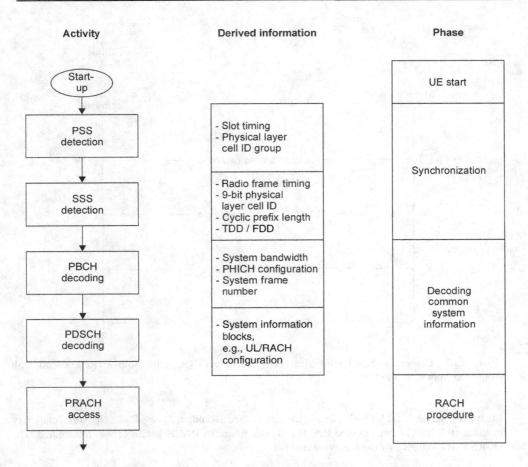

Figure 1.66   Initial cell access with level of retrieved information

Two random access procedures are defined to distinguish an eNB-initiated RACH procedure against a completely independent RACH access:

- **Non-contention based**: The eNB assignment of used RACH resources in cases of handover and DL data arrival. Collisions should not occur.
- **Contention based**: In previously fully unknown RACH resources because of random selection by UEs accessing a cell, for example, initial access or radio link failure. Collisions might occur.

The nature of the PRACH is described in Section 1.8.12. The contention-based RACH procedure is described in the following.

Figure 1.67 illustrates the contention-based RACH procedure with its four messages. A UE selects a random access preamble and transmits it in the time and frequency resources of the PRACH with initial RACH power settings derived from SIBs. This is depicted in Figure 1.67 as message 1.

Now the UE starts to monitor the PDCCH for a DL response from the eNB. This RAR is transmitted in the timewise RAR window (message 2). Within this window, the UE originating the RACH preamble

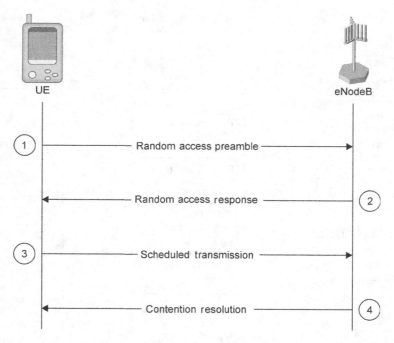

Figure 1.67   Contention-based random access procedure (TS36.300). (*Source*: Reproduced with permission from © 3GPP™.)

listens to the scheduled RAR with a special identifier called a Random Access Radio Network Temporary Identifier (RA-RNTI) matching the RACH time slot where its RACH preamble was transmitted. The RA-RNTI is calculated by the following equation:

```
RA-RNTI = 1 + t_id + 10*f_id
```

where t_id addresses the first subframe of the used PRACH, resulting in a possible range between 0 and 9, and f_id is the ascending frequency index within this subframe with a range between 0 and 5. Thus, the RA-RNTI has a range

```
1 ≤ RA-RNTI ≤ 60
```

in decimal or, in 16-bit hexadecimal,

```
0x0001 ≤ RA-RNTI ≤ 0x003c
```

All received preambles within one PRACH are answered by one MAC PDU containing multiple RAR messages addressing random access identifiers matching the preamble index used by a specific UE. Each preamble indicates a single UE random access attempt, which is addressed and answered by one MAC RAR as depicted in Figures 1.68 and 1.69.

```
***B7***  LTE-RLC/MAC  (MAC-RAR (DL))                     3GPP LTE-RLC/MAC Rel.8 (MAC TS 36.321
***B7***  1 MAC Random Access Response
01101110  1.1 MAC RAR Header Part
01101110  1.1.1 MAC Random Access ID Subheader
0-------  Extension field                                 Last MAC PDU subheader
-1------  Type Field                                      Random Access Identity
--101110  Random Access Identity                          46
***B6***  1.2 MAC RAR Payload Part
***B6***  1.2.1 MAC Random Access Response
0-------  Reserved                                        0
**b11***  Timing Advance Value                            2
**b20***  UpLink Grant Resources                          '00000001010010001100'B
***B2***  Temporary C-RNTI Value                          'a277'H
```

Figure 1.68   Example RAR message

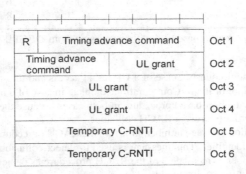

Figure 1.69   A 6-byte MAC RAR (TS36.321). (*Source*: Reproduced with permission from © 3GPP™.)

A new preamble with additional power is transmitted if there is no RAR message received within the RAR window by a UE that has sent a preamble. This random access power ramping procedure is repeated until a RAR message for its preamble is received or the maximum allowed transmit power is reached.

An initial timing advance command is calculated by the eNB in order to synchronize all UL traffic to the orthogonal UL slot structure. A first physical layer address, the temp C-RNTI, is assigned to the UE with the RAR message.

Furthermore, the RAR message carries a 20-bit UL grant field containing the following information:

- Hopping flag – 1 bit.
- Fixed-size UL resource block assignment for the first scheduled UL transmission in the PUSCH – 10 bits.
- Truncated MCS – 4 bits.
- TPC command for the scheduled UL transmission – 3 bits.
- UL delay – 1 bit.
- CQI request – 1 bit.

Other than TPC commands within an established connection (1-bit or 2-bit TPC commands), the TPC command within the MAC RAR message consists of 3 bits addressing the power adjustment range, as in Table 1.19.

The first scheduled UL transmission on the PUSCH (RACH contention-based procedure message 3) takes place after decoding the correct RAR message. This RACH procedure message number 3 is, for

Table 1.19   TPC command δ_{msg2} in RAR for scheduled PUSCH UL grant

TPC command	Value (in dB)
0	−6
1	−4
2	−2
3	0
4	2
5	4
6	6
7	8

Source: Reproduced with permission from © 3GPP™.

example, the rrcConnectionRequest in case of a UE initial access and the rrcReestablishment after, for example, a radio link failure or the rrcConfigurationComplete in case of a handover (HO). It is mandatory to use the RACH contention-based procedure in both cases because the eNB cannot assign RACH resources previously in order to prevent RACH collisions.

The contention resolution message number 4 mirrors the RACH procedure message number 3, which includes at least a NAS identifier. This message is used to detect collusions as each UE compares its transmitted message number 2 (message 3 of the RACH procedure as depicted in Figure 1.67) with the received contention resolution payload field.

A non-contention-based RACH procedure is defined as illustrated in Figure 1.70. This procedure is used in cases of handover and DL data arrival in the UL non-synchronized state.

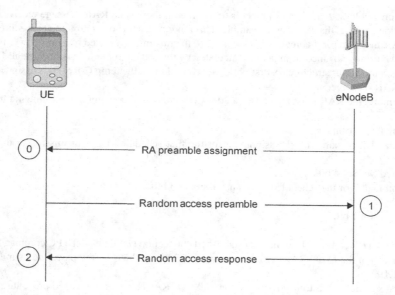

Figure 1.70   Non-contention-based random access procedure. (*Source*: Reproduced with permission from © 3GPP™.)

Message number 0 assigns a random access preamble to the UE, which is not broadcast for the contention-based RACH procedure. This is done on the source eNB triggered by a handover command via the X2 interface from the target eNB in the case of handover. It is signaled via the PDCCH on DL data arrival.

The assigned random access preamble is used within the PRACH resources of the target eNB or on PRACH in case of DL data arrival.

Message number 2 is the RAR message similar to the contention-based procedure but a Temp C-RNTI assignment is not mandatory as the UE might already have an RNTI assigned.

Regular UL and DL transmission can take place after the random access procedure as all mandatory parameters are known by both sides and UL synchronization is established.

1.9   Hybrid ARQ

By physics, wireless transmission of information always influences the wireless channel, leading to distortion of the transmitted information. The wireless channel is discussed in Section 1.8. This distortion, along with the receiver's hardware noise, results in wireless transmission errors of the transmitted information.

It is required to protect the information against errors due to the fact that false symbol detections are injected during transmission and reception of bits between sender and receiver. The first essential function is an entity to detect errors. The receiver is using a method called *Cyclic Redundancy Check* (CRC) to evaluate if a received transport block is correctly or incorrectly received.

A simple scheme to increase robustness and protect the transmission is to request a retransmission of the erroneous transport block. This is done by signaling this feedback information back to the transmitter by using an out-of-band feedback channel. The receiver sends an *Acknowledgement* (ACK) or a *Not Acknowledgement* (NACK) back to the transmitter to declare the correct or the incorrect reception of the transport block. The ACK or NACK feedback information is used by the transmitter to initiate a retransmission or a new transmission. Such basic procedure is known as Stop-and-Wait ARQ protocol.

Basically, three versions of ARQ procedures are described in literature:

- The simple *Stop-and-Wait* where the transmitter waits for feedback for each single packet before the next new transmission is initiated or a retransmission of the previous packet is conducted. No sequence numbers or buffering is required; the feedback includes only ACK or NACK.
- The *Go-back-N* scheme, which continues with transmissions even though no feedback, is yet received from the receiver. This scheme can be used to increase the throughput in cases of large propagation and processing delays between transmitter and receiver. Sequence numbers need to be introduced to signal ACK or NACK for backs received in the past. The transmitter will stop transmitting in case of a received NACK or time out for a packet, which was not ACK'd and will retransmit this packet and all other packets transmitted previously. In other words, the transmitter will jump back N sequence numbers and will continue transmission packets from this NACK'd sequence number. The Go-back-N scheme requires medium resource effort as sequence numbers, multiple timers and buffers on the transmitter are required.
- The Selective Repeat scheme, which is similar to the Go-back-N scheme, but it does not jump back N instances in case of a NACK and retransmits all packets already transmitted after the received NACK and only selectively repeats the NACK'd packets. This scheme additionally to the Go-back-N-type resource requirements demands a receive buffer for in-sequence delivery to high entities.

LTE uses two stacked processes for error correction. One is the outer-loop ARQ process embedded in the RLC layer. A faster inner-loop Hybrid ARQ process (described next) in the lower MAC layer close to PHY layer allows fast retransmissions. The outer-loop RLC ARQ process shall eliminate errors that could not be corrected by the inner-loop MAC Hybrid ARQ process. The subsequently mentioned ARQ entities are always meant to be understood as part of the lower loop Hybrid ARQ and not the outer-loop RLC ARQ process.

Standard ARQ procedures as indicated earlier are effective for transmissions with few error occurrences and can achieve in such scenarios good throughput results as the overhead for each transmission is very low. No additional redundancy is added to the transport block payload for error correction at the receiver side. The effort for implementation of both sides is low. On the other hand, the throughput decreases dramatically in transmission environments with high failure probabilities.

An error correction scheme at the receiver side is known as *Forward Error Correction* (FEC). FEC and source coding algorithms are discussed in various publications and are not scope of this book. In case FEC is used, the transmitter will apply source coding and transmit not only the original data but add redundancy to the original information as error protection. The redundancy is in literature referred to as parity (parity bits). The parity information is added to the source information symbols leading to a higher amount of symbols to be transmitted.

The relation between source data information and total amount of data after FEC source coding is called *code rate*. A code rate that equals one means no redundancy (no error protection) is added to the information; in other words, only net information is transmitted. A higher divisor (smaller quotient ratio) means more redundancy for robustness against transmission errors, but also a higher amount of overhead to be transmitted. The higher robustness and capability to correct errors is bought by less net information throughput.

A combination of both ARQ and FEC is known as *Hybrid ARQ* (HARQ). HARQ requires a close interworking of an ARQ entity and a FEC entity. The interworking functionality is required for both the transmitter and the receiver. It is a common scheme of most modern wireless standards such as HSxPA in 3G or WiMAX and LTE. HARQ dramatically increases reliability and performance of the overall system.

This interworking is essential as for each retransmission state (number of retransmission), a designated redundancy version of the information to be transmitted has to be generated. HARQ makes use of time diversity and increases the correct decoding probability by using different redundancy (code rates) for the first transmission and each following retransmission.

At the transmitter side, the ARQ entity informs the FEC entity about each retransmit state in order to acquire a new transport block with the appropriate code rate or parity bits for the current retransmission. In conjunction, the receiver ARQ entity has to provide the retransmission state to the FEC that the transport block can be decoded with the applying FEC settings. There are three different types of HARQ classifications known, which require a different amount of interworking between FEC and HARQ.

The first type transmits and retransmits the same data with the same FEC protection. The interaction between FEC and the lower ARQ process in this case is very little. All previous transmissions of the same transport block are discarded for decoding; thus, these do not make use of the valuable information of the first transmissions as the previously received information is not 100% waste even though it could not be completely decoded. It is much more effective to also consider previously received transport blocks in order to increase the decoding probability with each retransmission. The first type is barely used for the reason of wasting already received information. The next two types of HARQ are usually meant if one is talking about HARQ.

The second HARQ type is also known as *incremental redundancy* scheme. The FEC is transmitting the payload information with usually a low amount of parity bits for error protection. The packet is NACK'd if the first transmission could not be decoded correctly by the receiver and is stored in a buffer to be used with the decoding attempting at the upcoming retransmissions.

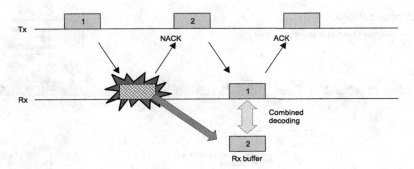

Figure 1.71   HARQ principle with reception error, NACK feedback, retransmission with successful combined decoding, and ACK feedback

The retransmissions then carry only parity bits of the first transmission generated by the source FEC. The additional parity bits are generated by punctuation. With each retransmission, the redundancy is increased (lower FEC code rate), which increases the decoding probability at the receiver. The receiver has to provide fast buffers for each transmission in order to combine them with each decoding attempt (see Figure 1.71). A drawback of this methodology is the high importance of the first transmission for the overall decoding even with a high amount of retransmissions as the native payload data is only carried in the first transmission and only parity bits are added with each retransmission.

The third HARQ principle is known as *case* or *soft combining* and overcomes the drawback of the incremental redundancy scheme where the first transmission holds a central role as the following retransmissions only bear parity information. The initial transmission and all retransmissions are individually decodable and are not dependent on each other. Essentially, each transmission is equivalent, that is, the same set of data with the same FEC code rate is transmitted.

This scheme also makes use of time diversity and uses previous transmissions to increase the decoding probability with each successive retransmission. The retransmission time diversity gain is achieved with case or soft combining. By using case or soft combining, the digital sampled "analogue soft" values are stored for each symbol within each (re-)transmission before the sampled soft values are applied to the bit decider unit.

The intention to store the soft values before the bit decisions are made is to increase the correct bit decision probability. Various soft combining and decision algorithms are discussed in other literature, and the general concept is to make the bit decision not only with one set of input symbols but also with all received HARQ transmissions for one set of data. The correct bit decision is increased by utilizing more HARQ retransmissions with sampled soft values in order to eliminate incorrect or unlikely values by adding the timely diverse soft values for each symbol.

Both principles, Incremental Redundancy and Soft Combining, increase the transmission performance and robustness against transmission errors but require additional processing resources, a close interaction between HARQ and FEC, and fast buffer for retransmissions or even soft values for each symbol of the retransmissions.

The LTE standard allows both HARQ methodologies of Incremental Redundancy and Soft or Case Combining. The number of maximum HARQ retransmissions can be configured as UE-dependent based on the UE capabilities signal on RRC layer.

Retransmission timing and scheduling can be handled in two different ways, Synchronous and Asynchronous. Both schemes are discussed in the next two sections.

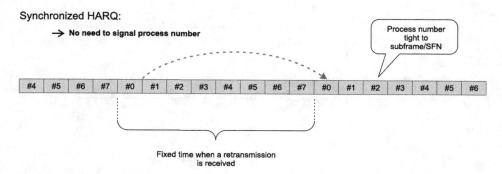

Figure 1.72   Synchronous HARQ with fixed time slots allocation to HARQ processes

1.9.1   Synchronous HARQ in LTE Uplink

LTE uses one uplink and one downlink HARQ entity, which has eight sub HARQ processes for each direction. Each HARQ process uses the simple stop and wait ARQ scheme. A pipelining effect across all eight HARQ processes is used in order to provide a continuous transmission for all transmission slots without wait stages.

A new HARQ process from the total of eight processes is utilized for each adjutant subframe in a round-robin pattern (see Figure 1.72). The assignment of which process is to be used has a fixed allocation to subframe numbers and system frame numbers. This way no additional signaling for the HARQ process number is required with the scheduling of UL transmissions as it is implicitly defined by the subframe that is used. This scheme is known as *Synchronous HARQ*.

The minimal duration between two transmissions of one HARQ process is 8 ms as the subframe duration is 1 ms and in between the residual 7 HARQ processes are served.

Figure 1.73 depicts the HARQ operation of multiple HARQ processes with a continuous transmission distributed over all eight HARQ processes. The HARQ feedback is sent after a fixed value of 4 ms. This gives the receiver 3 ms for packet propagation and to process the packet to generate a positive

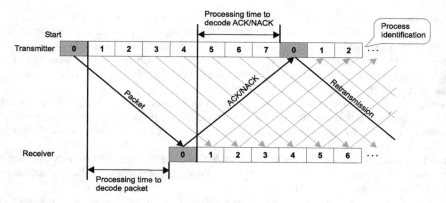

Figure 1.73   Synchronous HARQ transmission scheme with multiple parallel HARQ processes

acknowledge or a negative acknowledge. The ACK or NACK feedback is then transmitted four time slots (4 ms) after the payload packet started to be transmitted. The HARQ feedback is transmitted in downlink direction on a designated physical channel Physical Hybrid ARQ Channel (PHICH). Meanwhile, other HARQ processes have also received packets in parallel and process them individually.

Additional 4 ms later, the next packet for the current HARQ process is transmitted by the transmitter. Depending on the previous HARQ feedback, it will either transmit a new data or perform retransmission. Thus, the transmitter also has 3 ms to process the receiver's HARQ feedback and can generate the data to be either retransmitted in case a NACK is received or a new transmission in case of an ACK. The retransmission is either a new redundancy version (parity bits by punctuation) or a retransmission under the condition that case combining is used.

The synchronous HARQ scheme is defined in the LTE uplink direction. In LTE downlink, an asynchronous HARQ scheme is defined. The asynchronous HARQ downlink procedure is outlined in the next section.

1.9.2   Asynchronous HARQ in LTE Downlink

Similar to the uplink HARQ entity, the downlink HARQ entity utilizes multiple HARQ processes in order to guarantee a continuous transmission even when single HARQ processes are busy waiting for retransmissions or HARQ ACK/NACK feedback.

The downlink HARQ entity uses an *Asynchronous HARQ* scheme other than the uplink HARQ, which uses synchronous HARQ transmissions. The downlink asynchronous HARQ also schedules single transmissions to one of the eight HARQ processes. But other than in the synchronous uplink HARQ scheme, there is no fixed HARQ process pattern tied to the subframes. Each process number is signaled in with the scheduling information sent on the PDCCH.

Figure 1.74 depicts the asynchronous HARQ transmission scheme with multiple HARQ processes in downlink direction. The minimum response time is 4 ms; thus, the minimum delay for a repetitive transmission due to a missing HARQ feedback is 8 ms. The HARQ feedback is sent by the UEs either along with a data transmission on PUSCH or on the physical uplink control channel (PUCCH) if the UE is not scheduled 4 ms after the downlink transmission.

Figure 1.75 depicts the timings of a downlink HARQ transmission with a NACK and a retransmission. Each subframe has a duration of 1 ms on the horizontal dimension. The HARQ feedback is provided four subframes later, leaving 3 ms for signal propagation and time to decode the downlink transmission and

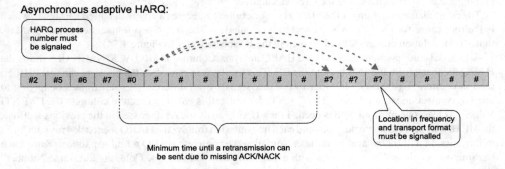

Figure 1.74   Asynchronous HARQ process with variable retransmission or next transmission slots

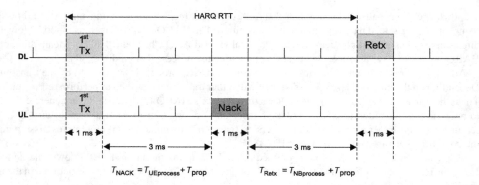

Figure 1.75   HARQ round-trip timing with NACK and retransmission

evaluate the CRC result (transmission error or correct reception). Same timing applies for the uplink reception: 3 ms for uplink propagation and eNB processing time. The downlink transmission will be scheduled again on PDCCH with the matching HARQ process number and can also be transmitted later (asynchronous principal).

The same timing scheme as depicted in Figure 1.75 also relates to the uplink HARQ timing.

The following section outlines some real-life HARQ process examples.

1.9.3   HARQ Example

This section illustrates two HARQ procedures with screenshots from an air interface monitoring probe. The air interface probe passively receives all uplink and downlink traffic in an LTE cell and decodes the full protocol stack from physical channels to RRC/NAS and user plane.

Figure 1.76 shows a screenshot of analysis software of an LTE Uu communication between an UE and an eNB. Each line in the upper frame of the window represents a transmission in uplink or downlink direction. The used physical channel is indicated in column "Last MSG". The lower frame of the screenshot shows the IQ constellation diagram of the highlighted uplink PUSCH transmission with frame ID 3655903. Column "CRC report for TB" shows the CRC result for each transport block, in case the highlighted PUSCH frame failed the CRC calculation.

The constellation diagram for the PUSCH in our example indicates a highly distorted RF signal, which is the root cause for the failed CRC checksum calculation. The frame was transmitted using QAM16; thus, 16 modulation clusters should be seen in the IQ sphere (refer to Figure 1.54).

Additional columns were selected for a HARQ analysis. Column "HARQ ACK/NACK UL" shows the HARQ feedback from the eNB back to the UE. This feedback is carried on the physical channel PHICH, seen in column "Last MSG." In order to not mix signals from different UEs, all frames belonging to one UE connection were selected. The C-RNTI for this call is e875H as seen in column "UE ID/RNTI Value." At least one more column is needed for a HARQ analysis. As discussed in the previous sections, the UL HARQ process is synchronous and each transmission follows the HARQ feedback 4 ms later and a retransmission will follow again 4 ms later, summing the complete HARQ round-trip time to 8 ms. Each transmission is aligned to subframes with a subframe duration of 1 ms. Column "Subframe Number" shows the subframe numbers 0–9 within a radio frame. The uplink and downlink frame structures are described in Sections 1.8.10 and 1.8.6, respectively.

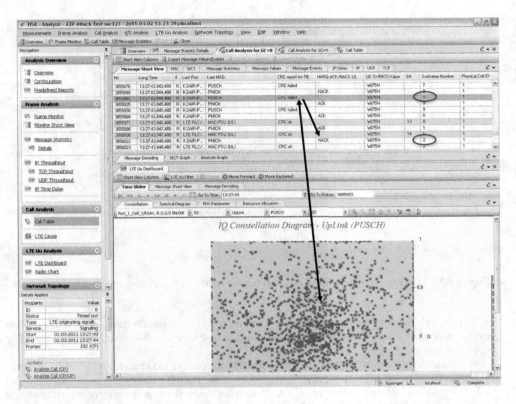

Figure 1.76   HARQ Example with IQ diagram on a noisy reception with failed decoding, NACK, and retransmission

The failed PUSCH transmission took place in subframe number 8. The HARQ feedback from the eNB that belongs to the example PUSCH transmission is transmitted in subframe 2 of the following radio frame and indicates the incorrect reception (by sending a NACK). The UE is retransmitting this frame in subframe 6 of the same radio frame (radio frame numbers are not depicted on example screenshot in Figure 1.76).

Another HARQ example is outlined in Figure 1.77. A screenshot of a protocol analyzer is shown in this example with a filter on one UE with C-RNTI "4cfb"H; thus, only frames from this particular UE are seen. The frames were recorded with a precise LTE air interface probe having a better digital RF processing than the eNodeB, which is indicated in this example.

The column "CRC report for TB" shows the UL reception at the eNodeB side but calculated by the air interface probe. The first circle in this column indicates three correctly received UL transport blocks transmitted from the UE to the eNodeB. The probe's calculated CRC result is "CRC ok." Even though the frames were decodable by the probe, the eNodeB was not able to correctly receive the frames and its HARQ processes are sending a NACK via the PHICH DL channel to the UE to indicate the incorrect reception and inquire a retransmission for all three frames. The first arrow shows the related three NACKs encircled in the next column.

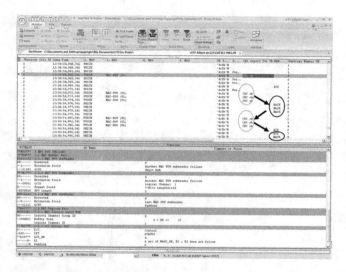

Figure 1.77   HARQ example with an eNodeB having issues with decoding received transmissions

The very right column outlines the subframe numbers within one radio frame as described in the previous example. Each NACK feedback is transmitted in four subframes (4 ms) after its corresponding transport block.

The UE correctly receives the three NACK on PHICH and each HARQ process retransmits the transport block. The eNodeB kept the previous transmission in the HARQ memory and combines the initial transmission and the first retransmission in order to increase the decoding probability of the retransmitted frames.

All three retransmissions are again correctly decoded by the air interface monitoring probe (CRC results in the protocol analyzer are generated by the probe in order to benchmark the eNodeB network decoding performance). The eNodeB could decode the first two transmissions that are indicated as ACK HARQ feedback in the HARQ feedback column, but the last retransmission could not be decoded by the eNodeB even with combining both transmissions. The NACK triggers the next retransmission of this transport block belonging to the third HARQ process, which could be finally correctly decoded (not seen in the screenshot of Figure 1.77).

1.10   LTE Advanced

New global mobile standards and their features are defined by the telecommunication institution International Telecommunications Union (ITU) based in Geneva, Switzerland. The ITU defines requirements and design targets but does not define specific technologies in order to leave the realization to the market and the completion to various technologies. That is one of the reasons why the third mobile generation 3G is realized as, for example, UMTS and WiMAX.

LTE is often already seen as a 4G standard, which is not true as it is defined by the 3G standardization groups and does not fulfill the requirements of the ITU 4G targets. The ITU formulated the requirements for the next generation mobile standard as IMT-Advanced (International Mobile Telecommunications-Advanced), which is formally the 4G standard. The 3GPPdesigns as standardization organization a new mobile communication standard under the name of LTE Advanced.

The first stable release of LTE Advanced is available with the publication of 3GPP Release 10 since the beginning of 2011. Main driver for LTE Advanced is to achieve higher data rates to keep up with the increasing global data traffic while using existing limited resources. Those goals are achieved with a higher spectral efficiency (transferring more information per give Hertz of spectral bandwidth), where higher modulation schemes and better code rates are applied. But LTE and LTE Advanced already operate at the natural limit of information theory at a given Signal-to-Noise-and-Interference Ratio (SNIR). As a result, LTE Advanced introduces techniques to reduce interference, for example, through spatial diversity or interference coordination. Especially, LTE Advanced decreases the performance spread between UEs close to the base station, compared to cell-edge users with low SNIR. Those techniques are discussed in the following sections, Inter-Cell Interference Coordination (ICIC) and Heterogeneous Networks.

The ITU delay requirements have not changed for the new standard; therefore, LTE Advanced has the same delay targets as LTE, a Control Plane delay of 10 ms, and a User Plane delay of 100 ms.

1.10.1   Increasing Spectral Efficiency

Obvious technique to increase spectral efficiency is to apply higher modulation rates as QAM64 and above along with more efficient code rates. However, there are limits for a given SINR and a design target of a 10% block error rate. LTE Advanced uses enhanced multiple smart antenna systems in order to further increase spectral efficiency. The spectral efficiency of LTE Advanced is up to 30 bps/Hz in downlink direction and up to 15 bps/Hz in uplink direction.

Release 10 introduces transmission modes with up to 8 antennas (8×8 MIMO) in downlink direction, which enables theoretically up to a factor of 8 of efficiency by transmitting 8 parallel data streams. The receiver can retrieve all eight individual data streams again. A MIMO gain is achieved with a spatial diverse multi-path channels. UEs inform the base station with the rank indicator about their DL reception conditions, so that the base station can apply the best pre-coding of the DL MIMO signals.

In uplink, a 4×4 MIMO scheme is defined, compared to single-antenna signal transmission in LTE. The drawback is a much large UL transmitter hardware in the UEs, which needs four individual transmitter hardware units, leading to higher cost and battery consumption.

1.10.2   Carrier Aggregation

As spectral efficiency is already close to the theoretical border, LTE Advanced defines a feature called Carrier Aggregation, allowing a UE to be scheduled across multiple Release 8 carriers of 20 MHz. Bundling up to five 20 MHz carriers allows a total of 100 MHz spectrum bandwidth for a single transmission. The individual 20 MHz carrier does not have to be adjacent to each other and could be even allocated in different frequency bands as most LTE bands do not host 100 MHz of contiguous spectrum.

Figure 1.78 depicts LTE Advanced terminals scheduled via multiple 20 MHz carriers backward compatible along with Release 8 LTE terminals. An LTE Advanced terminal always listens in one dedicated main 20 MHz carrier to the PDCCH, which also provides scheduling information about the other 20 MHz carriers where data is transmitted for the scheduled terminal as seen in the figure.

1.10.3   Heterogeneous Networks

The peak performance of LTE is only achieved close to the base station and in case of very good radio conditions. Compared to the best performance, on the other hand, cell-edge users can perceive more than 100 × weaker performance. Besides deploying centralized macro cells, operators have the option

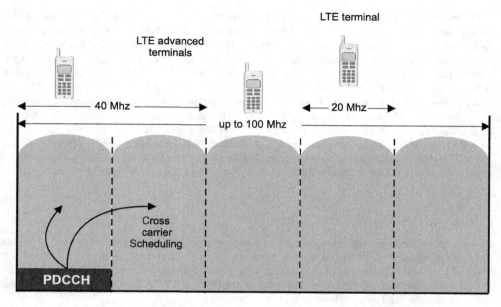

Figure 1.78 LTE Advanced Carrier Aggregation bundles multiple 20 MHz LTE carriers in order to use an increased bandwidth of up to 100 MHz for a single user

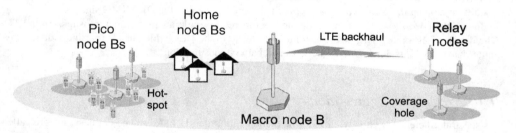

Figure 1.79 Heterogeneous Network with privately deployed Home eNBs, pico cells in hot spot areas, and Relay eNBs as range extension and to increase cell coverage

to overlay additional small cells and pico cells in order to minimize coverage holes or to enable cell range extensions on cell edges. Additionally, operators also can roll-out privately deployed Home eNBs to increase local indoor performance – a Heterogeneous Network (HetNet) is established.

Small and pico cells are usually deployed in hot spot metro areas with a high demand of capacity. Relay nodes are typically used as range extension of macro cells and to increase SNIR in cell-edge scenarios as outlined in Figure 1.79.

Relay eNodeBs are using the LTE air interface of the hosting macro cell as backhaul link. A frequency waste appears as a result, but overall average capacity can still be increased due to the poor performance of cell-edge users (>100x weaker performance compared to users close to the base station), which can be enhanced by certain factors.

Small and pico cells interfere heavily with the surrounding macro cells. The next section introduces LTE Advanced techniques to decrease this inter-cell interference.

1.10.4   Inter-Cell Interference Coordination

In the previous sections, various techniques were discussed to achieve better spectral efficiency and though higher peak data rates. Some of the applied methods are higher modulation schemes, code rate enhancements, carrier aggregation, and higher-order spatial multiplexing (MIMO).

Key to achieve an average high QoS for all cell users is to decrease the interference impact as most aforementioned methods require low interference to leverage their full potential. Increasing capacity in future networks requires utilization of HetNets (see Section 3.10.3) where it is essential to manage interference. LTE Advanced introduces an interference management scheme called Inter-Cell Interference Coordination (ICIC).

The X2 interface is used for communication of handovers between base stations and will also carry interference information and coordination messages. Deployed small and pico cells overlaying macro cells interfere with the macro cells leading to especially low SINRs for the cell-edge users of the pico cell. ICIC minimizes the interference with coordinated transmissions in specific subframes, where the macro cell will not schedule any traffic. This scheme seems to reduce capacity on the first glance, but with managing interference can increase throughput by the factors in the used subframes.

Figure 1.80 depicts a pico cell within an overlaid macro cell. With ICIC, either the cell-edge users of the pico cell receive a better service through interference minimization or the cell range of the pico cell can be extended. However, the standard does not provide additional protection on control channels, so that range extension is limited to keep control channel decoding errors in a reasonable range.

ICIC is based on two basic principles:

- Scheduling traffic in interfered pico cells and macro cells based on frequency multiplexing and carrier aggregation (see section 1.10.2)
- Scheduling traffic between both cells based on time multiplexing

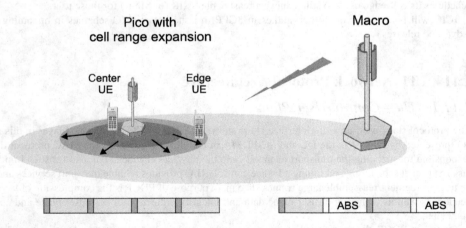

Figure 1.80   Pico cells interfere with surrounding macro cells. ICIC enables to enhance the SINR of cell-edge users of the pico cell by decreasing interference of the underlying macro cell. ICIC increases the performance of the cell-edge users of the pico cell and expands the pico cell's range

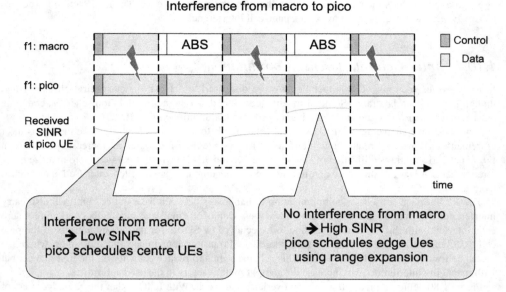

Figure 1.81 ICIC coordinates the interference from the macro cell to the pico cell

The so-called *Almost Blank Subframes* (ABS) are used for time multiplexing between both cells in order to have only one cell transmitting and the other cell injects an ABS frame. In case ABS is used, the cell transmitting will not interfere with the cell granting the ABS frame (see Figure 1.81). The cells fully interfere in subframes; both cells are transmitting, resulting in a low SNIR (or SINR) for users in the pico cell. Those subframes are used in the pico cell to schedule center users with a still fair SNIR (or SINR). In coordinated subframes where the macro cell is not transmitting (using ABS), the pico cell schedules its cell-edge users to still achieve a reasonable SNIR (or SINR) for those users.

ICIC will be a matter of further studies in 3GPP to issue better ICIC schemes in upcoming LTE Advanced releases.

1.11   LTE Network Protocol Architecture

1.11.1   Uu – Control/User Plane

The protocol stack used on radio interface Uu is shown in Figure 1.82. The physical layer in this stack is represented by OFDM in the DL and SC-FDMA in the UL. Then, we see the MAC protocol that is responsible for mapping the transport channels onto the physical channels, but also for such important tasks as packet scheduling and timing advance control. RLC provides reliable transport services and can be used to segment/reassemble large frames. The main purpose of PDCP is the compression of larger IP headers as well as ciphering of user plane data and integrity protection of both user plane and control plane data.

On top of PDCP, the stack is split into the user plane and control plane parts. On the control plane side, we see RRC protocol, that is, the expression for the communication between the UE and eNB.

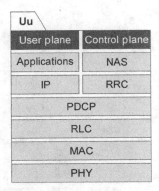

Figure 1.82 Protocol stack LTE Uu interface

RRC provides all the necessary functions to set up, maintain, and release a radio connection for a particular subscriber. Details of these functions are described in Section 1.12.7.

RRC also serves as a transport protocol for NAS signaling messages. NAS is the expression for the communication between the UE and MME in which MME represents the core network.

On the user plane side, we see IP as the transport layer for end-to-end applications. On the Uu stack, the IP is always end-to-end IP, which means that all these IP packets are transparently routed, often tunneled through the mobile network. The user plane IP frames we see on Uu are the same IP frames that can be monitored at SGi reference points before or behind the PDN-GW.

The IP version can be Internet Protocol version 4 (IPv4) or Internet Protocol version 6 (IPv6). In the case of VPN (Virtual Private Network) traffic, IPsec will be used.

The applications on top of IP in the user plane stack are all protocols of the TCP/IP suite, such as the File Transfer Protocol (FTP), HTTP (web browsing), and POP3/SMTP (for e-mail), but also Real-Time Transport Protocol (RTP) and SIP for real-time services like VoIP.

1.11.2 S1 – Control/User Plane

On the S1 reference point, the physical layer L1 will in most cases be realized by Gigabit Ethernet cables. L2 in this case will be Ethernet. On top of Ethernet, we find IP, but used as a transport protocol between two network nodes: eNB and MME. This lower layer IP does not represent the user plane frames.

Instead, the user plane IP frames (higher layer IP) are carried by the GTP Tunneling Packet Data Unit (T-PDU). The GTP is responsible for the transport of payload frames through the IP tunnels on S1-U. The transport layer for GTP-U is the User Datagram Protocol (UDP). As IP, this protocol may be found twice in the user plane stack: lower UDP for transport between the eNB and MME and higher UDP (not shown in Figure 1.83) that is transparently routed through the mobile network as the transport protocol for real-time application data. The higher layer IP on top of GTP-U and all application data on top of this higher layer IP are identical with the user plane information described in the previous section.

On the control plane side, the Streaming Control Transport Protocol (SCTP) provides reliable transport functionality for the very important signaling messages. S1AP is the communication expression between MME and S-GW while NAS – as already explained in the previous section – is used for communication between the UE and MME.

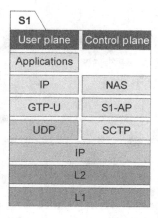

Figure 1.83   Protocol stack S1 control/user plane

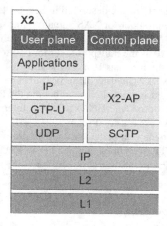

Figure 1.84   Protocol stack X2 control/user plane

1.11.3   X2 – User/Control Plane

On the X2 interface, the user plane protocol stack is identical to that of the S1 reference point. However, as shown in Figure 1.84, on the control plane, there is a different protocol: X2 Application Part (X2-AP).

The main purpose of X2-AP is to provide inter-eNB handover functionality. It is also used to exchange traffic-related and radio quality measurement reports between different eNBs.

1.11.4   S6a – Control Plane

There is no user plane at the S6a reference point due to the fact that we find here the connection between the MME and HSS, which is a plain signaling communication.

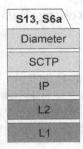

Figure 1.85   Protocol stack S6a

L1, L2, IP, and SCTP shown in Figure 1.85 provide the same functionality as explained in the section on the S1 protocol stack.

The new player in the S6a stack is the DIAMETER protocol. In the EPC network, DIAMETER has taken over the role of the Mobile Application Part (MAP). It is used to update the HSS about the current location of the subscriber and to provide crucial subscriber attributes stored in HSS databases to the MME so that network access can be granted and the subscriber's traffic can be routed according to these parameters. DIAMETER is also involved in the security functions of the network: subscriber authentication, integrity protection, and ciphering.

The protocol stack on S6a is identical to the protocol stack of the S13 reference point between MME and EIR – in case an EIR exists in the network.

1.11.5   S3/S4/S5/S8/S10/S11 – Control Plane/User Plane

At the S3, S4, S5, S8, S10, and S11 reference points, we find the same protocol stack as shown in Figure 1.86. The reason is that all these interfaces are used to tunnel IP payload transparently from one network node to the other. The tunnel management is provided by GTP-C, while the IP payload is transported using GTP-U T-PDUs. Indeed, the user plane protocol stack is the same as on the S1 reference point.

Simplifying the functionality of GTP-C, it can be said that this protocol is used to create, modify, and delete user plane tunnels. It also supports mobility of subscribers between core network nodes.

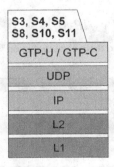

Figure 1.86   Protocol stack S3S4/S5/S8/S10/S11

1.12   Protocol Functions, Encoding, Basic Messages, and Information Elements

1.12.1   Ethernet

Ethernet is the typical transport layer protocol in IP networks. It is designed to transmit packets from a sender to a receiver, both identified on behalf of an address information element.

According to this limited functionality, Ethernet has a very small header. The header field (Figure 1.87) contains only the information elements:

- Destination address
- Source address
- Ethernet type.

Ethernet type is similar to a SAPI (Service Access Point Identifier). It contains information about which higher layer protocol information is transported by Ethernet frames.

The Ethernet addresses, often called MAC addresses (but with nothing in common with RLC/MAC!), consist of 6 bytes. These MAC addresses are fixed hardware addresses and, due to a defined numbering scheme, each address is unique worldwide.

If IP data is to be transmitted using Ethernet, the hardware MAC address of the receiver of IP packets is unknown when the connection starts. Only the target IP address is known. However, since Ethernet is the lowest layer of the connection, there must be a source and a destination MAC address included in each header. In other words, for each sender IP address, there is an appropriate sender hardware address, and for each destination IP address, there must be an appropriate target hardware address.

The target hardware address that is related to the target IP address is requested by the Address Resolution Protocol (ARP). Its sister protocol, the Reverse Address Resolution Protocol (RARP), can be used to find the target IP address (or, in terms of ARP/RARP, the target protocol address) to a known MAC address.

The address resolution procedure consists of two steps:

1. ARP request (req) message with Target Hardware Address = "0" is sent to *all(!)* IP clients in the network.
2. The client that has the target protocol address as set in the ARP req message sends ARP Replay (rpl). The sender hardware address in ARP rpl is the Ethernet MAC address related to the destination IP address that the sender of the ARP req is looking for. An example of the Ethernet address resolution procedure is shown in Figure 1.88.

1.12.2   Internet Protocol (IPv4/IPv6)

The IP frame is called the datagram and there exist two main versions of the IP: IPv4 and IPv6.

```
                                              Frame View
 BITMASK                      ID Name                                Comment or Value
Ethernet  IP6_2460   IPU4
Ethernet Link Layer (ETHER)   Ethernet (= Ethernet Packet)
Ethernet Packet
***B6***   destination addr                              01 00 5e 00 00 09
***B6***   source addr                                   00 07 84 b2 58 54
***B2***   ethernet type                                 Internet IP (IPv4)
```

Figure 1.87   Ethernet header example

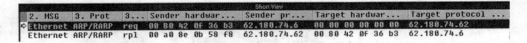

Figure 1.88   Ethernet address resolution

1.12.2.1   IPv4

The IP header has a minimum size of 20 bytes (if no options are used) and a maximum size of 64 bytes (including options and padding bits). Due to a set of different options that can be appended to the IPv4 header, these headers can become very large.

The included information elements shown in Figure 1.89 are:

- **Version**: IP protocol version, here IPv4.
- **Internet header length** (IHL length): The length of the header.
- **Type of Service**: The QoS parameters for IP.
- **Total Length**: The length of the IP frame including header and payload field.
- **Identification, Fragment Offset**: Both used in case of fragmentation/reassembly.
- **Time to Live**: A hop counter to prevent circular routing.
- **Protocol**: Indicates the higher layer protocol that uses IP as the transport layer; typical examples are ICMP, TCP, UDP.
- **Source Address**: IP address of the sender of the datagram.
- **Destination Address**: IP address of the receiver of the datagram.
- **Options**: For example, the timestamp of each router that the IP packet passed.
- **Padding**: Fill bits to align the header to a multiple of 32 bits.

Since the maximum packet size of an IP datagram can vary from one local network to the next, the IP is equipped with fragmentation/reassembly functionality that allows the transmission of larger frames in series of smaller portions. Figure 1.90 shows an example where a frame with 1600 bytes of data is fragmented into two smaller frames with 1480 and 120 bytes of data each. Fragmented frames do all have the same frame ID (in the example: 1234). As long as more fragments are following the first one,

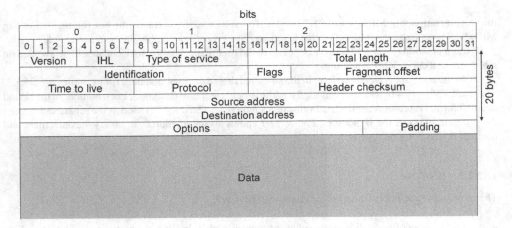

Figure 1.89   IP datagram structure

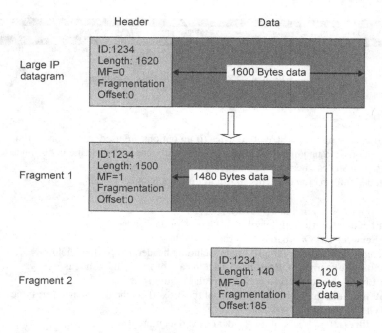

Figure 1.90 IP fragmentation

Dotted decimal notation: 195.24.1.2

0			31
1100 0011	00011000	0000 0001	0000 0010

Hexadecimal: 0x C3 18 01 02

Figure 1.91 Example of IPv4 address format

the fragmentation flag MF is set to "1." The last frame in a series of fragments has fragmentation flag MF = "0," but a fragmentation offset that is required for proper reassembly on the receiver side.

IP fragmentation (Figure 1.90) may be found in the user plane data streams, but should be avoided on interfaces that carry 3GPP signaling.

IPv4 addresses are typically written in the so-called dotted decimal notation, for example, 195.24.1.2. There are 32 bits (=4 bytes) reserved for the address fields in the IP datagram. Each number in the dotted decimal format represents the decimal value of a single byte. The dot "." is used as the separator between the different bytes of the IP address. Figure 1.91 shows a sample address in binary, hexadecimal, and decimal dotted notation format.

1.12.2.2 IPv6

The most important improvements that come with IPv6 are:

- A larger number of possible address values become available. In IPv4, the number of addresses is limited to 32 bits, which means in turn that 2^{32} (≈ 4.3 billion $= 4.3 \times 10^9$) possible values can be addressed.

IPv6 provides space for 2^{128} ($=3.4 \times 10^{38}$) possible address values. This is an improvement by a factor of 2^{96} and reached by a restructuring of the IP header. In the IPv6 header shown in Figure 1.92, 128 bits (16 bytes) is reserved for source and destination addresses. The larger address ranges available for IPv6 will also allow more direct end-to-end packet routing and, hence, less address translation in network nodes is required and the packet routing in the overall network is expected to be faster and more efficient.

- The automatic configuration of dynamically assigned IP addresses is improved and, in turn, legacy procedures like DHCP (Dynamic Host Configuration Protocol) become unnecessary.
- IPv6 supports Mobile IP, simplifies renumbering (change of dynamically assigned IP addresses), and allows multihoming of subscribers. The purpose of multihoming is to increase the reliability of Internet connections by using two different Internet service providers simultaneously. If the access to one of the providers is interrupted, a redirection of packets via the second connection is possible. Mobile IP means that the subscriber always gets the same IP address assigned, no matter if working at home or traveling around.
- IPsec is integrated into IPv6 to achieve a higher security of IP data transmission, while back in IPv4, no security functions were provided at all.
- All in all, the basic header of IPv6 has a simpler structure compared to the header of IPv4. Although the overall header size is larger than in IPv4 (40 bytes, most of them occupied by the longer IP addresses), there are less basic header fields.
- For the version, the decimal number 6 is encoded as binary bit sequence "0110."
- The IPv6 traffic class indicates the packet priority and should not be mistaken for the traffic class QoS element introduced in 3GPP standards that classifies the throughput sensitivity and delay sensitivity of application services. IPv6 traffic class priority values subdivide into two ranges: traffic, where the source provides congestion controlled and non-congestion controlled traffic.
- The flow label is used for QoS management and encoded in 20 bits. Packets having the same flow label value will be treated with the same priority and reliability. This is important for the routing of packets that contain real-time service data.
- The payload length indicates the size of the payload in octets and is encoded in 16 bits. When cleared to zero, the option is a "Jumbo Payload" (hop by hop). The size of the basic header is not counted by the payload length, but the optional header extensions are included. So payload length + 40 bytes (of basic header) = total length of the IPv6 packet.
- The next header information element specifies the next upper layer protocol of the transported payload such as UDP and TCP. The values are compatible with those specified for the IPv4 protocol field (8 bits). The next header information can also point to optional extension headers. In this case, the upper layer payload protocol is not indicated by this field.

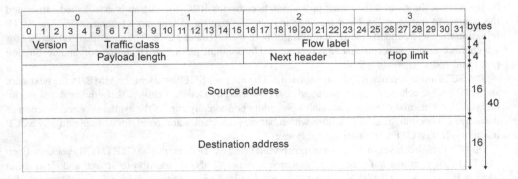

Figure 1.92   IPv6 header format

- The hop limit field (8 bits) indicates the maximum number of routers that are allowed to be involved in routing an IPv6 packet. It replaces the time to live field of IPv4. If the hop limit reaches the value "zero," the packet will be discarded by the router.
- Source and destination addresses, 128 bits each, represent the sender and receiver of the IPv6 datagram.

IPv6 addresses are normally written as eight groups of 16 bits, where each group is separated by a colon (:). For example, 2001:0db8:85a3:0000:0000:8a2e:0370:7334 is a valid IPv6 address.

To shorten the writing and presentation of addresses, several simplifications to the notation are permitted. Any leading zeros in a group may be omitted; thus, the given example becomes: 2001:db8:85a3:0:0:8a2e:370:7334.

Also, one or any number of consecutive groups of value 0 may be replaced with two colons (::): 2001:db8:85a3::8a2e:370:7334.

It is possible to use IPv6 addresses in the URL notation format. In this case, the IPv6 address information is enclosed in square brackets:

http://[2001:0db8:85a3:08d3:1319:8a2e:0370:7344]/.

The brackets prevent that part of the IPv6 address being misinterpreted as port number information. A URL including IPv6 address and port number looks like this:

http://[2001:0db8:85a3:08d3:1319:8a2e:0370:7344]:8080/.

1.12.3   Stream Control Transmission Protocol (SCTP)

Originally, the SCTP was defined as a transport protocol for SS7 messages to be transmitted over IP networks. As TCP and UDP, it is seen as a layer 4 transport protocol in the ISO OSI model.

The SCTP frames are called chunks. All chunks are associated to a connection that guarantees in-order delivery. However, within the same chunk, there might be data blocks of different connections transmitted simultaneously. In addition, it is also possible to send urgent packets "out of order" with a higher priority.

SCTP also supports multihoming scenarios where one host owns multiple valid IP addresses.

Besides the data streams, SCTP frequently sends heartbeat messages to test the state of connection.

How SCTP works will be demonstrated by means of an example. Figure 1.93 shows the message flow required to transport the NAS signaling message Attach Request from the eNB to MME across the S1 interface.

After setting up an RRC connection on the Uu interface between the UE and the eNB, the UE sends the attach request message. When the appropriate RRC transport container is received by the eNB, the establishment of a dedicated SCTP stream on the S1 interface as shown in Figure 1.93 is triggered.

The establishment of the SCTP stream starts with an SCTP initiation message. It will always be sent by the eNB in the case of the attach procedure, because the RRC connection is established earlier and the request to transport the NAS message triggers the request to have an S1 connection. The SCTP initiation message contains the IP addresses of both the eNB and MME. The individual subscriber for which this connection is established is represented by a unique pair of SCTP source port and destination port numbers.

The SCTP initiation needs to be acknowledged by the peer SCTP entity in the MME. In the next step, an SCTP cookie echo message is sent and acknowledged by Cookie Echo ACK. In the protocol world, this is called a heartbeat procedure. Such a procedure periodically checks the availability and function of the active connection. Similar functions with other message names are found, for example, in SS7 SCCP Inactivity Test or GTP Echo Request/Response.

On SCTP, higher layer messages are transported using SCTP datagram (SCTP DTGR) packets. Each SCTP DTGR contains a Transaction Sequence Number (TSN) in addition to source and destination address information. This TSN will later be used by the peer entity to acknowledge the successful

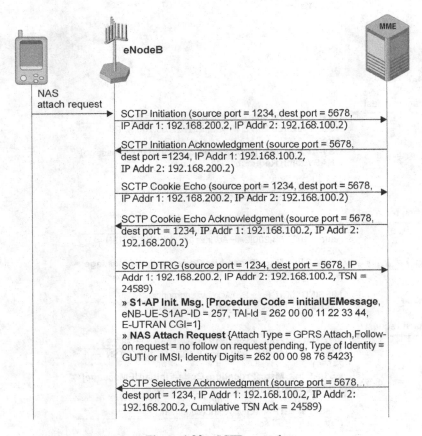

Figure 1.93   SCTP example

reception of the DTGR by sending an SCTP selective ACK message on S1 that confirms error-free reception of the SCTP DTGR that carried the attach request message. Further, S1AP and NAS messages of this connection will be transported in the same way and the Cookie Echo/Cookie Echo ACKs will be sent periodically as long as the connection remains active.

If the S1 signaling transport layer SCTP has problems in offering proper functionality, there will be no signaling transport on S1 if the problems are located in the eNB SCTP entity. If the MME suffers from congestion or protocol errors on the SCTP level as shown in Figure 1.94, the expected selective ACK messages will be missing (maybe not sent at all, maybe sent with a TSN out of the expected range). This malfunction may be detected by a NACK Cookie Echo, and as a result, the connection will be terminated, or the attach accept message expected to be received by the UE will be missed. The missing attach accept message will be recognized by the UE where a timer is guarding the NAS procedures. After the guard timer expires on the UE side, the attach request message will be repeatedly sent up to n times (the counter value of n is configurable and typically signaled on the broadcast channel SIBs; the default value recommended by 3GPP is $n = 5$). If neither an attach request message nor an attach reject message is received by the UE, the handset will go back to IDLE when the maximum number of attach request repetitions has been sent.

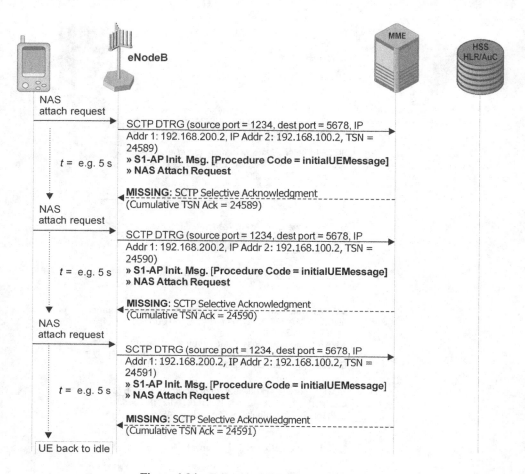

Figure 1.94   Failure in SCTP signaling transport

1.12.4   Radio Interface Layer 2 Protocols

LTE splits the layer 2 into three sublayers: MAC layer, RLC layer, and PDCP layer. Figure 1.95 illustrates the layer 2 architecture for DL and Figure 1.96 illustrates that for UL.

The sublayers (sublayers are called layers in the following) communicate via Service Access Points (SAPs). A lower layer provides services to the adjacent layer above through SAPs. From the point of view of the lower layer, the packets providing this service are Service Data Units (SDUs). This layer once more uses services from a lower layer, again by giving its packets through SAPs to the adjacent lower layer. Those packets are called Packet Data Units (PDUs). Hence, a PDU from a higher layer is an SDU from the point of view of the lower layer providing a service to the higher layer.

SAPs are depicted as ovals between the layers in Figures 1.95 and 1.96.

Each layer with its unique functionality is introduced in the next three sections.

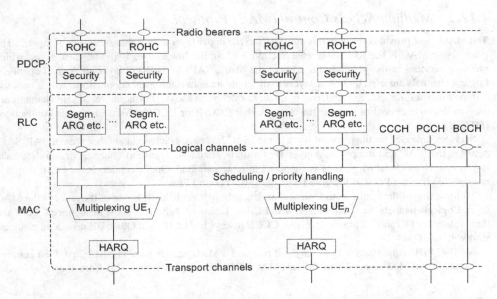

Figure 1.95 Layer 2 structure for DL (TS36.300). (*Source*: Reproduced with permission from ©
3GPP™.)

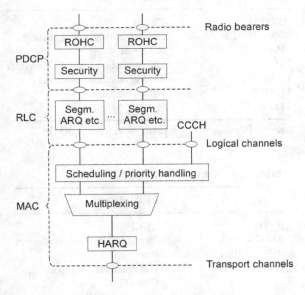

Figure 1.96 Layer 2 structure for UL (TS36.300). (*Source*: Reproduced with permission from ©
3GPP™.)

1.12.5 Medium Access Control (MAC) Protocol

The MAC layer provides services through SAPs to the upper layer as all other sublayers of layer 2. The layer above MAC is the RLC layer (see Section 1.12.6); the lower layer is the physical layer, which provides services to the MAC. In the case of the MAC, SAPs to the RLC layer are logical channels. Logical channels are used by higher layers to differentiate between logical connections, which may use different metrics, for example, in terms of quality or delay, and so on. Furthermore, logical channels are used to distinguish control plane connections, either CCCHs or DCCHs, from user plane connections (DTCHs).

Services provided by the physical layer to the MAC layer are granted via another type of SAP. SAPs between the MAC and the physical layer are transport channels. Transport channels match data units to physical channels in which data is supposed to be transmitted. One exception is the PCH, which is multiplexed into the PDSCH identified with the P-RNTI = 0xFFFE.

Multiplexing of data units from logical channels to transport channels is one of the tasks of the MAC layer. Logical channels are differentiated with LCIDs. Tables 1.20 and 1.21 show the defined LCIDs and their values for DL and UL, respectively. A CCCH always has LCID = 0. Other UE dedicated channels start with LCID = 1.

A MAC PDU consists of a MAC payload part and a MAC header part. The MAC payload conveys multiple units of MAC control elements and MAC SDUs from higher layers. Therefore, the MAC header

Table 1.20 Values of LCID for DL-SCH

Index	LCID values
00000	CCCH
00001–01010	Identity of the logical channel
01011–11011	Reserved
11100	UE contention resolution identity
11101	Timing advance command
11110	DRX command
11111	Padding

Source: Reproduced with permission from © 3GPP™.

Table 1.21 Values of LCID for UL-SCH

Index	LCID values
00000	CCCH
00001–01010	Identity of the logical channel
01011–11001	Reserved
11010	Power headroom report
11011	C-RNTI
11100	Truncated BSR
11101	Short BSR
11110	Long BSR
11111	Padding

Source: Reproduced with permission from © 3GPP™.

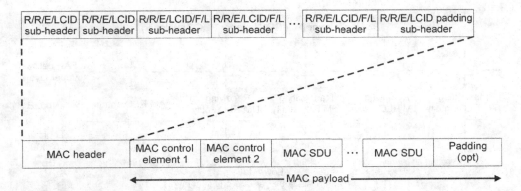

R/R/E/LCID sub-header	R/R/E/LCID sub-header	R/R/E/LCID/F/L sub-header	R/R/E/LCID/F/L sub-header	...	R/R/E/LCID/F/L sub-header	R/R/E/LCID padding sub-header

MAC header	MAC control element 1	MAC control element 2	MAC SDU	...	MAC SDU	Padding (opt)

◄──────────────────────────── MAC payload ────────────────────────────►

Figure 1.97 Example of MAC PDU consisting of MAC header, MAC control elements, MAC SDUs, and padding (TS36.321). (*Source*: Reproduced with permission from © 3GPP™.)

is also divided into sub-headers depending on the units carried in the MAC payload as MAC sub-headers describe the MAC payload units. There are various possible combinations of MAC control elements, MAC SDUs, and MAC padding derivatives. An example of a MAC PDU with a combination of MAC sub-headers, MAC control elements, and MAC SDUs in the payload section is depicted in Figure 1.97.

Logical channels, which are multiplexed to transport channels, are prioritized by the scheduling algorithm. The scheduling algorithm decides what to schedule on which physical resources as described in detail for DL scheduling in Section 1.8.6 and UL scheduling in Section 1.8.9. There is only one MAC entity per UE; thus, the UL within the UE has one MAC entity and the eNB executes multiple parallel MAC entities in the DL direction in case the eNB has to handle multiple UEs (as seen in Figures 1.95 and 1.96).

The MAC layer implements a soft combining N-process stop-and-wait FEC and detection mechanism or HARQ. Transport blocks are protected with a FEC algorithm known as turbo codes. Soft combining means blocks that are not correctly decoded are not acknowledged in order to conduct a retransmission, but the previously received block that is not decoded is held in a soft buffer to be recombined with the new retransmission. This process of soft combining two or more receptions increases the chance that the last received retransmission can be decoded error-free.

1.12.6   Radio Link Control (RLC) Protocol

The RLC layer uses SAPs of the MAC layer, which are logical channels. Those SAPs are used to induct its RLC PDUs from the RLC entities. The RLC layer, instead, provides basically three types of SAPs to the PDCP layer above the RLC layer. The SAPs provide access to the three operating modes to convey data PDUs, either the control plane or user plane: Transparent Mode (TM), Unacknowledged Mode (UM), and Acknowledged Mode (AM). Basic operating entities of the RLC layer are depicted in Figure 1.98.

1.12.6.1   RLC Transparent Mode (TM)

The TM transmits PDUs from the upper layer transparently, which means that the RLC layer is basically just storing and forwarding those PDUs. No feedback is requested on whether the data units were delivered or not. Furthermore, packets are transparently forwarded by keeping them at the same size.

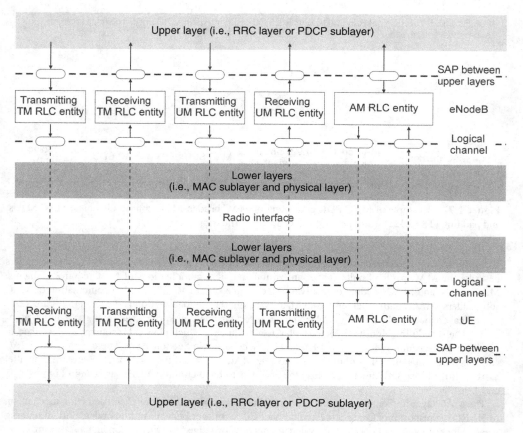

Figure 1.98 RLC layer overview with TM, UM, and AM RLC entities. (*Source*: Reproduced with permission from © 3GPP™.)

Figure 1.99 illustrates a transmitting and a receiving RLC TM entity.

The TM is used to transmit cell-wide channels as UE paging and the broadcast of system information. It is a typical mode for point-to-multipoint connections without any feedback from receivers. Additionally, CCCHs (LCID = 0) are sent by using the RLC TM.

1.12.6.2 RLC Unacknowledged Mode (UM)

In contrast to the TM, the UM adds segmentation and concatenation of PDUs. This processing step fits higher layer PDUs to the current available transport block size indicated by lower layers.

Segmentation reduces the size of PDUs by splitting them apart. Figure 1.100 illustrates the segmentation of PDCP SDUs.

Concatenation, instead, is merging PDCP SDUs to a larger RLC PDU. Figure 1.101 depicts this process of building longer PDUs, again in order to decrease padding (adding zeros at the end of a PDU) and use large transport blocks more efficiently.

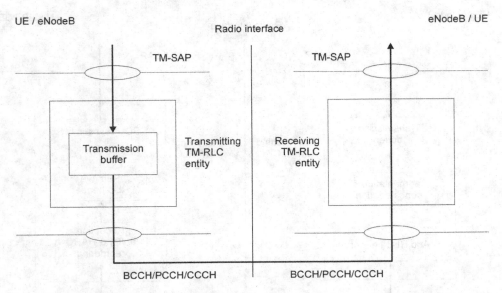

Figure 1.99   Model of two Transparent Mode peer entities. (*Source*: Reproduced with permission from © 3GPP™.)

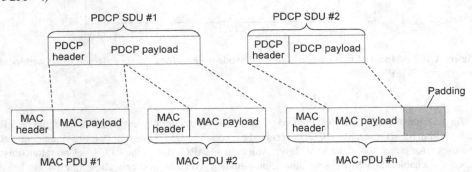

Figure 1.100   Example of RLC segmentation of PDCP SDUs into MAC PDUs. The MAC Payload is equivalent to RLC PDUs

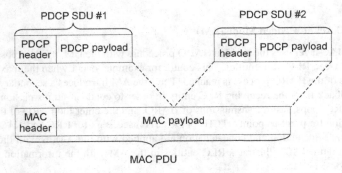

Figure 1.101   Example of RLC concatenation of several PDCP SDUs into one MAC PDU

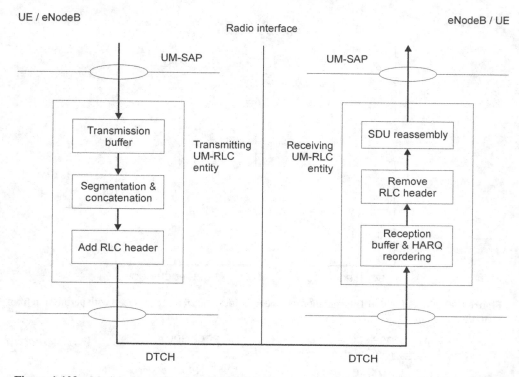

Figure 1.102   Model of two Unacknowledged Mode peer entities. (*Source*: Reproduced with permission from © 3GPP™.)

As the name already indicates, the UM (Figure 1.102) does not request feedback from the peer RLC entity within the receiver, like ACKs of received packets. The UM is used for point-to-point connections and uses error protection of the MAC layer, which uses HARQ. But on the RLC level, no retransmission process is enabled in UM transmissions. Some higher layer radio bearers allow rare packet loss, for example, IP/UDP traffic or IP/TCP traffic, which is taking care of lower layer packet loss with its own retransmissions on the application level.

1.12.6.3   RLC Acknowledged Mode (AM)

Although the MAC layer uses the enabled HARQ procedure with the UM, it is still possible that packets will get lost between RLC peer entities. This could, for example, occur when the maximum number of retransmissions of the HARQ process is reached. Thus, the AM introduces a secondary outer loop ARQ with ACK feedback from the receiving RLC entity in order to conduct retransmissions of lost packets. Two stacked ARQ loops, fast retransmissions of HARQ, and the outer loop ARQ of the RLC AM lead to good protection for point-to-point DTCHs or DCCHs (used, e.g., for RRC and NAS packets).

Furthermore, segmentation and concatenation are applied with the AM as with the UM in order to fit PDU sizes. Figure 1.103 illustrates RLC entities in the AM with the information flow through the different functions.

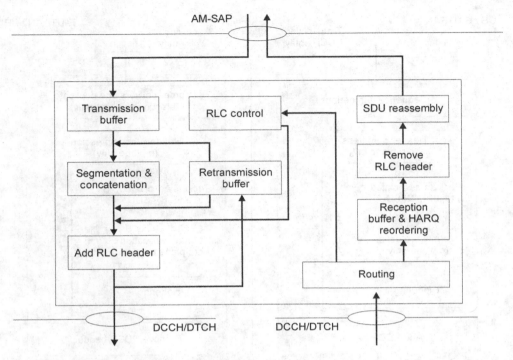

Figure 1.103 Model of an Acknowledged Mode entity. (*Source*: Reproduced with permission from ©️ 3GPP™.)

1.12.7 Packet Data Convergence Protocol (PDCP)

Located above the RLC layer is the PDCP layer. The PDCP layer provides direct data transport for control plane messages (RRC and NAS) and for user plane packets. User plane packets are application layer IP packets as LTE defines only packet-based (IP) data communication. Thus, voice or video service is always IP-based (VoIP).

SAPs above the PDCP layer are radio bearers, which are mapped to TM, UM, or AM SAPs of the RLC layer below. Figure 1.104 depicts the systematic functions and services of the PDCP layer.

Sequence numbering ensures in-sequence delivery of SDUs for the user plane and control plane of disordered delivered packets of lower layers. Different sequence number lengths are defined: 5, 7, and 12 bits. The 5-bit sequence numbers are reserved for Signaling Radio Bearers (SRBs) carrying control plane SDUs only. The 7-bit and 12-bit sequence numbers are dedicated to DRBs conveying user plane SDUs with application layer IP data.

Header compression is a feature for user plane transmission in order to decrease application header overhead and to increase overall efficiency. Application headers as IP headers are used for network routing and can be compressed on such air interface point-to-point transmission links. The header overhead is especially large when small packets are used, for example, with VoIP where, besides the IP and UDP header, also an RTP header is in front of the payload.

Only the Robust Header Compression (RoHC) algorithm is applied with LTE. RoHC is defined by the Internet Engineering Task Force (IETF). Several RFCs (describing profiles and protocols of the IETF) can be used with LTE. Table 1.22 lists the 10 defined RoHC profiles, which are permitted to be used.

UE / E-UTRAN   E-UTRAN / UE

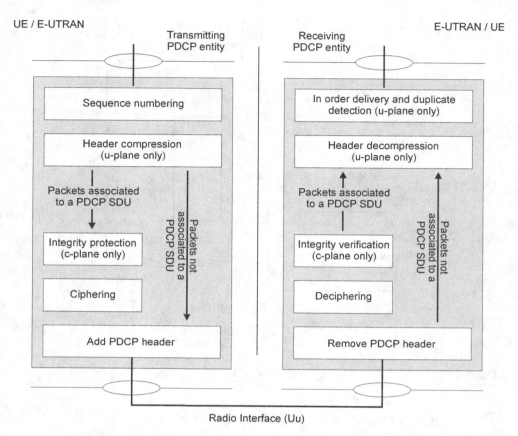

Figure 1.104 Functional overview of the PDCP layer (TS36.323). (*Source*: Reproduced with permission from © 3GPP™.)

Table 1.22 Supported header compression protocols and profiles

Profile identifier	Usage	Reference
0x0000	No compression	RFC 4995
0x0001	RTP/UDP/IP	RFC 3095, RFC 4815
0x0002	UDP/IP	RFC 3095, RFC 4815
0x0003	ESP/IP	RFC 3095, RFC 4815
0x0004	IP	RFC 3843, RFC 4815
0x0006	TCP/IP	RFC 4996
0x0101	RTP/UDP/IP	RFC 5225
0x0102	UDP/IP	RFC 5225
0x0103	ESP/IP	RFC 5225
0x0104	IP	RFC 5225

Source: Reproduced with permission from © 3GPP™.

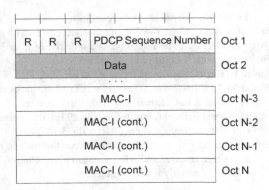

Figure 1.105 Format of PDCP data PDU for transport of signaling radio bearer information. (*Source*: Reproduced with permission from © 3GPP™.)

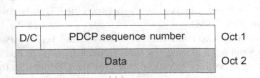

Figure 1.106 Format of PDCP data PDU for transport of user plane information. (*Source*: Reproduced with permission from © 3GPP™.)

Furthermore, integrity protection for the control plane, thus for SRBs, is a function of PDCP as well as ciphering and deciphering functionality.

The format of a PDCP Data PDU for the transport of SRB control plane information is shown in Figure 1.105. Besides the data field with higher layer protocol information, there is a PDCP sequence number, some reserved spare bits, and four octets for the integrity protection message authentication code (MAC-I).

For user plane transport, there are two different formats of PDCP PDUs defined: one with a long 12-bit PDPC sequence number and one with a short 7-bit sequence number (shown in Figure 1.106). The long sequence number format is used for transporting RLC AM and UM information, the short sequence number format for transporting RLC TM frames. The sequence number length for transporting SRB frames is 5 bits.

The D/C field is used to distinguish between user plane (D) and control plane (C) data carried by PDCP.

1.12.8 Radio Resource Control (RRC) Protocol

The RRC protocol is responsible for the setup, reconfiguration, and release of the radio interface connection. This includes the setup, modification, and release of SRBs, default and DRBs, along with the necessary QoS control and initial security activation. Also, the paging of UEs to request establishment of mobile terminated connections is a function of RRC. To provide network-specific access parameters, RRC system information is broadcasted by all cells of the network. The UE uses RRC to report a set of various measurements to the eNB. Some of these measurement reports can trigger intra-LTE or inter-RAT handover, and RRC is in charge of all these mobility procedures. As a special function that is required to support the best possible "always-on" scenario, RRC comes with an error recovery function that allows a dropped RRC connection to be re-established quickly. Similar functionality, called RRC re-establishment, was introduced with RRC used in 3G UTRAN.

Although the functions of LTE RRC are almost the same as for 3G UTRAN RRC, far fewer signaling messages have been defined for LTE RRC. So, at first sight, LTE RRC looks simpler. The trick used by the standard definition group is that it has defined only dedicated messages for RRC connection setup, release, reconfiguration, and re-establishment, but in fact, the LTE RRC reconfiguration procedure is a very complex process that combines all functions covered by the 3G UTRAN RRC protocol procedures, namely physical channel reconfiguration, transport channel reconfiguration, radio bearer setup, radio bearer reconfiguration, radio bearer release, and RRC measurement control. Consequently, it is now very difficult in LTE to find out what exactly is reconfigured.

1.12.8.1   RRC States

In contrast to 3G UMTS where four different RRC states have been defined, LTE recognizes only two RRC states, which means that the radio connection between the UE and network can be either active or not active (as was known from GSM).

In the RRC_IDLE state the radio connection is inactive. The UE mobility is not under control of the network and the UE does not need to send any measurement reports for updating, although it performs neighbor cell measurement for cell (re)selection. However, the UE monitors the PCH to detect incoming calls and it also monitors system information broadcast on the BCCH. This is the most important part of the system information, typically the MIB, since in LTE, the larger part of system information is not signaled on the BCH but on the DL-SCH.

In the RRC_CONNECTED state, the UE is able to send and receive data in the UL and DL direction. It measures the DL radio quality of neighbor cells and sends RRC measurement reports according to the measurement configuration received from the MME. However, it is the eNodeB and the MME that are respectively in charge of making handover decisions and triggering handover execution when necessary. The UE continues to monitor the PCH to detect incoming calls. In the RRC_CONNECTED mode, all system information sent on the DL-SCH, especially SIB 1, which contains information about change of system parameters, is readable by the UE.

When it is necessary to perform inter-RAT mobility, there will be a transition from LTE RRC states to UTRA or GSM states as illustrated in Figure 1.107.

When changing the RAT in the IDLE mode, this will always happen on account of reselection, which means the UE measures the radio quality of the available radio access technologies and selects the best suitable to log in and register to the network. This procedure also applies for UEs in the CELL_PCH and URA_PCH states in the 3G UTRAN. In the CELL_PCH or URA_PCH state, there is no active radio connection between the UE and network, but RRC context information stored in the RNC, and the state transition from CELL_PCH/URA_PCH to E-UTRA IDLE, mean that a UE in CELL_PCH is allowed without further notice to change the E-UTRAN.

A handover from E-UTRA RRC CONNECTED to 3G UMTS will see the UE enter 3G UMTS in the CELL_DCH state, and a handover to the GERAN starting from the E-UTRA RRC CONNECTED state will end up in the GSM_CONNECTED state for voice services after CS Fallback or GPRS packet transfer mode for non-real-time PS services. It is also possible that the UE is ordered to execute a Cell Change Order (CCO) from E-UTRA RRC CONNECTED to GSM_IDLE/GPRS Packet_IDLE or from GPRS packet transfer mode to E-UTRA RRC IDLE. In case of such a CCO, the UE must – as in the case

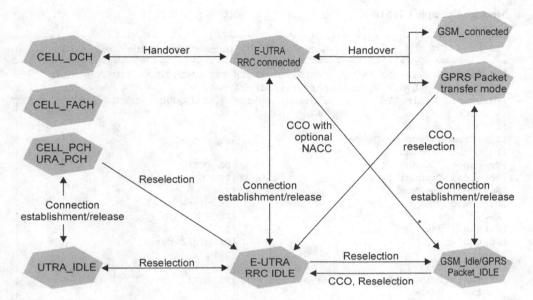

Figure 1.107 RRC state transitions in case of inter-RAT mobility. (*Source*: Reproduced with permission from © 3GPP™.)

of reselection – establish a radio connection and register at the network (e.g., with combined Routing/LA Update) before the user plane payload can be transferred under the umbrella of a still active PDP context. To minimize the delay for this registration procedure, which may take up to 10 seconds, the Network Assisted Cell Change (NACC) was introduced in 3GPP 44.901.

1.12.8.2 System Information

System information is divided into the MIB and a number of SIBs. The MIB includes the DL transmission bandwidth and the PHICH information of the broadcasting cell. The MIB is transmitted on the BCH (transport), that is, mapped onto the PBCH. The MIB is the only system information sent on the BCH. All other SIBs are transmitted using the DL-SCH.

To find the RBs that carry SIBs on the DL-SCH, an SI-RNTI is signaled on the PDCCH. The SI-RNTI indicates in which RBs the SIB *Type1* can be found.

The MIB and SIB 1 are sent periodically (MIB periodicity: 40 ms, SIB 1 periodicity: 80 ms); all other system information messages are flexibly scheduled.

SIB 1 (see Message Example 1.1) contains the PLMN identity, tracking area code, and CI of the broadcasting cell. It also contains Q-RxLevMin, which is the minimum RSRP threshold that a broadcasting cell should be measured with before initial cell selection, and later on, random access is performed by the UE. There is also SIB Mapping Info included to inform the UE which SIBs are transmitted and how they are scheduled. Furthermore, the MAC decoder output of Message Example

Message Example 1.1 shows the SI-RNTI and transport/physical channel used to transmit SIB 1.

Message Example 1.1: SIB 1

```
+---------------------------------------------+-----------------------------------+
|ID Name                                      |Comment or Value                   |
+---------------------------------------------+-----------------------------------+
|56 05:43:34,555,032 RRC-UU K2AIR-PHY PDSCH LTE-RLC/MAC MAC-TM-PDU (DL)
LTE-RRC_BCCH_DL_SCH systemInformationBlockType1            |
|Tektronix K2Air LTE PHY Data Message Header (K2AIR-PHY) PDSCH
(= PDSCH Message)                                                              |
|1 PDSCH Message                                                                |
|1.1 Common Message Header                                                      |
|Protocol Version                             |0                                |
|Transport Channel Type                       |DL-SCH                           |
|Physical Channel Type                        |PDSCH                            |
|System Frame Number                          |454                              |
|Direction                                    |Downlink                         |
|Radio Mode                                   |FDD                              |
|Internal use                                 |0                                |
|Status                                       |Original data                    |
|Reserved                                     |0                                |
|Physical Cell ID                             |0                                |
|UE ID/RNTI Type                              |SI-RNTI                          |
|Subframe Number                              |5                                |
|UE ID/RNTI Value                             |'ffff'H                          |
|1.2 PDSCH Header                                                               |
|CRC report                                   |CRC ok                           |
|HARQ process number                          |0                                |
|Reserved                                     |0                                |
|Transport Block Indicator                    |single TB info                   |
|Reserved                                     |0                                |
|1.2.1 Transport Block#1 Information                                            |
|Transport Block#1 Size                       |144                              |
|Modulation Order DL 1                        |QPSK                             |
|New Data Indicator DL 1                      |new data                         |
|Redundancy Version DL 1                      |1                                |
|Reserved                                     |0                                |
|Modulation Scheme Index DL 1                 |5                                |
|Reserved                                     |0                                |
|1.2.2 Transport Block Data                                                     |
|TB1 Mac-PDU Data                             |40 51 00 21 00 00 20 00 10       |
|                                              0c 14 01 10 21 00 68 22 b6      |
|Padding                                      |'0068'H                          |
|1.3 Additional Call related Info                                               |
|Number Of Logical Channel Informations       |1                                |
|1.3.1 Logical Channel Information                                              |
|LCID                                         |0                                |
|RLC Mode                                     |Transparent Mode                 |
|Radio Bearer ID                              |0                                |
|Radio Bearer Type                            |Control Plane (Signalling)       |
|Spare                                        |0                                |
|Spare                                        |0                                |
|Logical Channel Type                         |BCCH                             |
|Call ID                                      |'fffffff5'H                      |
```

```
|3GPP LTE-RLC/MAC Rel.8 (MAC TS 36.321 V8.5.0, 2009-03, RLC TS 36.322
|V8.5.0, 2009-03) (LTE-RLC/ MAC) MAC-TM-PDU (DL) (= MAC PDU (Transparent
|Content Downlink))              |
|1 MAC PDU (Transparent Content Downlink)                            |
|MAC Transparent Data                     |40 51 00 21 00 00 20 00 10
|                                           0c 14 01 10 21 00 68 22 b6    |
|RRC (BCCH DL SCH) 3GPP TS 36.331 V8.5.0 (2009-03) (LTE-RRC_BCCH_DL_SCH)
|systemInformationBlockType1 (= systemInformationBlockType1)         |
|bCCH-DL-SCH-Message                                                     |
|1 message                                                               |
|1.1 Standard                                                            |
|1.1.1 systemInformationBlockType1                                       |
|1.1.1.1 cellAccessRelatedInfo                                           |
|1.1.1.1.1 plmn-IdentityList                                             |
|1.1.1.1.1.1 pLMN-IdentityInfo                                           |
|1.1.1.1.1.1.1 plmn-Identity                                             |
|1.1.1.1.1.1.1.1 mcc                                                     |
|1.1.1.1.1.1.1.1.1 mCC-MNC-Digit           |2                           |
|1.1.1.1.1.1.1.1.2 mCC-MNC-Digit           |9                           |
|1.1.1.1.1.1.1.1.3 mCC-MNC-Digit           |9                           |
|1.1.1.1.1.1.1.2 mnc                                                     |
|1.1.1.1.1.1.1.2.1 mCC-MNC-Digit           |0                           |
|1.1.1.1.1.1.1.2.2 mCC-MNC-Digit           |0                           |
|1.1.1.1.1.2 cellReservedForOperatorUse    |notReserved                 |
|1.1.1.1.2 trackingAreaCode                |'0000'H                     |
|1.1.1.1.3 cellIdentity                    |'2000100'H                  |
|1.1.1.1.4 cellBarred                      |notBarred                   |
|1.1.1.1.5 intraFreqReselection            |notAllowed                  |
|1.1.1.1.6 csg-Indication                  |false                       |
|1.1.1.2 cellSelectionInfo                                               |
|1.1.1.2.1 q-RxLevMin                      |-65                         |
|1.1.1.3 freqBandIndicator                 |1                           |
|1.1.1.4 schedulingInfoList                                              |
|1.1.1.4.1 schedulingInfo                                                |
|1.1.1.4.1.1 si-Periodicity                |rf16                        |
|1.1.1.4.1.2 sib-MappingInfo                                             |
|1.1.1.4.2 schedulingInfo                                                |
|1.1.1.4.2.1 si-Periodicity                |rf32                        |
|1.1.1.4.2.2 sib-MappingInfo                                             |
|1.1.1.4.2.2.1 sIB-Typ                     |sibType3                    |
|1.1.1.4.2.2.2 sIB-Type                    |sibType6                    |
|1.1.1.4.3 schedulingInfo                                                |
|1.1.1.4.3.1 si-Periodicity                |rf32                        |
|1.1.1.4.3.2 sib-MappingInfo                                             |
|1.1.1.4.3.2.1 sIB-Type                    |sibType5                    |
|1.1.1.5 si-WindowLength                   |ms20                        |
|1.1.1.6 systemInfoValueTag                |22                          |
```

SIB 2 contains timers and constants, barring information, UL frequency information, and UL bandwidth information. SIB 3 contains parameters for the cell reselection procedure. SIB 4 contains neighbor cell information for intra-frequency cell reselection. SIB 5 contains information for inter-frequency cell reselection. SIB 6 contains information for inter-RAT cell reselection to the UTRAN. SIB 7 contains information for inter-RAT cell reselection to the GERAN. SIB 8 contains

Table 1.23   RRC measurement event IDs and description

Event ID	Description
A1	Serving becomes better than threshold
A2	Serving becomes worse than threshold
A3	Neighbor becomes offset better than serving
A4	Neighbor becomes better than threshold
A5	Serving becomes worse than threshold 1 and neighbor becomes better than threshold 2
A6	Neighbour becomes offset better than Secondary Cell (SCell)
B1	Inter-RAT neighbor becomes better than threshold
B2	Serving becomes worse than threshold 1 and inter-RAT neighbor becomes better than threshold 2

information for inter-RAT cell reselection to CDMA2000. SIB 9 is used to broadcast the home eNB name (HNB name). SIB 10 and SIB 11 can be used to broadcast warning information to subscribers (e.g., tsunami warnings).

1.12.8.3   RRC Measurements

As in 3G UTRAN, the RRC measurement reports are expected to be sent mostly as event-triggered reports. The events listed in Table 1.23 have been specified so far (3GPP 36.331v.11.5.0 2013-09).

The thresholds mentioned in the event description refer to either RSRP or RSRQ measurements. This is specified when the measurement is set up. To ensure that only significant changes of the radio quality are reported, the measurements are guarded with the additional parameters of hysteresis, offset, and time-to-trigger. The hysteresis parameter is used to eliminate ping-pong effects in measurement reporting as shown in Figure 1.108. Here, the events A1 and A2 will only be reported for the serving cell measured by UE 1, but not for the serving cell of UE 2.

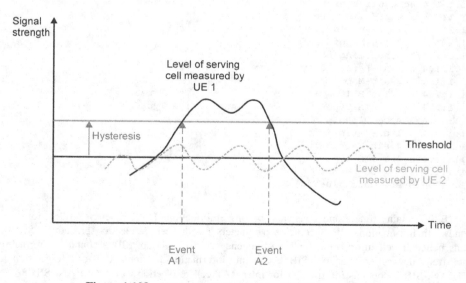

Figure 1.108   Hysteresis parameter for RRC measurements

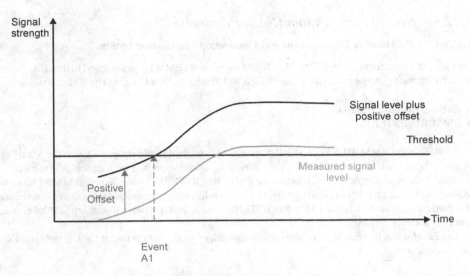

Figure 1.109   Offset parameter for RRC measurements

The offset parameter (see Figure 1.109) is not related to a predefined threshold, like hysteresis, but to the actual measurement result. The offset can have positive or negative values. The purpose of using the offset parameter is to speed up or slow down handover in/from strong/weak cells. Time-to-trigger should prevent short-time peaks of measured signals from triggering measurement reports and subsequent handover procedures. Looking at Figure 1.110, the setting of the time-to-trigger parameter prevents a handover to cell 2, where the radio quality would have dropped quickly back to below the required threshold. Instead, the call can be handed over to cell 1 after A1's report is sent, knowing that cell 1 offers not just a radio quality above the required threshold, but also stable radio conditions.

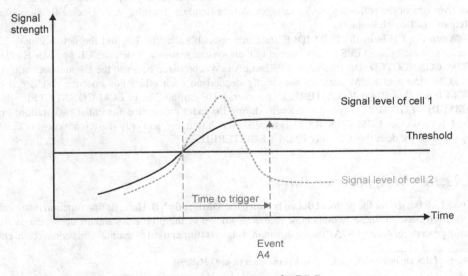

Figure 1.110   Time-to-trigger parameter for RRC measurements

1.12.9   Non-Access Stratum (NAS) Protocol

1.12.9.1   EPS Mobility Management and Connection Management States

On the NAS layer, there are two EPS Mobility Management (EMM) states defined to describe the results of mobility management procedures like Attach and Tracking Area Update. The EMM states are:

- EMM-DEREGISTERED
- EMM-REGISTERED

The UE in the EMM-DEREGISTERED state is not attached to the network. From the MME's point of view, there is no active context for a UE, no routing information, and no location information.

After a successful attach, the UE enters the EMM-REGISTERED state. Now the MME knows where to page the UE and the HSS is able to provide routing information for mobile terminating connections of this particular subscriber. In the EMM-REGISTERED state, the UE always has an active PDN connection and EPS security context.

The ECM states indicate if there is an active signaling connectivity between the UE and the EPC. The ECM states are:

- ECM-IDLE
- ECM-CONNECTED

The location of a UE in the ECM-IDLE state is known by the network on behalf of the current Tracking Area List. This list contains all tracking areas for which the UE performed successful initial registration (attach) to the network and subsequent successful tracking area update procedures. All tracking areas in a Tracking Area List to which a UE is registered must be connected to the same MME (according to 3GPP 23.401).

A UE is in the ECM-IDLE state when no active NAS signaling connection between the UE and network exists. In other words, there are no SRBs assigned to this mobile. If the UE changes its geographic position, it may perform cell selection/reselection when necessary.

The UE and MME should enter the ECM-CONNECTED state whenever the UE sends or MME receives one of the following NAS messages: Attach Request, Tracking Area Update Request, Service Request, or Detach Request.

When the UE is in the ECM-IDLE state, it is possible that the UE and the network have different sets of established EPS bearers stored in their system memory. When the UE and MME enter the ECM-CONNECTED state, the set of EPS bearers is synchronized between the UE and network.

ECM state and EMM state do not strictly depend on each other. For instance, a UE can be in ECM-IDLE, but EMM-REGISTERED. However, a UE can also be in ECM-CONNECTED while in EMM-DEREGISTERED, yet this happens during the radio connection for initial registration. Firstly, the UE must enter ECM-CONNECTED to send Attach Request, and only if Attach Accept is sent back by the network does this UE enter EMM-REGISTERED.

1.12.10   S1 Application Part (S1AP)[6]

The S1AP is used for the protocol dialog between the MME and eNB. Here, its first function is to establish a relation between these two network elements, known in the protocol standard document as the S1 setup procedure. A set of S1AP management messages is further used to maintain this connection, change

---

[6] Parts of this chapter are reproduced with permission from © 3GPP™.

configuration parameters, indicate possible errors and overload situations, balance the load, and reset, re-establish, or release the connection.

Once the relation between the eNB and MME is successfully established, UEs camping on the cells of this eNB can be paged using the S1AP to establish mobile terminating connections or can register and set up mobile originating calls by themselves. In the case of such a registration of connection setup procedure, the S1AP provides functions for transparent forwarding of NAS messages. During connection setup, the S1AP is in charge of the tunnel management of S1-U GTP tunnels. Using the S1AP signaling procedure, these S1 user plane tunnels can be established, modified according to changing QoS parameters, and released. The S1-U GTP tunnels are part of the E-RAB and in fact, it is the S1AP that provides full E-RAB management functions. Setup and modification of E-RABs are always triggered by the MME, while the release of E-RABs can be triggered by both the MME and eNB.

In using the S1AP during initial registration and connection setup, an S1 UE context is established in the eNB. Each S1AP connection related to a particular UE is identified by a pair of MME UE S1AP ID and eNB UE S1AP ID. These connection identifiers link all S1AP messages belonging to a single subscriber except paging messages (i.e., sent before the UE context is established). Table 1.24 shows the occurrence of the UE S1AP IDs in the different UE-related messages of this protocol.

Messages that are used to establish and maintain the relation between the MME and eNB on the network element level, like the messages of the S1 setup procedure, do not contain any UE ID.

As can be seen in Table 1.24, the S1AP is also responsible for different mobility functions. Belonging to this category is the path switch procedure, that is, a handover between eNBs connected to the same MME. Inter-MME handover and inter-RAT handover are executed using the set of S1AP handover messages – for more details on this topic, see Chapter 2.

E-RABs can be set up, modified, and released at any time during an active connection. Hence, separate E-RAB management messages are provided by the S1AP.

In addition, there are messages for the transmission of broadcast warning messages and for location reporting. The S1AP location report will keep the MME update about the ID of the serving cell currently used by the UE.

Looking a little deeper into the protocol details of the S1AP, it emerges that there are two classes of S1AP elementary procedures. Class 1 procedures are characterized by the fact that the initiating message should be answered with a successful or unsuccessful outcome message. The complete list of Class 1 elementary procedures and messages according to 3GPP 36.4313 is given in Table 1.25.

Class 1 elementary procedures are characterized by the fact that each request sent is positively or negatively acknowledged by an appropriate success or failure message.

As known from UTRAN RANAP, there is a difference between message names used in standard documents and encoding of these messages following ASN.1 encoding rules. An example will be given using the encoding of messages of the handover preparation procedure that is defined in the ASN.1 part of 3GPP 36.413 as follows:

```
handoverPreparation S1AP-ELEMENTARY-PROCEDURE ::= {
    INITIATING MESSAGE   HandoverRequired
    SUCCESSFUL OUTCOME   HandoverCommand
    UNSUCCESSFUL OUTCOME     HandoverPreparationFailure
    PROCEDURE CODE           id-HandoverPreparation
    CRITICALITY     reject
}
```

Translated into a written explanation, this ASN.1 notation means the handover required message will be encoded as Initiating Message with the procedure code = "id-HandoverPreparation," the handover command message will be encoded as Successful Outcome with the procedure

Table 1.24   S1AP UE IDs in UE-related messages

S1-AP UE related messages	MME UE S1AP ID	eNB UE S1AP ID
E-RAB setup Request	×	×
E-RAB setup Response	×	×
E-RAB Modify Request	×	×
E-RAB Modify Response	×	×
E-RAB Release Command	×	×
E-RAB Release Response	×	×
E-RAB Release Indication	×	×
Initial Context Setup Request	×	×
Initial Context Setup Response	×	×
Initial Context Setup Failure	×	×
UE Context Release Request	×	×
UE Context Release Command (optionally there is only the MME UE S1AP ID)	×	× (but optional)
UE Context Release Complete	×	×
UE Context Modification Request	×	×
UE Context Modification Response	×	×
UE Context Modification Failure	×	×
Handover Required	×	×
Handover Command	×	×
Handover Preparation Failure	×	×
Handover Request	×	No
Handover Request Acknowledge	×	×
Handover Failure	×	No
Handover Notify	×	×
Path Switch Request	No	×
Path Switch Request Acknowledge	×	×
Path Switch Request Failure	×	×
Handover Cancel	×	×
Handover Cancel Acknowledge	×	×
eNB Status Transfer	×	×
MME Status Transfer	×	×
Paging	No	No
Initial UE Message	No	×
Uplink NAS Transport	×	×
Downlink NAS Transport	×	×
NAS Non-Delivery Indication	×	×
Location Reporting Control	×	×
Location Report Failure Indication	×	×
Location Report	×	×

code = "id-HandoverPreparation," and Handover Preparation Failure will be encoded as Unsuccessful Outcome with the procedure code = "id-HandoverPreparation."

In contrast to the Class 1 message, the messages of Class 2 are unidirectional messages that are not explicitly acknowledged by the receiving entity. The complete list of Class 2 procedures and messages according to 3GPP 26.314 is given in Table 1.26.

Table 1.25   S1AP Class 1 elementary procedures

Elementary procedure	Initiating message	Successful outcome	Unsuccessful outcome
		Response message	Response message
Handover Preparation	HANDOVER REQUIRED	HANDOVER COMMAND	HANDOVER PREPARATION FAILURE
Handover Resource Allocation	HANDOVER REQUEST	HANDOVER REQUEST ACKNOWLEDGE	HANDOVER FAILURE
Path switch REquest	PATH SWITCH REQUEST	PATH SWITCH REQUEST ACKNOWLEDGE	PATH SWITCH REQUEST FAILURE
Handover Cancellation	HANDOVER CANCEL	HANDOVER CANCEL ACKNOWLEDGE	–
E-RAB SETup	E-RAB SETUP REQUEST	E-RAB SETUP RESPONSE	–
E-RAB MODify	E-RAB MODIFY REQUEST	E-RAB MODIFY RESPONSE	–
E-RAB RELease	E-RAB RELEASE COMMAND	E-RAB RELEASE RESPONSE	–
Initial Context Setup	INITIAL CONTEXT SETUP REQUEST	INITIAL CONTEXT SETUP RESPONSE	INITIAL CONTEXT SETUP FAILURE
Reset	RESET	RESET ACKNOWLEDGE	–
S1 SETUP	S1 SETUP REQUEST	S1 SETUP RESPONSE	S1 SETUP FAILURE
UE Context Release	UE CONTEXT RELEASE COMMAND	UE CONTEXT RELEASE COMPLETE	–
UE Context Modification	UE CONTEXT MODIFICATION REQUEST	UE CONTEXT MODIFICATION RESPONSE	UE CONTEXT MODIFICATION FAILURE
eNB Configuration Update	ENB CONFIGURATION UPDATE	ENB UPDATE CONFIGURATION ACKNOWLEDGE	ENB CONFIGURATION UPDATE FAILURE
MME Configuration Update	MME CONFIGURATION UPDATE	MME CONFIGURATION UPDATE ACKNOWLEDGE	MME CONFIGURATION UPDATE FAILURE
Write-Replace Warning	WRITE-REPLACE WARNING REQUEST	WRITE-REPLACE WARNING RESPONSE	–

Source: Reproduced with permission from © 3GPP™.

Table 1.26   S1AP Class 2 elementary procedures

Elementary procedure	Message
Handover Notification	HANDOVER NOTIFY
E-RAB Release Request	E-RAB RELEASE REQUEST
Paging	PAGING
Initial UE Message	INITIAL UE MESSAGE
Downlink NAS Transport	DOWNLINK NAS TRANSPORT
Uplink NAS Transport	UPLINK NAS TRANSPORT
NAS Non-Delivery Indication	NAS NON DELIVERY INDICATION
Error Indication	ERROR INDICATION
UE Context Release Request	UE CONTEXT RELEASE REQUEST
Downlink S1 CDMA2000 Tunneling	DOWNLINK S1 CDMA2000 TUNNELING
Uplink S1 CDMA2000 Tunneling	UPLINK S1 CDMA2000 TUNNELING
UE Capability Info Indication	UE CAPABILITY INFO INDICATION
eNB Status Transfer	eNB STATUS TRANSFER
MME Status Transfer	MME STATUS TRANSFER
Deactivate Trace	DEACTIVATE TRACE
Trace Start	TRACE START
Trace Failure Indication	TRACE FAILURE INDICATION
Location Reporting Control	LOCATION REPORTING CONTROL
Location Reporting Failure Indication	LOCATION REPORTING FAILURE INDICATION
Location Report	LOCATION REPORT
Overload Start	OVERLOAD START
Overload Stop	OVERLOAD STOP
eNB Direct Information Transfer	eNB DIRECT INFORMATION TRANSFER
MME Direct Information Transfer	MME DIRECT INFORMATION TRANSFER

Source: Reproduced with permission from © 3GPP™.

Call scenarios, message examples, and the most important parameters of S1AP messages are described in Chapter 2.

1.12.11   User Datagram Protocol (UDP)

UDP is a connectionless transport protocol (ISO OSI Model Layer 4), that is, a member of the IP family. UDP is used to transmit higher data quickly from one end of a connection to the other. The source and destination port information elements in the UDP datagram header identify the applications on top of UDP. In the example shown in Figure 1.111, the higher layer application is GTP, so this sample UDP

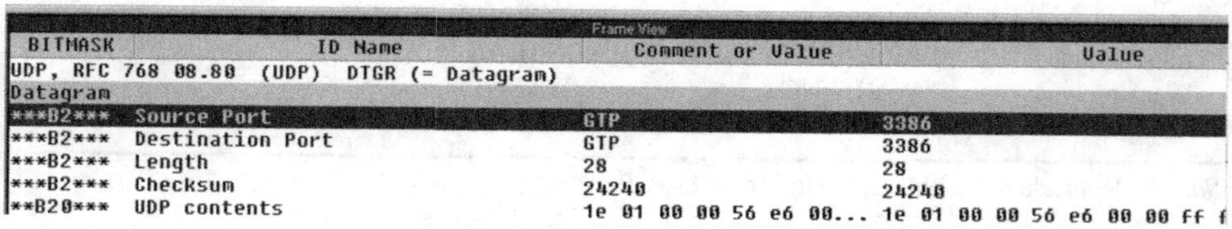

BITMASK	ID Name	Comment or Value	Value
UDP, RFC 768 08.80 (UDP)	DTGR (= Datagram)		
Datagram			
B2	Source Port	GTP	3386
B2	Destination Port	GTP	3386
B2	Length	28	28
B2	Checksum	24240	24240
B20*	UDP contents	1e 01 00 00 56 e6 00...	1e 01 00 00 56 e6 00 00 ff f

Figure 1.111   UDP datagram

frame is lower layer UDP as can be found on all interfaces of the mobile network that allow tunneling of IP payload.

1.12.12 GPRS Tunneling Protocol (GTP)[7]

GTP is used on various EPC interfaces. Its main functionality is to create, modify, and delete tunnels for IP payload transport. However, besides the tunnel management functionality, the GTP control plane protocol also offers functions and messages for:

- Path management
- Mobility management
- CS fallback
- Non-3GPP-related access
- Trace management

Table 1.27 gives an overview of all GTP messages. As in the S1AP, we see message pairs of connection-oriented procedures, for instance, Create Session Request/Create Session Response. However, there is also a set of connectionless unidirectional messages like Modify Bearer Command.

It should further be noticed that the name of a message often indicates who initiates a connection setup or change of connection parameters. If there is a connection set up by the mobile, the create session procedure will be used to activate the default bearer. If the network requests a bearer establishment, it will perform the create bearer procedure. The same differentiation can be seen in the case of bearer modification: a modification requested by the UE side will be performed using the modify bearer procedure; a modification requested by the network is executed using the update bearer procedure.

All in all, the use of GTP on various interfaces in the EPC has the advantage of simplified protocol stacks. This is also an advantage for NEMs, because the implementation effort for network element protocol entity software can be reduced due to the fact that the same or very similar code can be used to program the software, for example, S-GW and PDN-GW. However, for measurement equipment manufacturers like Tektronix Communications, the simplified protocol stacks make it harder to track messages of a single connection across multiple interfaces. Additional algorithms in the software are necessary to program call trace functions (filter and sequence messages belonging to a single connection) and to correlate performance measurement statistical counters and measurement with the different network elements. In other words, in the past, it was quite clear that the GTP-C protocol was only used on the Gn interface between SGSN and GGSN. Now, in EPC, the GTP-C protocol is found on many different interfaces and a bunch of different network elements are using it. Further, a state-of-the-art protocol tester or network monitoring system is required to detect automatically which interfaces and network elements are monitored.

1.12.12.1 Path Management

The path management messages are used to check the activity status of a GTP tunnel on behalf of a heartbeat check mechanism. The messages involved in this procedure are GTP Echo Request and GTP Echo Response as shown in Figure 1.112.

GTP Echo Request is sent with a sequence number that is expected to be received again in a GTP Echo Response sent by the peer entity. The sending of Echo Request is triggered by a timer, but should not happen more often than one Echo Request in 60 seconds for each GTP tunnel.

The functions of the non-3GPP-related access ensure interworking (handover) between E-UTRAN and the CDMA2000 HRDP radio access network.

---

[7] Parts of this chapter are reproduced with permission from © 3GPP™.

Table 1.27 GTP messages

Message type value (decimal)	Message	Reference	GTP-C	GTP-U
0	Reserved	–	–	–
1	Echo Request	–	×	×
2	Echo Response	–	×	×
3	Version Not Supported Indication	–	×	–
4–24	Reserved for S101 interface	TS 29.276 [14]	–	–
25–31	Reserved for Sv interface	TS 29.280 [15]	–	–
SGSN/MME to PGW (S4/S11, S5/S8)				
32	Create Session Request	–	×	–
33	Create Session Response	–	×	–
34	Modify Bearer Request	–	×	–
35	Modify Bearer Response	–	×	–
36	Delete Session Request	–	×	–
37	Delete Session Response	–	×	–
38	Change Notification Request	–	×	–
39	Change Notification Response	–	×	–
40–63	For future use	–	–	–
Messages without explicit response				
64	Modify Bearer Command (MME/SGSN to PGW – S11/S4, S5/S8)	–	×	–
65	Modify Bearer Failure Indication (PGW to MME/SGSN – S5/S8, S11/S4)	–	×	–
66	Delete Bearer Command (MME to PGW – S11, S5/S8)	–	×	–
67	Delete Bearer Failure Indication (PGW to MME – S5/S8, S11)	–	×	–
68	Bearer Resource Command (MME/SGSN to PGW – S11/S4, S5/S8)	–	×	–
69	Bearer Resource Failure Indication (PGW to MME/SGSN – S5/S8, S11/S4)	–	×	–
70	Downlink Data Notification Failure Indication (SGSN/MME to S-GW – S4/S11)	–	×	–
71	Trace Session Activation	–	×	–
72	Trace Session Deactivation	–	×	–
73	Stop Paging Indication	–	×	–
74–94	For future use	–	–	–
PGW to SGSN/MME (S5/S8, S4/S11)				
95	Create Bearer Request	–	×	–
96	Create Bearer Response	–	×	–
97	Update Bearer Request	–	×	–
98	Update Bearer Response	–	×	–
99	Delete Bearer Request	–	×	–
100	Delete Bearer Response	–	×	–
101–127	For future use	–	–	–

Table 1.27 (*continued*)

Message type value (decimal)	Message	Reference	GTP-C	GTP-U
	MME to MME, SGSN to MME, MME to SGSN, SGSN to SGSN (S3/10/S16)			
128	Identification Request	–	×	–
129	Identification Response	–	×	–
130	Context Request	–	×	–
131	Context Response	–	×	–
132	Context Acknowledge	–	×	–
133	Forward Relocation Request	–	×	–
134	Forward Relocation Response	–	×	–
135	Forward Relocation Complete Notification	–	×	–
136	Forward Relocation Complete Acknowledge	–	×	–
137	Forward Access Context Notification	–	×	–
138	Forward Access Context Acknowledge	–	×	–
139	Relocation Cancel Request	–	×	–
140	Relocation Cancel Response	–	×	–
141	Configuration Transfer Tunnel	–	×	–
142–148	For future use	–	–	–
	SGSN to MME, MME to SGSN (S3)			
149	Detach Notification	–	×	–
150	Detach Acknowledge	–	×	–
151	CS Paging Indication	–	×	–
152	RAN Information Relay	–	–	–
153–159	For future use	–	–	–
	MME to S-GW (S11)			
160	Create Forwarding Tunnel Request	–	×	–
161	Create Forwarding Tunnel Response	–	×	–
162	Suspend Notification	–	×	–
163	Suspend Acknowledge	–	×	–
164	Resume Notification	–	×	–
165	Resume Acknowledge	–	×	–
166	Create Indirect Data Forwarding Tunnel Request	–	×	–
167	Create Indirect Data Forwarding Tunnel Response	–	×	–
168	Delete Indirect Data Forwarding Tunnel Request	–	×	–
169	Delete Indirect Data Forwarding Tunnel Response	–	×	–
170	Release Access Bearers Request	–	×	–
171	Release Access Bearers Response	–	×	–
172–175	For future use	–	–	–
	S-GW to SGSN/MME (S4/S11)			
176	Downlink Data Notification	–	×	–
177	Downlink Data Notification Acknowledgement	–	×	–
178	Update Bearer Complete	–	×	–
179–191	For future use	–	–	–
	Other			
192–255	For future use	–	–	–

Source: Reproduced with permission from © 3GPP™.

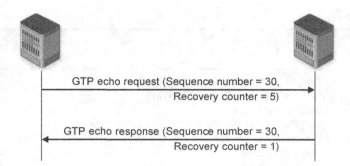

GTP echo request (Sequence number = 30,
Recovery counter = 5)

GTP echo response (Sequence number = 30,
Recovery counter = 1)

Figure 1.112   GTP path management

Trace management is a term that covers all necessary functions to track the activities of selected subscribers. The trace functionality can be used for network element troubleshooting or legal interception.

The path management heartbeat messages are found on both GTP-C and GTP-U connections.

1.12.12.2   Tunnel Management

The create session request message should be sent on the S11 interface by the MME to the S-GW and on the S5/S8 interface by the S-GW to the PDN GW as part of the procedure:

- E-UTRAN initial attach
- UE requested PDN connectivity

The message should also be sent on the S11 interface by the MME to the S-GW in case the S-GW is changed due to mobility. In such cases, the create session procedure is embedded in a tracking area update or handover/relocation procedure.

In the create session request message, the IMSI is included as a mandatory information element. The network elements on both ends of the GTP-C connection will be identified by their Tunnel Endpoint Identifiers (TEIDs) that during initial assignment are signaled together with the IP addresses of the appropriate network element (see Section 2.2 for more details). Since there will always be a default bearer established when a session is activated, the EPS Bearer ID (EBI) is included as a mandatory information element as well. Together with the TEIDs for the GTP-C signaling connection, a unique triplet of information elements is formed that can always be used to identify a single subscriber's session or bearer unambiguously on any GTP-C interface. The TEIDs are not only used to identify the network elements on both sides of the GTP tunnel, but also used to determine the direction in which the message is sent. This is especially important for monitoring and analyzing user plane traffic.

Figure 1.113 shows the key parameters involved in tunnel management procedures. TEID a and TEID b are used to identify the participating core network elements on the control plane level. EBI X and EBI Y represent two different bearers for the same UE. A maximum of up to 256 bearers per UE is supported. Now, within each bearer, there is a unique pair of TEIDs that is used to route the traffic in the UL and DL directions. Since the EBI is only signaled on the control plane, these TEIDs are also used to identify the tunnel established for a particular bearer.

In addition to the differences in TEID parameter values, the IP addresses configured on the transport network layer for user plane and control plane transport are often different as well. However, this is not a standardized setting.

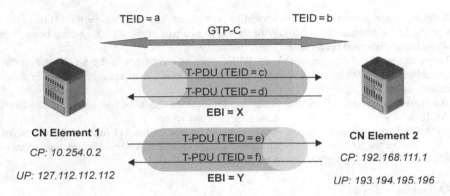

Figure 1.113 GTP tunnel management

The create bearer request message should be sent on the S5/S8 interface by the PGW (Packet Gateway) to the S-GW and on the S11 interface by the S-GW to the MME as part of the dedicated bearer activation procedure. It contains the EBI as a mandatory information element that links the new dedicated bearer with the previously established session.

If the UE wishes to modify an established bearer, a bearer resource command message should be sent from the MME to the S-GW and forwarded to the PDN-GW. The Bearer Resource Command is triggered by a NAS message with the same name received by the MME. Again, it is the EBI information element and the TEIDs that link this.

A bearer resource failure indication should be sent by the PGW to an S-GW and forwarded to the MME to indicate failure of the UE requested bearer resource modification procedure.

The modify bearer request message should only be sent on the S11 interface by the MME to the S-GW and on the S5/S8 interfaces by the S-GW to the PDN GW in case of Tracking Area Update without changing the S-GW, UE-triggered Service Request, or UE.

A DL data notification procedure is used on the S11 interface between the S-GW and the MME to execute the network-triggered service request procedure (mobile terminated PS connection request).

The delete indirect data forwarding tunnel request message is sent on the S11 interface by the MME to the S-GW as part of handover procedures.

The Modify Bearer Command should be sent on the S11 interface by the MME to the S-GW, on the S5/S8 interface by the S-GW to the P-GW, and on the S4 interface by the SGSN to the S-GW whenever the subscribed QoS needs to be modified according to QoS information stored in the HSS.

For GTP-based S5/S8, the Update Bearer Request should be sent by the PGW to the S-GW and forwarded to the MME by the following procedures:

- PGW-initiated bearer modification with bearer QoS update
- HSS-initiated subscribed QoS modification
- PGW-initiated bearer modification without bearer QoS update
- UE request bearer resource modification procedure.

The message should also be sent on the S5/S8 interface by the PGW to the S-GW and on the S4 interface by the S-GW to the SGSN as part of the following procedures:

- PGW-initiated EPS bearer modification
- Execution part of MS-initiated EPS bearer modification
- SGSN-initiated EPS bearer modification procedure using S4.

A delete bearer command message should be sent on the S11 interface by the MME to the S-GW and on the S5/S8 interface by the S-GW to the PGW as a part of the eNB requested bearer release or MME-initiated dedicated bearer deactivation procedure.

The message should also be sent on the S4 interface by the SGSN to the S-GW and on the S5/S8 interface by the S-GW to the PGW as part of the MS and SGSN-initiated non-default bearer deactivation procedure using S4.

The create indirect data forwarding tunnel request message should be sent on the S11/S4 interface by the MME/SGSN to the S-GW as part of the handover procedures.

The release access bearers request message should be sent on the S11 interface by the MME to the S-GW as part of the S1 release procedure. The message should also be sent on the S4 interface by the SGSN to the S-GW as part of the following procedures:

- RAB release using S4
- Iu release using S4

A stop paging indication message should be sent on the S11/S4 interface by the S-GW to the MME/SGSN as part of the network-triggered service request procedure.

1.12.12.3 Mobility Management

A forward relocation request message should be sent from the source MME to the target MME over the S10 interface as part of S1-based handover relocation procedure from the source MME to the target SGSN, or from the source SGSN to the target MME over the S3 interface as part of inter-RAT handover and combined hard handover and SRNS relocation procedures, or from the source SGSN to the target SGSN over the S16 interface as part of the SRNS relocation and PS handover procedures.

The new MME/SGSN should send the context request message to the old MME/SGSN on the S3/S16/S10 interface as part of the TAU/RAU procedure to get the MM and EPS bearer contexts for the UE. If the UE identifies itself with a temporary identity and it has changed SGSN/MME since detaching in the attach procedure, the new MME/SGSN should send an identification request message to the old SGSN/MME over the S3, S16, or S10 interface to request IMSI.

A forward access context notification message should be sent from the old SGSN to the new SGSN over the S16 interface to forward the RNC contexts to the target system, or sent from the old MME to the new MME over the S10 interface to forward the RNC/eNB contexts to the target system.

A detach notification message should be sent from an MME to the associated SGSN, or from an SGSN to the associated MME, as part of the detach procedure if the ISR (Idle Mode Signaling Reduction) is activated between the MME and SGSN for the UE. Possible cause values are:

- "Local Detach"
- "Complete Detach"

"Local Detach" indicates that this detach is local to the MME/SGSN and so the associated SGSN/MME registration where the ISR is activated should not be detached. The MME/SGSN that receives this message including this cause value of "Local Detach" only deactivates the ISR. This cause value should be included in the following procedures:

- MME/SGSN-initiated detach procedure in case of implicit detach.
- HSS-initiated detach procedure.

"Complete Detach" indicates that both the MME registration and the SGSN registration that the ISR is activated for should be detached. This "Complete Detach" cause value should be included in the following procedures:

- UE-initiated detach procedure.
- MME/SGSN-initiated detach procedure in case of explicit detach.

The change notification request message is sent on the S4 interface by the SGSN to the S-GW and on the S5/S8interface by the S-GW to the PGW as part of location-dependent charging-related procedures.

A relocation cancel request message should be sent from the source MME/SGSN to the target MME/SGSN on the S3/S10/S16 interface as part of the inter-RAT handover cancel procedure and on the S16 interface as part of the SRNS relocation cancel procedure. A relocation cancellation typically happens if:

- a relocation timer expires in the source MME/RNC – this means the relocation/handover cannot be performed within an expected time;
- the UE cannot hand over on the radio interface either because of defects in the UE software (as seen with early releases of HSDPA data cards) or because of the wrong radio interface parameters assigned by the target system (e.g., wrong BCCH ARFCN of the GSM cell).

A configuration transfer tunnel message should be used to tunnel eNB configuration transfer messages from a source MME to a target MME over the S10 interface. The purpose of the eNB direct configuration transfer is to transfer information from an eNB to another eNB in UM. This is a possibility to exchange important information (e.g., UL interference status) between eNBs that are not physically connected using the X2 interface.

The RAN information relay message should be sent on the S3 interface between the SGSN and MME to transfer the RAN information received by an SGSN from BSS or RNS (GERAN Iu mode) or by an MME from eNB.

1.12.12.4   CS Fallback

The purpose of CS fallback is to send service requests for voice calls from LTE/EPC to UMTS/GSM. Network operators running both GSM/UMTS and 4G networks in parallel may optimize their resource usage by reserving the high-speed PS transmission capabilities of LTE and EPC for PS traffic, while voice calls can be delegated by using CS fallback functions to the reliable and profitable legacy GSM and UMTS network. The following GTP messages support this functionality.

The suspend notification message should be sent on the S11 interface by the MME to the S-GW as part of the CS fallback from E-UTRAN access to UTRAN/GERAN CS domain access-related procedures. It is answered with a suspend acknowledge message by the receiving entity. To suspend the call means to stop the call proceeding for a short time, here as long as is necessary to switch the signaling connection from 4G to UMTS/GSM.

The resume notification message should be sent on the S11 interface by the MME to the S-GW as part of the resume procedure returning from CS fallback to the E-UTRAN. With the resume notification message, a previously suspended call is taken back into service. Resume Notification is answered by the receiving entity with Resume Acknowledge.

The CS paging indication should be sent on the S3 interface by the MME to the associated SGSN when ISR is activated as part of mobile-terminated CS services. This will happen if a UE was redirected to UTRAN/GERAN radio access by CS fallback, but paging is still received over the EPS.

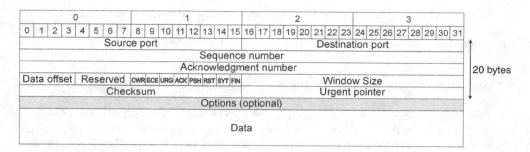

Figure 1.114   TCP header format

1.12.12.5   Non-3GPP-Related Access

The messages for non-3GPP-related access are used to support mobility between E-UTRAN and CDMA2000 radio access networks.

A create forwarding tunnel request message should be sent by an MME to an S-GW as part of the MME configuration resources for indirect data forwarding during the active handover procedure from E-UTRAN to CDMA2000 HRPD access. This message is answered with a create forwarding tunnel response message sent back by the receiving entity.

1.12.12.6   Trace Management

The trace session activation message should be sent on S11 by the MME to the S-GW, and on S5/S8 by the S-GW to the PGW, when session trace is activated for a particular IMSI or IMEI for a UE that is attached and active or attached and idle.

The trace session deactivation message should be sent on S11 by the MME to the S-GW, and on S5/S8 by the S-GW to the PGW, when session trace is deactivated for a particular subscriber that is attached and active or attached and idle.

The report data of the trace functionality has a proprietary structure defined by the individual NEMs.

1.12.13   Transmission Control Protocol (TCP)

Like its sister protocol UDP, TCP is a member of the IP family. However, TCP is a reliable, connection-oriented protocol that has capabilities for error detection, error correction, and in-sequence delivery of data streams. Figure 1.114 shows the generic header format of TCP frames.

As in case of UDP, the source and destination ports identify the application on top of the transport layer. Each TCP packet is sent with a unique sequence number that is acknowledged by the receiver after error-free reception of the packet. There is no separate ACK message. Instead, the ACK is transmitted using the next available frame in the opposite direction.

The meaning of the other TCP header information elements is as follows:

- Data offset: The length of the TCP header without the data field (payload).
- Flags (8 bits) (aka control bits): These are eight 1-bit flags with the following meaning:
    - Congestion Window Reduced (CWR) (1 bit): This flag is set by the sending host to indicate that it received a TCP segment with the ECN-Echo (ECE) flag set and had responded in the congestion control mechanism.

- ECE (1 bit): Value (1) indicates that the TCP peer is ECN capable during three-way handshake, and value (2) indicates that a packet with the congestion experienced flag in the IP header set is received during normal transmission.
- URG (1 bit): This is the URGent pointer field for datagrams with high priority.
- ACK (1 bit): This indicates that the ACK field is significant.
- PSH (1 bit): The push function.
- RST (1 bit): Reset the connection.
- SYN (1 bit): Synchronize sequence numbers.
- FIN (1 bit): No more data from sender. This flag should trigger the TCP FIN procedure to terminate the TCP connection.

- Window (16 bits): This is the size of the receive window, which specifies the number of bytes (beyond the sequence number in the ACK field) that the receiver is currently willing to receive.
- Checksum (16 bits): This 16-bit field is used for error checking of the header *and* the payload data.
- Urgent pointer (16 bits): If the URG flag is set, then this 16-bit field is an offset from the sequence number indicating the last urgent data byte.
- Options (variable bits): This is an appendix with optional additional information. The total length of the option field must be a multiple of a 32-bit word and the data offset field adjusted appropriately.

Each TCP connection starts with a so-called three-way handshake, as shown in the first three lines of Figure 1.115.

Firstly, the side that wants to establish a TCP connection sends a TCP SYN message. The sequence number in this SYN message is a random number. The ACK number is always 0, because no received frame needs to be acknowledged.

When the peer entity receives the SYN message, it sends back a SYN-ACK including the previous sequence number + 1 as the ACK number. The originating party of the connection will do the same to acknowledge the SYN-ACK with the first payload packet to be sent. Starting with this first payload packet, the size of the payload data field is encoded in the sequence number and the SYN-ACK is again acknowledged with its sequence number value + 1.

Due to the fact that the initial round-trip time must be calculated on each side of the connection, the TCP startup procedure may take quite a while, especially if there is a long distance or a long chain of network nodes between the endpoints. Based on delay measurement results, it has become common to describe this typical behavior as "TCP slow start."

TCP detects missing packets on behalf of timer expiry. The initial timer values on both sides of the connection are calculated during the startup procedure as the initial round-trip time. If the expected ACK for a sent packet does not arrive before the timer in the sending entity expires, the previously sent frame is retransmitted as shown in Figure 1.116.

During the ongoing connection, the round-trip time on both endpoints of the connection is adjusted multiple times. However, due to outliers in end-to-end latency, it is also possible that duplicated messages

4. Prot	4. MSG	5. Prot	5. MSG	Source Port	Destination Port	Sequence Number	Acknowledgment Number
TCP	syn			2867 : esps-portal	21 : ftp - File Transfer [Control]	4062981465	0
TCP	syn-ack			21 : ftp - File Transfer [Control]	2867 : esps-portal	1886411453	4062981466
TCP	ack			2867 : esps-portal	21 : ftp - File Transfer [Control]	4062981466	1886411454
TCP	ack	FTP	repl	21 : ftp - File Transfer [Control]	2867 : esps-portal	1886411454	4062981466
TCP	ack	FTP	cmd	2867 : esps-portal	21 : ftp - File Transfer [Control]	4062981466	1886411474
TCP	ack	FTP	repl	21 : ftp - File Transfer [Control]	2867 : esps-portal	1886411474	4062981479
TCP	ack	FTP	cmd	2867 : esps-portal	21 : ftp - File Transfer [Control]	4062981479	1886411508
TCP	ack	FTP	repl	21 : ftp - File Transfer [Control]	2867 : esps-portal	1886411508	4062981495
TCP	ack	FTP	cmd	2867 : esps-portal	21 : ftp - File Transfer [Control]	4062981495	1886411531
TCP	ack	FTP	repl	21 : ftp - File Transfer [Control]	2867 : esps-portal	1886411531	4062981503
TCP	ack	FTP	cmd	2867 : esps-portal	21 : ftp - File Transfer [Control]	4062981503	1886411562
TCP	ack	FTP	repl	21 : ftp - File Transfer [Control]	2867 : esps-portal	1886411562	4062981511
TCP	ack	FTP	cmd	2867 : esps-portal	21 : ftp - File Transfer [Control]	4062981511	1886411582
TCP	ack	FTP	repl	21 : ftp - File Transfer [Control]	2867 : esps-portal	1886411582	4062981519

Figure 1.115 TCP startup for an FTP service

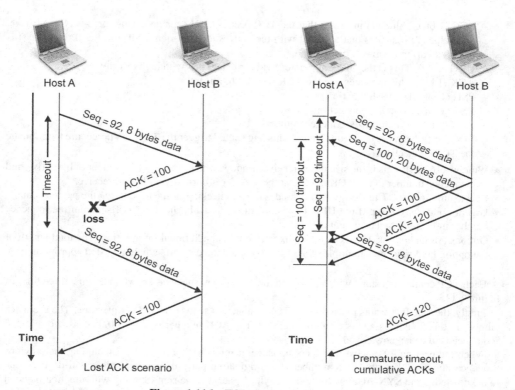

Figure 1.116   TCP retransmission scenarios

are sent in case the receiving entity detects timeout and hence a necessary retransmission, but the expected frame is indeed received – after the retransmission was already sent.

Malfunctions, lost frames, and duplicated ACKs have an impact on the network QoS. They slow down the transmission of payload data. Duplicated ACKs are a waste of precious bandwidth on the radio interface and necessary retransmissions increase the payload packet delay.

1.12.14   Session Initiation Protocol (SIP)

SIP is used to establish VoIP calls using the E-UTRAN as the access network. On E-UTRAN interfaces, the SIP signaling will run as user plane in-band signaling, which means it can only be monitored inside the user plane bearers.

As all protocols, SIP has its specific terms. In SIP, we do not see plain messages, but "methods" that are invoked by clients. Important SIP methods are:

- **REGISTER**: Used to bind a SIPURI address to an IP address.
- **INVITE**: Sent by the originating party to initiate a SIP session. It can be also used to modify an existing session, for instance, to add or remove a data stream.
- **BYE**: Used to terminate a session initiated with INVITE.

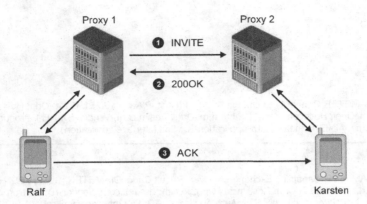

Figure 1.117   SIP session initiation

- **CANCEL**: Used to cancel a pending request, for example, if INVITE was answered with 200OK, but final ACK is missed.
- **OPTIONS**: Can be used by a SIP agent to request a capability list from another SIP agent or from a SIP proxy.
- **MESSAGE**: Used for instant messaging.

In return to a method invoked, the receiving SIP party sends back a response code to the originating party. Basically, these response codes are identical to the ones used in HTTP/1.1. For a full list, refer to RFC 3261.

The most common response code is 200OK, which is typically sent to acknowledge the reception of a SIP INVITE. Figure 1.117 shows how a SIP session between two users is set up. The terminals in this scenario are called user agents – the term used for a SIP subscriber. The SIP proxies have to make sure that all SIP messages are properly routed. In the architecture of the EPC, the SIP proxy functions can be found, for example, in the IMS.

In step 1, the user agent of Ralf sends an INVITE that contains Karsten's URI. This message is routed through the proxies to Karsten's terminal. As Karsten accepts the session invitation, his user agent sends a 200OK response code back to Ralf (step 2). Finally, Ralf's user agent sends the final ACK that may be transmitted on a different route not involving the proxies. This is possible because, after the previous messages have been sent, each user agent knows the address of its connection peer.

This SIP call setup procedure is also known as the SIP trapezoid.

SIP by itself does not provide any media format description or QoS indication. For this reason, it is generally used in combination with the Session Description Protocol (SDP) that is used to negotiate, for example, the audio codec of the VoIP connection.

1.12.15   DIAMETER on EPC Interfaces

On the EPC network interfaces that are used to connect the HSS with other network elements such as the MME, a specific version of the DIAMETER protocol is used. DIAMETER is a protocol designed for authentication, authorization, and accounting. In the IP world, it is the successor to the RADIUS protocol. To express the step forward in words, the name DIAMETER was invented as a word play: in geometry, the diameter is double the radius. So, in contrast to many other protocols, DIAMETER is not an acronym.

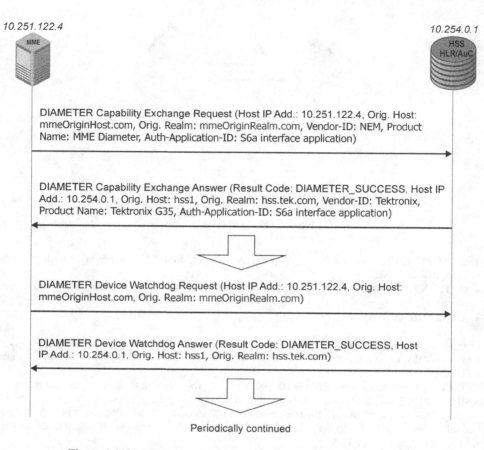

Figure 1.118 MME startup and activity check on S6a reference point

When looking at EPC DIAMETER decoder output on the S6a interface, there are two groups of messages that can be seen very often: messages for Mobility Service as specified in 3GPP 29.272 and DIAMETER Common Messages according to RFC 3588.

While the mobility service messages will be discussed in the core network call scenarios later in the book, the four most important common messages are shown in the call flow of Figure 1.118. This figure illustrates the message flow that is exchanged between the MME and HSS if a new MME is taken into service.

Whenever a new MME is set up in the network and its DIAMETER protocol entity is starting, this MME will send a DIAMETER Capability Exchange Request to the HSS. The address of the HSS was previously configured in the MME by the network operator. With the capability exchange request message, the MME informs the HSS about its IP address as well as its host name and realm name in DNS format. The home realm is an administrative domain. The vendor of the MME is included together with the product name. In these fields, we expect very proprietary information. Finally, the specific DIAMETER application running on the new network element is indicated: S6a interface applications for Mobility Services.

Message Example 1.2 shows the details of the capability exchange request message. Besides the already mentioned parameters, there is some header information. The hop-by-hop identifier allows tracking of DIAMETER messages on their point-to-point transport connections across the IP transport network, for example, from router to router. The end-to-end identifier allows identification of a DIAMETER message unambiguously by two DIAMETER peer entities. Error bit and retransmission bit have self-explanatory names and underline that DIAMETER is a very reliable protocol.

Message Example 1.2: DIAMETER Capabilities Exchange Request

```
|DIAMETER CER (= Capabilities-Exchange-Request (RFC3588))            |
|1 Capabilities-Exchange-Request (RFC3588)                           |
|Diameter Version                    |1                              |
|Message Length                      |180                            |
|Request-Bit                         |The message is a request       |
|Proxiable-Bit                       |Message MUST be locally        |
|                                    |  processed                    |
|Error-Bit                           |Message contains no error      |
|T(re-transmission)-Bit              |No re-transmission indication  |
|Spare                               |'0000'B                        |
|Command Code                        |Capabilities-Exchange-Request  |
|Application-ID                      |Diameter Common Messages       |
|Hop-by-Hop Identifier               |'653740ce'H                    |
|End-to-End Identifier               |'3e927b1f'H                    |
|1.1 Diameter AVPs                                                   |
|1.1.1 Host-IP-Address                                               |
|1.1.1.1 Host-IP-Address (RFC 3588)                                  |
|AVP Code                            |Host-IP-Address                |
|1.1.1.1.1.1 Address                                                 |
|Address-Type                        |IP (IP version 4)              |
|Address                             |10.251.122.4                   |
|1.1.2 Origin-Host                                                   |
|1.1.2.1 Origin-Host (RFC 3588)                                      |
|AVP Code                            |Origin-Host                    |
|Origin-Host                         |"mmeOriginHost.com"            |
|1.1.3 Origin-Realm                                                  |
|1.1.3.1 Origin-Realm (RFC 3588)                                     |
|AVP Code                            |Origin-Realm                   |
|Origin-Realm                        |"mmeOriginRealm.com"           |
|1.1.4 Vendor-Id                                                     |
|Vendor-Id                           |NEM                            |
|1.1.5 Product-Name                                                  |
|1.1.5.1 Product-Name (RFC 3588)                                     |
|Product-Name                        |"MME Diameter"                 |
|1.1.6 Auth-Application-Id                                           |
|1.1.6.1 Auth-Application-Id (RFC 3588)                              |
|Auth-Application-Id                 |S6a interface application      |
|1.1.8 Vendor-Specific-Application-Id                                |
| |1.1.8.1 Vendor-Specific-Application-Id (RFC 3588)                 |
|Auth-Application-Id                 |S6a interface application      |
|1.1.8.1.1.2 Vendor-Id                                               |
| |1.1.8.1.1.2.1 Vendor-Id (RFC 3588)                                |
|Vendor-Id                           |3GPP                           |
```

In response to the capability exchange request message, the HSS will send a capability exchange answer message, as shown in Message Example 1.3. This message contains the IP address, host name and realm name of the HSS, and the DIAMETER Result Code. The Result Code indicates a successful procedure on the HSS side or, in the case of failure, it contains the appropriate error cause.

Message Example 1.3: DIAMETER Capabilities Exchange Answer

```
|DIAMETER CEA (= Capabilities-Exchange-Answer (RFC3588)         |
|1 Capabilities-Exchange-Answer (RFC3588)                       |
|Diameter Version                    |1                         |
|Message Length                      |160                       |
|Request-Bit                         |The message is an answer  |
|Proxiable-Bit                       |Message MUST be locally   |
|                                       processed               |
|Error-Bit                           |Message contains no error |
|T(re-transmission)-Bit              |No re-transmission indication|
|Command Code                        |Capabilities-Exchange-Answer|
|Application-ID                      |Diameter Common Messages  |
|Hop-by-Hop Identifier               |'653740ce'H               |
|End-to-End Identifier               |'3e927b1f'H               |
|1.1 Diameter AVPs                                              |
|1.1.1 Result-Code                                              |
|1.1.1.1 Result-Code (RFC 3588)                                 |
|AVP Code                            |Result-Code               |
|Result-Code                         |DIAMETER_SUCCESS          |
|1.1.2 Origin-Host                                              |
|1.1.2.1 Origin-Host (RFC 3588)                                 |
|Origin-Host                         |"hss1"                    |
|1.1.3 Origin-Realm                                             |
|1.1.3.1 Origin-Realm (RFC 3588)                                |
|Origin-Realm                        |"hss.tek.com"             |
|Padding                             |'00'H                     |
|1.1.4 Host-IP-Address                                          |
|1.1.4.1 Host-IP-Address (RFC 3588)                             |
|AVP Code                            |Host-IP-Address           |
|Address-Type                        |IP (IP version 4)         |
|Address                             |10.254.0.1                |
|1.1.5 Vendor-Id                                                |
|1.1.5.1 Vendor-Id (RFC 3588)                                   |
|Vendor-Id                           |Tektronix Communications  |
|1.1.6 Product-Name                                             |
|1.1.6.1 Product-Name (RFC 3588)                                |
|Product-Name                        |"Tektronix G35"           |
|1.1.7 Vendor-Specific-Application-Id                           |
|1.1.7.1 Vendor-Specific-Application-Id (RFC 3588)              |
|Vendor-Id                           |3GPP                      |
|1.1.7.1.1.2 Auth-Application-Id                                |
|Auth-Application-Id                 |S6a interface application |
```

Now, after the two peer entities have exchanged their address information and capabilities, they will check the availability of their connection periodically using a heartbeat check procedure. The DIAMETER-specific name for this period check is the device watchdog procedure and it works as follows.

One entity sends a DIAMETER device watchdog request message (as shown in Message Example 1.4) containing the host name and real name of the sender.

Message Example 1.4: DIAMETER Device Watchdog Request

```
|DIAMETER DWR (= Device-Watchdog-Request (RFC3588))           |
|1 Device-Watchdog-Request (RFC3588)                          |
|Command Code                        |Device-Watchdog-Request |
|Application-ID                      |Diameter Common Messages|
|Hop-by-Hop Identifier               |'653740d0'H             |
|End-to-End Identifier               |'3f3db3e6'H             |
|1.1 Diameter AVPs                                            |
|1.1.1 Origin-Host                                            |
|1.1.1.1 Origin-Host (RFC 3588)                               |
|AVP Code                            |Origin-Host             |
|Origin-Host                         |"mmeOriginHost.com"     |
|Padding                             |'000000'H               |
|1.1.2 Origin-Realm                                           |
|1.1.2.1 Origin-Realm (RFC 3588)                              |
|AVP Code                            |Origin-Realm            |
|Origin-Realm                        |"mmeOriginRealm.com"    |
```

On receiving the Device Watchdog Request, the peer entity sends back a Device Watchdog Answer (Message Example 1.5) including the address information and Result Code. Again, the Result Code indicates a successful procedure or, in case of failure, it contains the appropriate error cause.

Message Example 1.5: DIAMETER Device Watchdog Answer

```
|DIAMETER DWA (= Device-Watchdog-Answer (RFC3588))            |
|1 Device-Watchdog-Answer (RFC3588)                           |
|Command Code                        |Device-Watchdog-Answer  |
|1.1.1.1 Result-Code (RFC 3588)                               |
|AVP Code                            |Result-Code             |
|Result-Code                         |DIAMETER_SUCCESS        |
|1.1.2 Origin-Host                                            |
|Origin-Host                         |"hss1"                  |
|1.1.3 Origin-Realm                                           |
|1.1.3.1 Origin-Realm (RFC 3588)                              |
|AVP Code                            |Origin-Realm            |
|Origin-Realm                        |"hss.tek.com"           |
```

2

E-UTRAN/EPC Signaling

In this second chapter of the book, we take a closer look at the signaling procedures across the wired or fibered transport links of the E-UTRAN (Evolved Universal Terrestrial Radio Access Network) and the EPC (Evolved Packet Core) network. This will give some insights into the details of session and bearer management. Also, we explain how the mobile communicates with the network on the Non-Access Stratum (NAS) layer that is responsible for subscriber registration and macro-mobility.

Also in this chapter, the signaling procedures on the S1 interface between eNodeB (eNB) and the (Mobility Management Entity) MME, on S6a between the MME and Home Subscriber Server (HSS), on S11 between the MME and Serving Gateway (S-GW), and on the S5 reference point between S-GW and the Packet Data Network Gateway (PDN-GW) are discussed in detail. This will also provide insights into how the user plane packets are tunneled and routed through the EPC network.

2.1 S1 Setup

The S1 Application Part (S1AP) S1 setup procedure is used to take a new eNB into service. Basically, this is the same functionality as the NodeB setup on the Iub interface of 3G Universal Terrestrial Radio Access Network (UTRAN). However, in comparison to the Iub signaling procedure, S1 Setup is simpler. S1 Setup is directly triggered by the eNB while the 3G NodeB only requests to be audited by the RNC. This means that in the 3G UTRAN, the RNC is the master of all NodeB configuration parameters, but in the E-UTRAN, the eNB is already configured when it is taken into service, and on behalf of the S1 setup procedure, the MME is just informed about the main parameters configured in the eNB that is launched.

2.1.1 S1 Setup: Message Flow

Figure 2.1 shows how the S1 setup procedure is executed for two different eNBs. Each eNB sends an S1AP initiating Message for the S1 setup (Third Generation Partnership Project (3GPP) message name: S1 Setup Request – see Message Example 2.1) to the MME. Each eNB is unambiguously identified by its Global-ENB-ID that consists of the Mobile Country Code (MCC) + Mobile Network Code (MNC) + macroENB-ID. There is also a unique Transaction Area Code (TAC) configured for each eNB in the example. Furthermore, the Source IP address (Scr IP) of the IP transport layer (the IP layer below the Stream Control Transmission Protocol (SCTP)) reveals that each eNB has its own IP address.

LTE Signaling, Troubleshooting and Performance Measurement, Second Edition. Ralf Kreher and Karsten Gaenger.
© 2016 John Wiley & Sons, Ltd. Published 2016 by John Wiley & Sons, Ltd.

5. Prot	5. MSG	Procedure Code	tAC	Src IP	Dst IP	macroENB-ID	mME-Group-ID	mME-Code
S1-AP	initiatingMessage	id-S1Setup	'0001'H	10.254.0.3	10.251.122.3	'00001'H		
S1-AP	initiatingMessage	id-S1Setup	'00c8'H	10.254.0.4	10.251.122.3	'0000c'H		
S1-AP	successfulOutcome	id-S1Setup		10.251.122.3	10.254.0.3		'0001'H	'01'H
S1-AP	initiatingMessage	id-Reset		10.254.0.3	10.251.122.3			
S1-AP	successfulOutcome	id-S1Setup		10.251.122.3	10.254.0.4		'0001'H	'01'H
S1-AP	initiatingMessage	id-Reset		10.254.0.4	10.251.122.3			
S1-AP	successfulOutcome	id-Reset		10.251.122.3	10.254.0.3			
S1-AP	successfulOutcome	id-Reset		10.251.122.3	10.254.0.4			

Figure 2.1 Combined S1 setup/reset for two eNodeBs connected to the same MME. (*Source*: Tektronix Communications.)

With this triplet of configuration parameters (macroENB-ID, TAC, and IP address) the main configuration parameters of these eNBs in the overall E-UTRAN topology have been identified.

Message Example 2.1: S1AP S1 Setup Request

```
|S1 Application Protocol TS 36.413 V8.4.0 (2008-12) (S1-AP) initiatingMessage|
|s1apPDU
|1 initiatingMessage                                                        |
|1.1 procedureCode                |id-S1Setup                               |
|1.2 criticality                  |reject                                   |
|1.3 value                                                                  |
|1.3.1 protocolIEs                                                          |
|1.3.1.1 sequence                                                           |
|1.3.1.1.1 id                     |id-Global-ENB-ID                         |
|1.3.1.1.2 criticality            |reject                                   |
|1.3.1.1.3 value                                                            |
|1.3.1.1.3.1 TBCD String          |'299010'                                 |
|1.3.1.1.3.2 eNB-ID                                                         |
|1.3.1.1.3.2.1 macroENB-ID        |'0000c'H (12 dec.)                       |
|1.3.1.2 sequence                                                           |
|1.3.1.2.1 id                     |id-eNBname                               |
|1.3.1.2.2 criticality            |ignore                                   |
|1.3.1.2.3 value                  |'0000c0'H (12 dec.)                      |
|1.3.1.3 sequence                                                           |
|1.3.1.3.1 id                     |id-SupportedTAs                          |
|1.3.1.3.2 criticality            |reject                                   |
|1.3.1.3.3 value                                                            |
|1.3.1.3.3.1 supportedTAs-Item                                              |
|1.3.1.3.3.1.1 tAC                |'00c8'H (200 dec.)                       |
|1.3.1.3.3.1.2 broadcastPLMNs                                               |
|1.3.1.3.3.1.2.1 TBCD String      |'299010'                                 |
```

Looking at the details of the decoder output shown in Message Example 2.2, it should be mentioned that multiple TACs per eNB are possible; the theoretical maximum number of TACs per eNB based on the size of the TAC information element in the S1AP specification is 256 (8-bit TAC field). Also, the same eNB can serve more than one Public Land Mobile Network (PLMN). The theoretical maximum number of broadcast PLMNs per eNB is six.

In return, the MME sends an S1AP successful outcome message for S1 Setup to each eNB using the previously identified IP address and updating the base stations with the Globally Unique MME Identifier (GUMMEI). Figure 2.1 only shows the MME group ID and MME code that are part of the GUMMEI (see Section 1.4.4).

Message Example 2.2: S1AP S1 Setup Response

```
|S1 Application Protocol TS 36.413 V8.4.0 (2008-12) (S1-AP) successfulOutcome|
|s1apPDU                                                                       |
|1 successfulOutcome                                                           |
|1.1 procedureCode                      |id-S1Setup                           |
|1.2 criticality                        |reject                               |
|1.3 value                                                                     |
|1.3.1 protocolIEs                                                             |
|1.3.1.1 sequence                                                              |
|1.3.1.1.1 id                           |id-MMEname                           |
|1.3.1.1.2 criticality                  |ignore                               |
|1.3.1.1.3 value                        |6a 6a 62 48 54 43 67 44              |
|1.3.1.2 sequence                                                              |
|1.3.1.2.1 id                           |id-ServedPLMNs                       |
|1.3.1.2.2 criticality                  |ignore                               |
|1.3.1.2.3 value                                                               |
|1.3.1.2.3.1 TBCD String                |'299010'                             |
|1.3.1.3 sequence                                                              |
|1.3.1.3.1 id                           |id-ServedGUMMEIs                     |
|1.3.1.3.2 criticality                  |ignore                               |
|1.3.1.3.3 value                                                               |
|1.3.1.3.3.1 gUMMEI                                                            |
|1.3.1.3.3.1.1 TBCD String              |'299010'                             |
|1.3.1.3.3.1.2 mME-Group-ID             |'0001'H                              |
|1.3.1.3.3.1.3 mME-Code                 |'01'H                                |
|1.3.1.4 sequence                                                              |
|1.3.1.4.1 id                           |id-RelativeMMECapacity               |
|1.3.1.4.2 criticality                  |ignore                               |
|1.3.1.4.3 value                        |50                                   |
```

In Figure 2.2, the message flow of S1 Setup is shown together with the network topology. The relative MME capacity found in the successful outcome message indicates the relative processing capacity of an MME with respect to the other MMEs in an MME pool. The parameter value ranges from 0 to 255 and is used for balancing the load within an MME pool. It represents the weight factor defined in 3GPP 23.401. A higher weight factor (= higher relative MME capacity) means that this MME is able to handle more User Equipment (UE) connections compared to other MMEs in the same pool. So, its load can be increased. If an eNB is connected to multiple MMEs, the MME with the highest weight factor should be selected to set up a new UE connection.

After successful S1 Setup, the S1 connection between the new eNB and the MME is reset using the S1AP reset procedure (not shown in Figure 2.2 but in Figure 2.1). This reset is done to synchronize the S1AP state machines in the peer entities after S1 Setup. Also, here the difference from the NodeB setup/audit procedure in 3G UTRAN should be noted: in 3G, the audit of a NodeB is triggered by a previous NBAP Reset and new NodeB/cell parameters are assigned by the RNC. In E-UTRAN, S1AP Reset is executed to complete the setup of a new eNB and now parameter changes become effective.

2.1.2 S1 Setup: Failure Analysis

If the MME is not able to perform the S1 setup procedure, it will send back an unsuccessful outcome message (S1 Setup Failure) to the eNB as shown in Figure 2.3. This failure message includes a cause value.

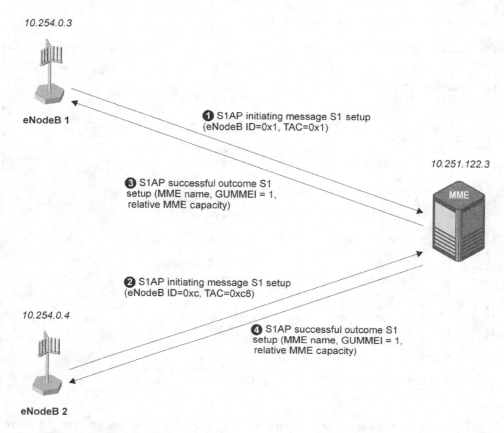

Figure 2.2   S1 Setup for two different eNodeBs. (*Source*: Tektronix Communications.)

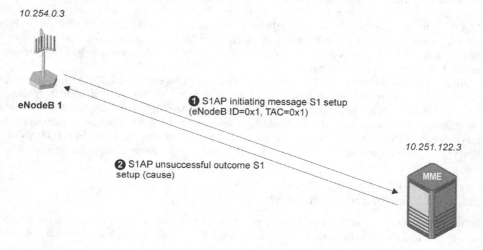

Figure 2.3   S1 Setup failure. (*Source*: Tektronix Communications.)

Typical cause values that can be expected are found in the protocol cause group and miscellaneous cause group of S1 AP:

- Semantic error.
- Abstract syntax error.
- Control processing overload.
- Hardware failure.
- Unspecified.
- Unknown PLMN.

Cause values from the NAS cause group and the radio network layer cause group are not expected to be used in the S1 setup failure message.

2.2   Initial Attach

As known from 2G and 3G mobile networks, subscribers need to register to the network. Once the registration is successfully completed, the services offered by the network operator are available for the subscribers.

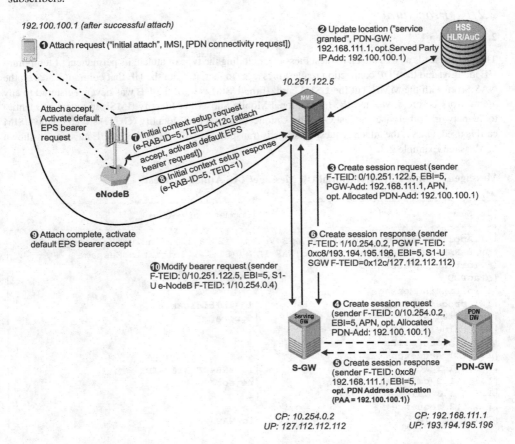

Figure 2.4   E-UTRAN attach procedure. (*Source*: Tektronix Communications.)

Typical subscriber actions behind the signaling procedure of attach is to switch on a handset. In the case of data cards or USB sticks commonly used for connecting laptop PCs to the mobile networks, clicking on the CONNECT button of the mobile connection administration software installed together with the device will trigger an attach of the UE.

In addition, as described in 3GPP 23.401, any change in the UE's E-UTRAN capabilities requires the UE to detach and then reattach to the network.

The main difference between an attach to E-UTRAN and one to 3G or 2G RAN is that in E-UTRAN, a default bearer for user plane transport will immediately be assigned to the subscriber before the attach is completed. The purpose of this immediate assignment is to reduce the access latency. Indeed, there is not just the S1-U bearer established, but all bearers in the core network on the radio interface that are required to transport payload packets between the UE and the PDN-GW or vice versa. The bearer setup on E-UTRAN and EPC interfaces is embraced by setting up a PDP connection, in core network terminology also called a session. If the subscriber has access to the network, if this access is limited or not, and what kind of service and Quality of Service (QoS) are allowed for this particular subscriber – all these are controlled by the HSS.

As shown in Figure 2.4, the main steps of the attach procedure are as follows.

2.2.1   Procedure

2.2.1.1   Step 1

The mobile sends a NAS attach request message including the type of attach, its permanent or temporary UE identity, and the PDP connection request message. In fact, it is not the UE that communicates on the NAS layer with the MME, but the USIM card. If the USIM card of the UE was never registered to any 4G network before, it will use the International Mobile Subscriber Identity (IMSI) to reveal its identity to the network; otherwise, the last Globally Unique Temporary UE Identity (GUTI) stored on the USIM card is used. On S1, the attach request message is transported by the S1AP initial UE message as shown in Message Example 2.3.

Message Example 2.3: S1AP Initial UE Message/NAS Attach Request

```
+-------------------------------------+----------------------------------------+
|ID Name                              |Comment or Value                        |
+-------------------------------------+----------------------------------------+
|S1 Application Protocol TS 36.413 V8.4.0 (2008-12) (including Changes
for E-RABToBeSetupItemCtxtSUReq:NAS PDU; (S1-AP) initiatingMessage            |
(= initiatingMessage)
|s1apPDU                                                                       |
|1 initiatingMessage                                                           |
|1.1 procedureCode                    |id-initialUEMessage                     |
|1.2 criticality                      |ignore                                  |
|1.3 value                                                                     |
|1.3.1 protocolIEs                                                             |
|1.3.1.1 sequence                                                              |
|1.3.1.1.1 id                         |id-eNB-UE-S1AP-ID                       |
|1.3.1.1.2 criticality                |reject                                  |
|1.3.1.1.3 value                      |1                                       |
|1.3.1.2 sequence                                                              |
|1.3.1.2.1 id                         |id-NAS-PDU                              |
```

```
|1.3.1.2.2 criticality                    |reject                          |
|1.3.1.2.3 value                          |07 41 01 08 39 01 10 02 10 00 01 10 |
|                                         |05 00 00...                     |
|                                         |                                |
|1.3.1.3 sequence                         |                                |
|1.3.1.3.1 id                             |id-TAI                          |
|1.3.1.3.2 criticality                    |reject                          |
|1.3.1.3.3 value                          |                                |
|1.3.1.3.3.1 TBCD String                  |'299000'                        |
|1.3.1.3.3.2 tAC                          |'00c8'H                         |
|1.3.1.4 sequence                         |                                |
|1.3.1.4.1 id                             |id-EUTRAN-CGI                   |
|1.3.1.4.2 criticality                    |ignore                          |
|1.3.1.4.3 value                          |                                |
|1.3.1.4.3.1 TBCD String                  |'299000'                        |
|1.3.1.4.3.2 cell-ID                      |'0000c00'H                      |
|NAS LTE TS24.301 V8.0.0 (2008-12) (Secured NAS) ATRQ (= Attach Request) |
|1 Attach Request                         |                                |
|Security header type                     |No security protection          |
|Protocol discriminator                   |EPS mobility management messages|
|Type of Management                       |EPS mobility management         |
|Message type                             |ATRQ                            |
|1.1 NAS key set                          |                                |
|identifier                               |                                |
|Type of security context flag (TSC)      |cached security context         |
|NAS key set identifier                   |0 (Possible values for the NAS key |
|                                         |set identi...                   |
|                                         |                                |
|1.2 EPS attach type                      |                                |
|Spare                                    |0                               |
|EPS attach type                          |EPS attach                      |
|1.3 Old GUTI or IMSI                     |                                |
|Length                                   |8                               |
|1.3.1 IMSI                               |                                |
|Type of identity                         |IMSI                            |
|Odd/Even indicator                       |Odd                             |
|Identity Digits                          |'299000001001001'               |
|1.4 UE network capability                |                                |
|Length                                   |5                               |
|EEA0                                     |EPS encryption algorithm supported |
|Spare                                    |0                               |
|EIA1                                     |EPS integrity algorithm supported |
|1xSRVCC                                  |SRVCC from E-UTRAN to 3GPP2 1xCS not |
|                                         |supported                       |
|ISR                                      |The UE does not support ISR     |
|1.5 ESM Message Container                |                                |
|Length                                   |32                              |
|ESM Message Container                    |02 01 d0 11 27 1a 80 21 0a 03 60 00 |
|                                         |0a 81 06...                     |
|1.6 MS network capability                |                                |
|Tag                                      |MS network capability           |
|NAS LTE TS24.301 V8.0.0 (2008-12) (NAS ESM) PCONRQ |
|(= PDN connectivity request)             |                                |
|1 PDN connectivity request               |                                |
|EPS bearer id                            |0                               |
```

```
|Protocol discriminator               |EPS session management messages    |
|Procedure transaction identity       |1                                  |
|Type of Management                   |EPS session management             |
|Message type                         |PCONRQ                             |
|1.1 PDN type                         |                                   |
|Spare                                |0                                  |
|PDN type                             |IPv4                               |
|1.2 Request type                     |                                   |
|Spare                                |0                                  |
|Request type                         |initial attach                     |
|1.3 Protocol configuration options   |                                   |
|Tag                                  |Protocol configuration options     |
|Length                               |26                                 |
|Extension bit                        |1                                  |
|Spare                                |'0000'B                            |
|Configuration protocol               |PPP for use with IP PDP type       |
|Protocol configuration options       |21 0a 03 60 00 0a 81 06 00 00 00 00 |
                                       80 21 0a...
```

The S1AP initial UE message starts the NAS transport between eNB and the MME. The procedure code information element identifies the message itself. The e-NodeB-UE-S1AP-ID is used during the entire S1AP connection between this eNB and the MME to identify the mobile station unambiguously within the eNB. The first S1AP Downlink (DL) NAS transport message sent by the MME will contain a MME-UE-S1AP-ID value that is the unique identifier of the UE on the S1AP entity of the MME. Both UE-S1AP-IDs form a unique context used to identify the logical S1AP signaling connection for a single UE. Hence, these parameters are key to the design of S1AP call trace functionality.

Together with the NAS PDU (Packet Data Unit – this is the embedded NAS message in raw data format), the eNB sends the current location of the subscriber represented by the tracking area ID and the E-UTRAN Cell Group Identifier (CGI) to the MME. These parameters will later be reported to the HSS during the DIAMETER update location procedure. The TBCD string shown in Message Example 2.3 contains the MCC and MNC of the E-UTRAN. This information is especially important in the case of network sharing, which means the same eNB hardware including antennas can be shared by different network operators.

When the eNB has received the S-TMSI from the UE during Radio Resource Control (RRC) connection establishment, this parameter will also be included in the S1AP initial UE message.

In Message Example 2.3, the TAC and cell-ID are still decoded in hexadecimal format. Next-generation decoders will translate this into decimal values and performance measurement counters will also use decimal values of these parameters to correlate call-related protocol events such as the attempt counter of the attach procedure with the location of the subscriber.

In the attach request itself, the subscriber's identity can be found. In the case of Message Example 2.3, we see the IMSI here, but attach request messages monitored in live networks will contain the GUTI instead. The security header is a specific information element for the new Long-Term Evolution (LTE) NAS protocol that indicates in Message Example 2.3 that security protection is not yet active for this NAS signaling connection. Indeed, this will be mostly the case for attach request messages, because, typically, activation of security functions is triggered by the network after the initial UE message is received. However, as stated in 3GPP 24.301, an already existing NAS security context does not need to be deleted if the UE detaches from the network. In such a case, if valid NAS security exists on the UE side, the attach request message will be integrity protected.

In the attach request message, the Evolved Packet System (EPS) attach type is also found. Possible values for this information element are "EPS attach" and "combined EPS/IMSI attach." "EPS attach" means the UE is allowed to use LTE radio access and the transport services of the EPC network.

"IMSI attach" is the protocol term used to describe registration and access to non-EPS services provided by GERAN/UTRAN or other radio access technologies and their appropriate core network transport. A UE that has successfully performed "combined EPS/IMSI attach" can seamlessly be handed over to UTRAN/GERAN, while a UE that has only performed "EPS attach" must remain in E-UTRAN or attach to UTRAN/GERAN before a handover is allowed. It should be noted that a "combined EPS/IMSI attach" may partially fail in such a way that EPS attach is successful, but IMSI attach is not. In such a case, an attach accept message is sent back to the UE and the EPS Mobility Management (EMM) cause value will indicate the reason for the failed IMSI attach.

In the UE network capability section, the LTE and UMTS integrity protection and ciphering algorithms supported by the handset that sent the attach request message are indicated. In another section called MS Network Capability, the GERAN-related information is transmitted, for example, the list of supported General Packet Radio Service (GPRS) encryption algorithms.

The PDN connectivity request information is embedded in the ESM message container. As the name of the container already suggests, the PDN Connectivity Request is an EPS session management message while the Attach Request is an EMM message, and these messages are sent in parallel due to the requirement that, in LTE and EPC, for each UE a default bearer needs to be established when the mobile attaches.

The Request Type in the PDN Connectivity Request indicates "initial attach" (as shown in Message Example 2.3) or "handover." In the Protocol Configuration Options, typically the dial-in information for the subscriber's Internet access such as user name and password is seen. Addresses of primary and secondary DNS servers can also be found there.

2.2.1.2   Step 2

The Attach Request is received by the MME. This reception triggers the DIAMETER update location procedure on the S6a interface. With this procedure, the geographic position of the freshly detected UE is stored in the HSS database and in return, the MME receives the subscribed QoS parameters stored in the HSS together with a confirmation that the subscriber is allowed to use the service of the network.

Message Example 2.4 shows an example of the DIAMETER update location request message. The message type is encoded as the DIAMETER Command Code. The Application-ID indicates that this message was monitored on the S6a interface between the MME and HSS. The sender is identified by Origin Host and Origin Realm, the receiver by its Destination Realm and Destination Host (at the end of the decoded message).

The IMSI of the subscriber who sent the Attach Request is found as "User Name" in the DIAMETER protocol. RAT Type indicates that the E-UTRAN is the radio access technology used by the subscriber and the visited PLMN is identified by the already well-known combination of MCC and MNC.

Message Example 2.4: DIAMETER Update Location Request

```
+---------------------------------------+---------------------------------------+
|ID Name                                |Comment or Value                       |
+---------------------------------------+---------------------------------------+
|DIAMETER based on RFC3588/4005/4006/4072, 3GPP
 29.061/.109/.140/.209/.210/.234/.229/.329/32.299 (DIAMETER)                     |
ULR (= Update-Location-Request)
|1 Update-Location-Request                                                        |
|Diameter Version                       |1                                        |
|Message Length                         |224                                      |
|Request-Bit                            |The message is a request                 |
```

```
|Proxiable-Bit                          |Message MAY be proxied, relayed or  |
                                        redirected
|Error-Bit                              |Message contains no error           |
|T(re-transmission)-Bit                 |No re-transmission indication       |
|Spare                                  |'0000'B                             |
|Command Code                           |Update-Location-Request             |
|Application-ID                         |S6a interface application           |
|1.1.3 Origin-Host                                                           |
|1.1.3.1 Origin-Host (RFC 3588)                                              |
|Origin-Host                            |"mmeOriginHost.com"                 |
|1.1.4 Origin-Realm                                                          |
|Origin-Realm                           |"mmeOriginRealm.com"                |
|1.1.5 Destination-Realm                                                     |
|1.1.5.1 Destination-Realm (RFC 3588)                                        |
|Destination-Realm                      |"hss.tek.com"                       |
|1.1.6 User-Name                                                             |
|1.1.6.1 User-Name (RFC 3588)                                                |
|User-Name                              |"299000001001001"                   |
|1.1.7 RAT-Type                                                              |
|Vendor-Id                              |3GPP                                |
|RAT-Type                               |EUTRAN                              |
|1.1.9 Visited-PLMN-Id                                                       |
|Vendor-Id                              |3GPP                                |
|MCC Digit 1                            |2                                   |
|MCC Digit 2                            |9                                   |
|MCC Digit 3                            |9                                   |
|MNC Digit 3                            |0                                   |
|MNC Digit 1                            |0                                   |
|MNC Digit 2                            |0                                   |
|1.1.10 Destination-Host                                                     |
|Destination-Host                       |"hss1"                              |
```

Message Example 2.5 shows the response of the HSS, the DIAMETER update location answer message. It is the Request bit in the DIAMETER header that marks this message as an "Answer" message and again the Application-ID reveals the monitored interface while Origin Real/Host and Destination Realm/Host are the senders/receivers of this message.

The next interesting parameter is the DIAMETER Result Code. This information element distinguishes between a successful and failed update location procedure. In case the value is "Failure," an appropriate cause value will be delivered together with the result code.

The subscriber status information element contains information on whether the desired access to the network is granted to the subscriber or not. Independently of the DIAMETER Result Code, this field may also indicate a failed location update, for example, in the case of roaming subscribers that are not allowed to access the network.

Message Example 2.5: DIAMETER Update Location Answer

```
+-------------------------------------+--------------------------------------+
|ID Name                              |Comment or Value                      |
+-------------------------------------+--------------------------------------+
|DIAMETER based on RFC3588/4005/4006/4072, 3GPP                              |
| 29.061/.109/.140/.209/.210/.234/.229/.329/32.299 (DIAMETER) ULA            |
|(= Update-Location-Answer )                                                 |
|1 Update-Location-Answer                                                    |
```

```
|Diameter Version                          |1                                    |
|Message Length                            |648                                  |
|Request-Bit                               |The message is an answer             |
|Proxiable-Bit                             |Message MAY be proxied, relayed or   |
|                                          |redirected                           |
|Error-Bit                                 |Message contains no error            |
|T(re-transmission)-Bit                    |No re-transmission indication        |
|Spare                                     |'0000'B                              |
|Command Code                              |Update-Location-Answer               |
|Application-ID                            |S6a interface application            |
|1.1.3.1 Origin-Host (RFC 3588)            |                                     |
|AVP Code                                  |Origin-Host                          |
|Origin-Host                               |"hss1"                               |
|1.1.4 Origin-Realm                        |                                     |
|Origin-Realm                              |"hss.tek.com"                        |
|1.1.5 Result-Code                         |                                     |
|1.1.5.1 Result-Code (RFC 3588)            |                                     |
|AVP Code                                  |Result-Code                          |
|Result-Code                               |DIAMETER_SUCCESS                     |
|1.1.7.1.1 Subscriber-Status               |                                     |
|Subscriber-Status                         |SERVICE_GRANTED                      |
|1.1.7.1.2 AMBR                            |                                     |
|1.1.7.1.2.1.1 Max-Requested-Bandwidth-UL  |                                     |
|AVP Code                                  |Max-Requested-Bandwidth-UL           |
|Max-Requested-Bandwidth-UL                |10000000                             |
|1.1.7.1.2.1.2 Max-Requested-Bandwidth-DL  |                                     |
|Max-Requested-Bandwidth-DL                |100000000                            |
|1.1.7.1.3 3GPP-Charging-Characteristics   |                                     |
||.1.7.1.4 Access-Restriction-Data         |                                     |
|HO-To-Non-3GPP-Access Not Allowed         |0                                    |
|E-UTRAN Not Allowed                       |0                                    |
|I-HSPA-Evolution Not Allowed              |0                                    |
|GAN Not Allowed                           |0                                    |
|GERAN Not Allowed                         |1                                    |
|UTRAN Not Allowed                         |1                                    |
|1.1.7.1.5 MSISDN                          |                                     |
|MSISDN                                    |'491999792001'                       |
|1.1.7.1.6 APN-Configuration-Profile       |                                     |
|Called-Station-Id                         |"rhinophone.com"                     |
|1.1.7.1.6.1.3.1.3 PDN Type                |                                     |
|AVP Code                                  |PDN Type                             |
|PDN-Type                                  |IPv4                                 |
|1.1.7.1.6.1.3.1.4 VPLMN-Dynamic-Address-Allowed |                               |
|VPLMN-Dynamic-Address-Allowed             |NOT ALLOWED                          |
|1.1.7.1.6.1.3.1.5 PDN-GW-Identity         |                                     |
|1.1.7.1.6.1.3.1.5.1.1.1 PDN-GW-Address    |                                     |
|1.1.7.1.6.1.3.1.5.1.1.1.1 Address         |                                     |
|Address-Type                              |IP (IP version 4)                    |
|Address                                   |192.168.111.1                        |
|1.1.7.1.6.1.3.1.5.1.2 PDN-GW-Name         |                                     |
|AVP Code                                  |PDN-GW-Name                          |
|PDN-GW-Name                               |"rhinophone.com"                     |
|1.1.7.1.6.1.3.1.6 PDN-GW-Allocation-Type  |                                     |
```

```
|PDN-GW-Allocation-Type                    |STATIC                            |
|1.1.7.1.6.1.3.1.7 EPS-Subscribed-QoS-Profile                                 |
|1.1.7.1.6.1.3.1.7.1.1 QoS-Class-Identifier                                   |
|AVP Code                                  |QoS-Class-Identifier              |
|QoS-Class-Identifier                      |9                                 |
|1.1.7.1.6.1.3.1.7.1.2 ARP                                                    |
|1.1.7.1.6.1.3.1.8.1 Served-Party-IP-Address                                  |
|1.1.7.1.6.1.3.1.8.1.1 Address                                                |
|Address-Type                              |IP (IP version 4)                 |
|Address                                   |192.100.100.1                     |
|1.1.7.1.6.1.3.1.9 3GPP-Charging-Characteristics                              |
|1.1.7.1.6.1.3.1.10.1 AMBR AVP-List                                           |
|1.1.7.1.6.1.3.1.10.1.1 Max-Requested-Bandwidth-UL                            |
|Max-Requested-Bandwidth-UL                |12000000                          |
|1.1.7.1.6.1.3.1.10.1.2 Max-Requested-Bandwidth-DL                            |
|Max-Requested-Bandwidth-DL                |120000000                         |
```

The Aggregated Maximum Bit Rate (AMBR) for uplink (UL) and DL traffic occurs twice in the message: first related to the subscriber, then related to the Access Point Name (APN). The APN – which is clearly visible in the decoder output – is the PDN gateway itself. The difference between the two AMBR values that are also specified as Maximum Requested Bandwidth is that in the first section where the values 10 Mbps on UL and 100 Mbps on DL have been allocated, this is the maximum bit rate granted to all payload traffic carried by the default bearer. In the second section (below the charging characteristics section), the limits for the aggregated bandwidth of *all bearers* of this single subscriber within the single APN (PDN-GW) are defined. This means that if 10 Mbps is reserved for the default bearer on the UL and 12 Mbps set as Maximum Requested UL Bandwidth for this subscriber on the APN side, then only 2 Mbps is left for other bearers running in parallel with the default bearer.

In general, the QoS granted to the subscriber during the update location procedure is defined by the Quality of Service Class Identifier (QCI) that was already described in Chapter 1. In the message example, the QCI has the value 9, which means a nonguaranteed bit rate handled with best effort by the network.

In addition to QoS parameters in the sample call flow and message example, the HSS provides the IP address to this particular UE that will be used as long as it is attached to this network (in the message example: 192.100.100.1). This IP address provided by the HSS is a fixed address stored together with the subscriber's QoS according to the service level agreement in its HSS. However, the most common scenario of IP address allocation that is expected to be seen later in live networks is the assignment of dynamic IP addresses, and the network element responsible for dynamic IP address management and assignment is the PDN-GW. Thus, the Served Party IP Address value shown in Figure 2.4 and Message Example 2.5 is only an optional parameter.

The HSS also provides the IP address of the PDN-GW together with the APN itself. Having this information, the default route for all IP traffic between this mobile and external IP networks such as the Internet is determined.

2.2.1.3 Step 3

In step 3, this default route needs to be established and the first step is that the MME sends a GPRS Tunneling Protocol (GTP) Create Session Request to the S-GW. This messages reveals the Fully Qualified Tunnel Endpoint Identifier (F-TEID) of the sender that is a combination of a simple Tunnel Endpoint Identifier (TEID) value and the IP address of the MME. Also, the IP address of the UE (if provided by MME in step 2, otherwise default value 0.0.0.0 is used in this message), its IMSI, the PDN-GW address, and APN (all allocated earlier by the HSS) are included. The EPS Bearer ID (EBI) identifies the particular bearer in case there are more bearers established for the same UE later on.

Message Example 2.6: S11 GTP Create Session Request

```
+-------------------------------------------+-------------------------------------------+
|ID Name                                    |Comment or Value                           |
+-------------------------------------------+-------------------------------------------+
|Tunnelling Protocol for Control plane (GTPv2-C) 3GPP TS 29.274 V8.0.0
|(2008-12) (GTP_C) CSREQ (= Create Session Request) |
|1 Create Session Request                                                               |
|Version                                    |GTPv2                                      |
|Spare                                      |0                                          |
|T-Bit                                      |TEID is present in the GTP-C header        |
|Spare                                      |0                                          |
|Message Type                               |Create Session Request                     |
|Message Length                             |247                                        |
|TEID                                       |'00000000'H                                |
|Sequence Number                            |1                                          |
|Spare                                      |0                                          |
|1.1 ieSetCsReq                                                                         |
|1.1.1 International Mobile Subscriber Identity (IMSI)                                   |
|Type                                       |International Mobile Subscriber            |
|                                            Identity (IMSI)                            |
|Identity Digits                            |'299000001001001'                          |
|1.1.2 RAT Type                                                                         |
|Type                                       |RAT Type                                   |
|RAT Type                                   |EUTRAN                                     |
|1.1.3 Indication Flags                                                                 |
|Type                                       |Indication                                 |
|Length                                     |2                                          |
|CR                                         |1                                          |
|Spare                                      |0                                          |
|Instance                                   |0                                          |
|DAF (Dual Address Bearer)                  |0                                          |
|DTF (Direct Tunnel)                        |0                                          |
|HI (Handover Indication)                   |0                                          |
|DFI (Direct Forwarding Indication)         |0                                          |
|OI (Operation Indication)                  |0                                          |
|ISRSI (Idle Mode Signalling                |0                                          |
|Reduction..                                                                            |
|ISRAI (Idle Mode Signalling                |0                                          |
|Reduction..                                                                            |
|SGWCI (SGW Change Indication)              |0                                          |
|Spare                                      |0                                          |
|PT (Protocol Type)                         |1                                          |
|TDI (Teardown Indication)                  |0                                          |
|SI (Scope Indication)                      |0                                          |
|MSV (MS Validated)                         |0                                          |
|1.1.4 Sender F-TEID for Control Plane                                                  |
|Type                                       |Fully Qualified Tunnel Endpoint           |
|                                            Identifier (F-...                          |
|Interface Type                             |S11 MME GTP-C interface                    |
|TEID/GRE Key                               |'00000000'H                                |
|IPv4 Address                               |10.251.122.5                               |
|Spare                                      |0                                          |
|EPS Bearer ID (EBI)                        |5                                          |
```

```
|1.1.5 PDN Type
|Type                                      |PDN Type
|PDN type                                  |IPv4
|1.1.6 Maximum APN Restriction
|Type                                      |APN Restriction
|Restriction Type                          |No Existing Contexts or Restrictions
|1.1.7 Bearer Context to be created
|Type                                      |Bearer Context
|1.1.7.1 Information Elements
|1.1.7.1.1 EPS Bearer ID (EBI)
|Type                                      |EPS Bearer ID (EBI)
|EPS Bearer ID (EBI)                       |5
|1.1.7.1.2 Bearer Level Quality of Service (Bearer QoS)
|Type                                      |Bearer Level Quality of Service
|                                          |(Bearer QoS)
|ARP                                       |3
|QCI                                       |Network selects the QCI 9 [UL]
|Maximum bit rate for uplink               |00 00 00 00 00
|Maximum bit rate for downlink             |00 00 00 00 00
|Guaranteed bit rate for uplink            |00 00 00 00 00
|Guaranteed bit rate for downlink          |00 00 00 00 00
|1.1.8 MSISDN
|Type                                      |MSISDN
|MSISDN                                    |'491999792001'
|1.1.9 User Location Info (ULI)
|Type                                      |User Location Info (ULI)
|E-UTRAN Cell Global Identification        |is included
|(E..
|Tracking Area Identity (TAI)              |is included
|Routing Area Identification (RAI)         |is not included
|Service Area Identity (SAI)               |is not included
|Cell Global Identification (CGI)          |is not included
|1.1.9.1 TAI Contents
|MCC                                       |'299'
|MNC                                       |'000'
|TAC                                       |'00c8'H
|1.1.9.2 ECGI Contents
|MCC                                       |'299'
|MNC                                       |'000'
|Spare                                     |0
|ECI                                       |'00c0000'H
|1.1.10 Serving Network
|Type                                      |Serving Network
|MCC                                       |'299'
|MNC                                       |'000'
|1.1.11 PGW S5/S8 Address for Control Plane or PMIP
|Type                                      |Fully Qualified Tunnel Endpoint
|                                          |Identifier (F-...
|Length                                    |9
|CR                                        |1
|Spare                                     |0
|Instance                                  |1
|IPv4 Address                              |is included
```

```
|IPv6 Address                              |is not included                          |
|EPS Bearer Id                             |is not included                          |
|Interface Type                            |S5/S8 PGW PMIPv6 interface               |
|TEID/GRE Key                              |'00000000'H                              |
|IPv4 Address                              |192.168.111.1                            |
|1.1.12 Access Point Name (APN)            |                                         |
|Type                                      |Access Point Name (APN)                  |
|1.1.12.1 Access Point Name                |                                         |
|Length Of Label                           |49                                       |
|Accesspoint name label                    |"rhinophone.com"                         |
|1.1.14 PDN Address Allocation (PAA)        |                                         |
|Type                                      |PDN Address Allocation (PAA)             |
|PDN type                                  |IPv4                                     |
|PDN address (IPv4)                        |192.100.100.1                            |
|1.1.15 Aggregate Maximum Bit Rate (AMBR)  |                                         |
|Type                                      |Aggregate Maximum Bit Rate (AMBR)        |
|APN-AMBR for uplink [kbps]                |12000000                                 |
|APN-AMBR for downlink [kbps]              |120000000                                |
|1.1.16 Protocol Configuration Options (PCO)|                                        |
|Type                                      |Protocol Configuration Options (PCO)     |
```

A look at Message Example 2.6 shows the detailed contents and structure of the GTP Create Session Request message. It is quite interesting, because it reveals many different addresses. There are the IMSI and MSISDN of the subscriber, the IP address of the PDN-GW/APN, the IP address assigned to the subscriber, and the IP address of the MME. There are the EBIs (=5) together with the previously discussed QCI and the maximum and guaranteed bit rate values (they are all set to zero, because QCI value "9" indicates "best effort"). Also, Tracking Area Identity (TAI) and E-UTRAN Cell Group Identifier (ECGI) are included and allow the determination of the current location of the subscriber. Finally, the protocol configuration options (e.g., user name and password of the Internet service provider) are also included.

In the RAT Type, there are furthermore a couple of flags that signal if the UE supports functions like idle mode signaling reduction, a mechanism to limit signaling during inter-RAT cell reselection in idle mode. It is also indicated if the create session request message is sent as part of the handover or data forwarding procedure.

2.2.1.4   Step 4

The Create Session Request received by the S-GW triggers another create session procedure, this time between S-GW and PDN-GW. Again, the IP addresses of the involved network elements and the EBI can be found in the GTP signaling messages. The arrows of these messages in Figure 2.4 are dashed, because the S5 interface was not monitored in the real-world scenario discussed in this chapter. The PDN-GW also receives the IP address of the UE and the APN value.

2.2.1.5   Step 5

With the GTP Create Session Response message, the PDN-GW confirms reception of all UE-related information and completes the setup of the S5 bearer. In case of dynamic IP address allocation by the APN the PDN Address Allocation information element of this message will be the first signaling message containing the UE's IP address.

2.2.1.6   Step 6

Now, the S-GW signals completion of the session and bearer setup to the MME across the S11 interface. Since the payload on the way from the UE to the S-GW will be routed on a different interface (S1-U), a special F-TEID for the S1-U user plane entity of the S-GW is signaled to the MME. Also, the routing information for the PDN-GW is sent to the MME. This gives the MME full control of all user plane packet routing and it is later up to the MME to select new routes due to the mobility of subscribers or due to changing load conditions in the packet core entities.

2.2.1.7   Step 7

After the bearers in the core network have been established, it is now necessary to establish user plane transport functions on the radio interface as well as on S1. A S1AP initial context setup request message is sent by the MME to the eNB. It includes all UE-specific parameters to be stored in the eNB, the user plane TEID of the S-GW for GTP tunnel establishment on S1-U, and the NAS messages Attach Accept and Activate Default Bearer Request that are transparently forwarded to the UE over the radio interface. An example of this important, but very complex signaling message is given in Message Example 2.7 (where all transport layer and redundant decoder outputs have been deleted to highlight the essential parts).

Message Example 2.7: S1AP Initial Context Setup Request/NAS Attach Accept/Activate Default Bearer Request

```
+-------------------------------------+--------------------------------------+
|ID Name                              |Comment or Value                      |
+-------------------------------------+--------------------------------------+
|S1 Application Protocol TS 36.413 V8.4.0 (2008-12) (including Changes
for E-RABToBeSetupItemCtxtSUReq:NAS PDU) (S1-AP) initiatingMessage
(= initiatingMessage)                                                        |
|s1apPDU                                                                     |
|1 initiatingMessage                                                         |
|1.1 procedureCode                    |id-InitialContextSetup                |
|1.2 criticality                      |reject                                |
|1.3 value                                                                   |
|1.3.1 protocolIEs                                                           |
|1.3.1.1 sequence                                                            |
|1.3.1.1.1 id                         |id-MME-UE-S1AP-ID                     |
|1.3.1.1.2 criticality                |reject                                |
|1.3.1.1.3 value                      |0                                     |
|1.3.1.2 sequence                                                            |
|1.3.1.2.1 id                         |id-eNB-UE-S1AP-ID                     |
|1.3.1.2.2 criticality                |reject                                |
|1.3.1.2.3 value                      |1                                     |
|1.3.1.3 sequence                                                            |
|1.3.1.3.1 id                         |id-uEaggregateMaximumBitrate          |
|1.3.1.3.2 criticality                |reject                                |
|1.3.1.3.3 value                                                             |
|1.3.1.3.3.1                          |'05f5e100'H                           |
uEaggregateMaximumBitRateDL                                                  |
|1.3.1.3.3.2                          |'989680'H                             |
uEaggregateMaximumBitRateUL                                                  |
|1.3.1.4 sequence                                                            |
```

```
|1.3.1.4.1 id                      |id-E-RABToBeSetupListCtxtSUReq  |
|1.3.1.4.2 criticality             |reject                          |
|1.3.1.4.3 value                                                    |
|1.3.1.4.3.1 sequenceOf                                             |
|1.3.1.4.3.1.1 id                  |id-E-RABToBeSetupItemCtxtSUReq  |
|1.3.1.4.3.1.2 criticality         |reject                          |
|1.3.1.4.3.1.3 value                                                |
|1.3.1.4.3.1.3.1 e-RAB-ID          |5                               |
|1.3.1.4.3.1.3.2                                                    |
e-RABlevelQoSParameters
|1.3.1.4.3.1.3.2.1 qCI             |9                               |
|1.3.1.4.3.1.3.2.2 allocationRetentionPriority                      |
|1.3.1.4.3.1.3.2.2.1 priorityLevel |spare                           |
|1.3.1.4.3.1.3.2.2.2               |shall-not-trigger-pre-emption   |
pre-emptionCapabi..
|1.3.1.4.3.1.3.2.2.3               |not-pre-emptable                |
pre-emptionVulner..
|1.3.1.4.3.1.3.2.3 gbrQosInformation                                |
|1.3.1.4.3.1.3.2.3.1               |                                |
e-RAB-MaximumBitr..
                             232 kbps
|1.3.1.4.3.1.3.2.3.2               |                                |
e-RAB-MaximumBitr..
                             368 kbps
|1.3.1.4.3.1.3.2.3.3               |                                |
e-RAB-GuaranteedB..
                             504 kbps
|1.3.1.4.3.1.3.2.3.4               |                                |
e-RAB-GuaranteedB..
                             1088 kbps
|1.3.1.4.3.1.3.3                   |7f 70 70 70 00 00 00 00 00 00 00 00 |
transportLayerAddress             00 00 00 00
|                                  |00 00 00 00                     |
|1.3.1.4.3.1.3.4 gTP-TEID          |'0000012c'H                     |
|1.3.1.4.3.1.3.5 nAS-PDU           |27 00 00 00 00 01 07 42 01 54 06 20 |
                                   13 20 10...

|1.3.1.5 sequence                                                   |
|1.3.1.5.1 id                      |id-UESecurityCapabilities       |
|1.3.1.5.2 criticality             |reject                          |
|1.3.1.5.3 value                                                    |
|1.3.1.5.3.1 encryptionAlgorithms  |'0000'H                         |
|1.3.1.5.3.2                       |'0000'H                         |
integrityProtectionAlgori..
|1.3.1.6 sequence                                                   |
|1.3.1.6.1 id                      |id-SecurityKey                  |
|NAS LTE TS24.301 V8.0.0 (2008-12) (Secured NAS) IPCMSG (= Integrity
protected and ciphered NAS message) |
|1 Integrity protected and ciphered NAS message                     |
|Security header type              |Integrity protected and ciphered |
|Protocol discriminator            |EPS mobility management messages |
|Message authentication code       |'00000000'H                     |
|Sequence number                   |1                               |
|NAS message(s) FFS                |07 42 01 54 06 20 13 20 10 00 c8 00 |
                                   5f 52 01...
```

```
|NAS LTE TS24.301 V8.0.0 (2008-12) (Plain NAS) ATAC (= Attach Accept)      |
|1 Attach Accept                                                           |
|Security header type            |No security protection                   |
|Protocol discriminator          |EPS mobility management messages         |
|Type of Management               |EPS mobility management                  |
|Message type                     |ATAC                                     |
|1.1 Spare Half Octet                                                      |
|Spare                            |0                                        |
|1.2 EPS attach result                                                     |
|Spare                            |0                                        |
|EPS attach result                |EPS only                                 |
|1.3 T3412 value                                                           |
|Timer value unit                 |value is incremented in multiples of     |
|                                 decihours                                |
|Timer value                      |20                                       |
|1.4 TAI list                                                              |
|Length                           |6                                        |
|Spare                            |0                                        |
|Type of list                     |one PLMN, with consecutive TAC           |
|                                 values                                   |
|Number of elements               |0                                        |
|1.4.1 Contents                                                            |
|1.4.1.1 Partial tracking area identity list (001)                        |
|MCC Digit 1                      |2                                        |
|MCC Digit 2                      |9                                        |
|MCC Digit 3                      |9                                        |
|MNC Digit 3                      |0                                        |
|MNC Digit 1                      |0                                        |
|MNC Digit 2                      |0                                        |
|TAC                              |200                                      |
|1.5 ESM Message Container                                                 |
|Length                           |95                                       |
|ESM Message Container            |52 01 c1 05 09 55 66 77 88 32 31 76      |
|                                 65 72 69...                              |
|1.6 GUTI                                                                  |
|Tag                              |EPS mobile identity                      |
|Length                           |11                                       |
|1.6.1 GUTI                                                                |
|Type of identity                 |GUTI                                     |
|Odd/Even indicator               |Even                                     |
|Spare                            |15                                       |
|MCC Digit 1                      |2                                        |
|MCC Digit 2                      |9                                        |
|MCC Digit 3                      |9                                        |
|MNC Digit 3                      |0                                        |
|MNC Digit 1                      |0                                        |
|MNC Digit 2                      |0                                        |
|MME Group ID (MMEGI)             |1                                        |
|MME Group Code (MMEC)            |1                                        |
|M-TMSI                           |'c0000000'H                              |
|1.7 T3402 value                                                           |
|Tag                              |GPRS timer                               |
|Timer value unit                 |value is incremented in multiples of     |
|                                 1 minute                                 |
```

```
|Timer value                                 |12                                               |
|NAS LTE TS24.301 V8.0.0 (2008-12) (NAS ESM) ACTDEFRQ (= Activate
default EPS bearer context request) |
|1 Activate default EPS bearer context request                                              |
|EPS bearer id                               |5                                                |
|Protocol discriminator                      |EPS session management messages                  |
|Procedure transaction identity              |1                                                |
|Type of Management                          |EPS session management                           |
|Message type                                |ACTDEFRQ                                         |
|1.1 EPS QoS                                                                                  |
|Length                                      |5                                                |
|QCI                                         |Network selects the QCI 9 [UL]                   |
|Maximum bit rate for uplink                 |232 kbps                                         |
|Maximum bit rate for downlink               |368 kbps                                         |
|Guaranteed bit rate for uplink              |504 kbps                                         |
|Guaranteed bit rate for downlink            |1088 kbps                                        |
|1.2 Access point name                                                                        |
|Length                                      |50                                               |
|1.2.1 Access Point Name                                                                      |
|Length Of Label                             |49                                               |
|Accesspoint name label                      |"rhinophone.com"                                 |
|                                            |" "                                              |
|1.3 PDN address                                                                              |
|PDN address (IPv4)                          |192.100.100.1                                    |
|1.4 Protocol configuration options                                                           |
|Tag                                         |Protocol configuration options                  |
```

Looking closer at the parameters, the eNB-UE-S1AP-ID is the same parameter that was received by the MME in the initial UE message. Now, the MME-UE-S1AP-ID is sent back to the eNB to complete the pair of unique context identifiers. All NAS messages of this particular UE connection transported from now on between eNB and the MME or vice versa will have the same ID values.

The UE AMBR values for UL and DL are the same ones that were monitored on the S6a interface (DIAMETER Location Update Response) before. However, in Message Example 2.7, they are still encoded in hexadecimal format.

The next part is the e-RAB-to-be-setup-List. This list contains all information for at least one e-RAB that should be established. If multiple e-RABs are to be established, the list will contain multiple sections of e-RAB-IDs and individual QoS parameters for each Radio Access Bearer (RAB). In the message example, there is just one e-RAB to be established. This e-RAB has e-RAB-ID = 5. The value "5" corresponds to the smallest possible NSAPI value used in GPRS Session Management (defined in 3GPP 24.008). The identical value is chosen to ensure the best possible inter-operability between the SGSN and MME.

The transport layer address is a bit string of 32 bits in the case of an IPv4 address and 128 bits in the case of an IPv6 address. In the message example, we see an IPv4 address in hexadecimal format (dotted decimal notation of this string: 127.112.112.112). Thus, all bits reserved for the IPv6 address format are set to "0." The transport layer address identifies the user plane entity of the S-GW. Also, the following GTP TEID belongs to the S-GW side of the user plane tunnel. The UE security capabilities and the security key are sent by the MME to eNB to enable ciphering and integrity protection for the radio interface transmission as described in Section 1.7.

Now comes the NAS part, starting with the attach accept message. This message belongs to EMM as the Protocol Discriminator reveals. The EPS Attach Result shows that an "EPS only" attach was performed – the UE was not registered in the SGSN simultaneously.

The timer T 3412 is used to control periodic Tracking Area Updates. In the message example, a value of 20 decihours is signaled to the UE; 1 decihour is 6 minutes. Hence, the UE is requested to perform period tracking area updates every two hours.

The next section of the message contains the Tracking Area List. We see here only one TAC, but indeed the UE can be registered in multiple tracking areas simultaneously.

Finally, we see the GUTI, the temporary identity of the UE within the E-UTRAN and a GPRS timer value of 12 minutes. The purpose of the GPRS timer information element is to specify GPRS specific timer values, for example, for the READY timer that is used to send the UE to the STANDBY state when no data is transmitted during the time of the timer value.

The second NAS message transported by the S1AP Initial Context Setup Request is the Activate Default EPS Bearer Context Request, a session management message sent by the MME to the UE. This message will trigger the setup of a radio bearer on the Uu interface for the defined QoS (Channel Quality Indicator, CQI = 9), the APN, and the UE's IP PDN address assigned earlier by the HSS.

2.2.1.8   Step 8

With the Initial Context Setup Response, the eNB confirms the establishment of the GTP tunnel on the S1-U that is part of the RAB. As shown in Message Example 2.8, this message contains an ERAB-Setup-List in which for each successfully established e-RAB-ID, the user plane transport layer address (here IPv4, dotted decimal notation: 10.254.0.4) and the appropriate GTP TEID of the user plane tunnel endpoint on the eNodeB side are transmitted.

Message Example 2.8: S1AP Initial Context Setup Response

```
|S1 Application Protocol TS 36.413 V8.4.0 (2008-12) (including Changes for
E-RABToBeSetupItemCtxtSUReq:NAS PDU) (S1-AP) successfulOutcome            |
(= successfulOutcome)
|s1apPDU                                                                  |
|1 successfulOutcome                                                      |
|1.1 procedureCode                        |id-InitialContextSetup         |
|1.2 criticality                          |reject                         |
|1.3 value                                                                |
|1.3.1 protocolIEs                                                        |
|1.3.1.1 sequence                                                         |
|1.3.1.1.1 id                             |id-MME-UE-S1AP-ID              |
|1.3.1.1.2 criticality                    |ignore                         |
|1.3.1.1.3 value                          |0                              |
|1.3.1.2 sequence                                                         |
|1.3.1.2.1 id                             |id-eNB-UE-S1AP-ID              |
|1.3.1.2.2 criticality                    |ignore                         |
|1.3.1.2.3 value                          |1                              |
|1.3.1.3 sequence                                                         |
|1.3.1.3.1 id                             |id-E-RABSetupListCtxtSURes     |
|1.3.1.3.2 criticality                    |ignore                         |
|1.3.1.3.3 value                                                          |
|1.3.1.3.3.1 sequenceOf                                                   |
|1.3.1.3.3.1.1 id                         |id-E-RABSetupItemCtxtSURes     |
```

```
|1.3.1.3.3.1.2 criticality          |ignore                                      |
|1.3.1.3.3.1.3 value                |                                            |
|1.3.1.3.3.1.3.1 e-RAB-ID           |5                                           |
|1.3.1.3.3.1.3.2                    |0a fe 00 04 00 00 00 00 00 00 00 00        |
transportLayerAddress               00 00 00 00                                 
|                                   |00 00 00 00                                 |
|1.3.1.3.3.1.3.3 gTP-TEID           |'00000001'H                                 |
```

2.2.1.9   Step 9

The UE sends the NAS message Attach Complete in case a new GUTI was assigned with Attach Accept. The assignment of a new GUTI during attach is optional, but is expected to be typically seen in live networks. The other NAS message in this transaction is Activate Default EPS Bearer Accept. With this message, the UE confirms that from now on it is ready to send and receive user plane data using the QoS profile assigned to the default bearer.

2.2.1.10   Step 10

After the UE and eNodeB are ready to send/receive payload data, there is only one piece of important information missing to start this data transfer: that is, the S-GW has not yet been informed about the eNodeB's user plane IP address and GTP TEID to be used on S1-U. To close this last open link in the activation chain, a GTP Modify Bearer Request is sent by the MME to the S-GW. This message contains the required information encoded as the S1-U eNodeB F-TEID information element sequence. A GTP modify bearer response message is sent back from the S-GW to the MME (not shown in Figure 2.4).

After this flood of addresses and identifiers, Figure 2.5 shows a summary overview of the involved network elements and identifiers for a particular UE connection and how the signaling and user plane packets are routed through the network.

For the signaling connection between the UE and MME, the UE is now identified by its GUTI. The GUTI is linked to the IMSI that we encountered during the location update procedure on S6a between the MME and HSS, but also on during the create session procedure on S11 between the MME and S-GW and on S5 between S-GW and PDN-GW. These GTP signaling connections for a single UE are now unambiguously marked by the IP addresses of the involved network elements (for S-GW and PDN-GW, the IP address for control plane traffic) and on each interface by a pair of TEIDs for the control plane. The same statement is true for the user plane transport tunnels on S1-U and S5. On S5 between S-GW and PDN-GW, we can also see how the user plane is separated from the control plane traffic by defining intelligent numbering plans. Indeed, there is no standard document that requires different IP addresses for control plane and user plane traffic in the core network, but from experience gathered in numerous lab and field trial scenarios, it can be postulated that this is a kind of de facto standard followed by all Network Equipment Manufacturers (NEMs).

The logical connection for one UE on the S1 control plane is embraced by the pair of S1AP-IDs. On the radio interface, the same function is provided by the Cell Radio Network Temporary Identifier (C-RNTI) that is signaled on the Medium Access Control (MAC) layer. The C-RNTI is valid for both the user plane and control plane, but eNodeB acts like a switch that sends the control plane traffic across one way to the MME and user plane packets across another way directly to the S-GW.

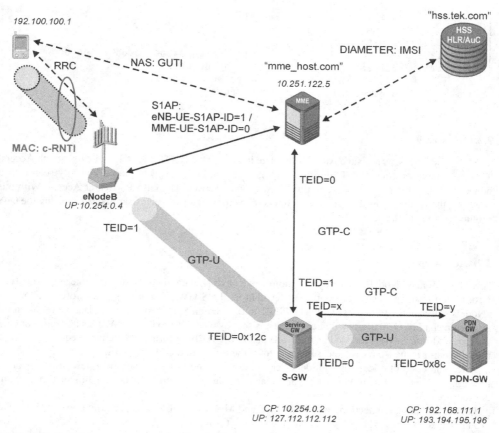

Figure 2.5 Signaling and user plane connection after successful attach. (*Source*: Tektronix Communications.)

2.3 UE Context Release Requested by eNodeB

When user inactivity is detected by the network on behalf of timer expiry, the default bearer and the UE context between e-NodeB and MME may be deleted at any time when the S1AP UE context release procedure is triggered by eNodeB as illustrated in Figure 2.6.

Note especially that the UE context release request message is also sent in some other cases, for instance, if the radio connection with the UE is lost and the call drops or if the timer for MME relocation after X2 handover expires.

2.3.1 Procedure

2.3.1.1 Step 1

After the user inactivity timer expires in eNodeB, it sends a S1AP UE Context Release Request with the cause value "user inactivity."

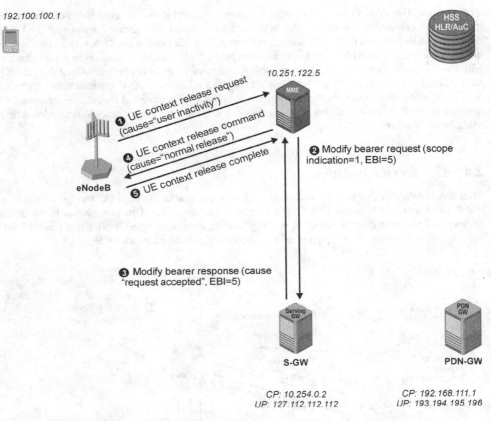

Figure 2.6 UE context release due to user inactivity. (*Source*: Tektronix Communications.)

2.3.1.2 Step 2

When the MME receives the UE context release request message, it triggers release of the S1-U bearer that is executed by sending a GTP modify bearer request message to the S-GW. In this modify bearer request message, the Scope Indication flag is set to "1." This is the indication that the associated EPS bearer (in the example identified by EBI = 5) should be deleted and the transport resources related to the bearer, especially the TEID on eNodeB and S-GW side should be released. However, the GTP user plane tunnel between S-GW and PDN-GW remains active and is not deleted during this procedure!

2.3.1.3 Step 3

The S-GW confirms the deletion of the S1-U bearer by sending GTP Modify Bearer Response with the bearer identity (EBI) and the cause "request accepted."

2.3.1.4 Steps 4 and 5

The MME must now also order eNodeB to release the resources for this particular UE connection. This is invoked by sending a S1AP UE context release command to eNodeB. Also, this message contains a cause value. However, since the UE context release command starts the termination of the complete logical S1AP connection, the cause value in this message should not indicate the status of the UE, but rather give an indication if the S1AP connection is normally terminated or not. Hence, this cause message indicates the state of the S1AP state machine in the MME. This state is "normal" even if the connection with the UE on the radio interface has been lost. Unfortunately, it cannot be guaranteed that all NEMs have this understanding. In 3G RANAP implementations of different RNC manufacturers, there have been cases where the cause values of the Release Request were repeated in the release command. This makes it difficult to count performance counters properly and compare statistical results of different MME/RNC vendors.

2.4   UE Service Request

After the S1AP context release procedure, the UE remains attached to the EPC and the default bearer is still active as a logical connection. Only the radio and channel card element resources in eNodeB are released to be used in the most efficient way. By sending a service request message on the NAS layer, the connection is resumed very quickly as described in the call flow scenario of Figure 2.7.

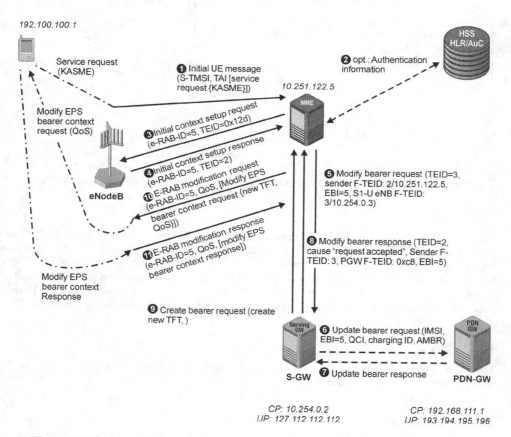

Figure 2.7   Service request after UE context release. (*Source*: Tektronix Communications.)

2.4.1   Procedure

2.4.1.1   Step 1

A service request message is sent on the NAS layer by the UE to the MME. At eNodeB, the NAS PDU is piggybacked by the S1AP transport function and forwarded to the MME using an S1AP initial UE message. The decoder output of both messages tied to each other is shown in Message Example 2.9.

In the S1AP part, the current location of the UE identified by the tracking area and E-UTRAN cell is found. The UE itself is identified by its S-TMSI that is typically a part of the GUTI. However, for the initial UE message, the GUTI is not defined as a mandatory parameter and to transmit the S-TMSI that consists of MME code and M-TMSI is sufficient.

Message Example 2.9: S1AP Initial UE Message/NAS Service Request

```
+----------------------------------+-------------------------------------+
|ID Name                           |Comment or Value                     |
+----------------------------------+-------------------------------------+
|S1 Application Protocol TS 36.413 V8.4.0 (2008-12) (including Changes  |
|for ERABToBeSetupItemCtxtSUReq:NAS PDU) (S1-AP) initiatingMessage       |
|(= initiatingMessage)                                                   |
|s1apPDU                                                                 |
|1 initiatingMessage                                                     |
|1.1 procedureCode                 |id-InitialContextSetup               |
|1.2 criticality                   |ignore                               |
|1.3 value                                                               |
|1.3.1 protocolIEs                                                       |
|1.3.1.1 sequence                                                        |
|1.3.1.1.1 id                      |id-MME-UE-S1AP-ID                    |
|1.3.1.1.2 criticality             |reject                               |
|1.3.1.1.3 value                   |2                                    |
|1.3.1.2 sequence                                                        |
|1.3.1.2.1 id                      |id-NAS-PDU                           |
|1.3.1.2.2 criticality             |reject                               |
|1.3.1.2.3 value                   |'c7020000'H                          |
|1.3.1.3 sequence                                                        |
|1.3.1.3.1 id                      |id-TAI                               |
|1.3.1.3.2 criticality             |reject                               |
|1.3.1.3.3 value                                                         |
|1.3.1.3.3.1 TBCD String           |'299000'                             |
|1.3.1.3.3.2 tAC                   |'0001'H                              |
|1.3.1.4 sequence                                                        |
|1.3.1.4.1 id                      |id-EUTRAN-CGI                        |
|1.3.1.4.2 criticality             |ignore                               |
|1.3.1.4.3 value                                                         |
|1.3.1.4.3.1 TBCD String           |'299000'                             |
|1.3.1.4.3.2 cell-ID               |'0000100'H                           |
|1.3.1.5 sequence                                                        |
|1.3.1.5.1 id                      |id-S-TMSI                            |
|1.3.1.5.2 criticality             |reject                               |
|1.3.1.5.3 value                                                         |
|1.3.1.5.3.1 mMEC                  |'01'H                                |
|1.3.1.5.3.2 m-TMSI                |'c0000001'H                          |
|NAS LTE TS24.301 V8.0.0 (2008-12) (Plain NAS) SREQ (= Service request) |
```

```
|1 Service request                                                     |
|Security header type                    |Security header for the SERVICE |
|                                         REQUEST message
|Protocol discriminator                  |EPS mobility management messages |
|1.1 KSI and sequence number                                           |
|KSI (ASME)                              |0                               |
|Sequence number (short)                 |2                               |
|1.2 Message authentication code                                       |
|(short)
|Message authentication code (short)  |0                                 |
```

In the NAS service request message, the used security key identifier KSI (ASME, Access Security Management Identity) is found together with a Sequence Number (SN) that should be used in the algorithm to restart the NAS ciphering. In addition, the NAS message authentication code for integrity protection is included.

2.4.1.2 Step 2

The reception of the NAS Service Request and its included security parameters may trigger an authentication information procedure on the S6a interface between the MME and HSS using the DIAMETER protocol.

Message Example 2.10 shows the DIAMETER authentication information request message that is sent by the MME to request one or more authentication vectors to be provided by the HSS.

Message Example 2.10: DIAMETER Authentication Information Request

```
+-----------------------------------+------------------------------------+
|ID Name                            |Comment or Value                    |
+-----------------------------------+------------------------------------+
|DIAMETER based on RFC3588/4005/4006/4072, 3GPP
29.061/.109/.140/.209/.210/.234/.229/.329/32.299
(DIAMETER) AIR (= Authentication-Information-Request ) |
|1 Authentication-Information-Request                                    |
|Origin-Host                        |"mmeOriginHost.com"                 |
|Origin-Realm                       |"mmeOriginRealm.com"                |
|Destination-Realm                  |"hss.tek.com"                       |
|User-Name                          |"299000001001001"                   |
|Destination-Host                   |"hss1"                              |
|1.1.9 Requested-EUTRAN-Authentication-Info                             |
|Vendor-Id                          |3GPP                                |
|Number-Of-Requested-Vectors        |1                                   |
|Requesting-Node-Type               |MME                                 |
```

Besides the address information of the sender and receiver of the DIAMETER message, there is also the IMSI of the user included in the "user name" field, and information is given that one vector is requested by the MME and the vector should follow the 3GPP standards.

The HSS sends back the DIAMETER authentication information answer message shown in Message Example 2.11 in compressed format. Here the different parameters of the Authentication Vector can be seen:

- Random number RAND.
- Expected Response (XRES).

- Authentication Token (AUTN).
- Security Key from ASME (KASME).

Message Example 2.11: DIAMETER Authentication Information Answer

```
+-------------------------------------------+-------------------------------------------------+
|ID Name                                    |Comment or Value                                 |
+-------------------------------------------+-------------------------------------------------+
|DIAMETER based on RFC3588/4005/4006/4072, 3GPP
29.061/.109/.140/.209/.210/.234/.229/.329/32.299
(DIAMETER) AIR (= Authentication-Information-Request ) |
|1 Authentication-Information-Answer                                                          |
|Origin-Host                                |"hss1"                                           |
|Origin-Realm                               |"hss.tek.com"                                    |
|Result-Code                                |"DIAMETER_SUCCESS                                |
|1.1.6 Authentication-Info                                                                    |
|Vendor-Id                                  |3GPP                                             |
|1.1.6.1 Authentication-Info AVP-list                                                         |
|1.1.6.1.1 E-UTRAN-Vector                                                                     |
|Vendor-Id                                  |3GPP                                             |
|Item-Number                                |1                                                |
|RAND                                       |3f 18 eb dd db 84 29 3d 64 79 ca d7              |
                                            81 1e 26 58
|XRES                                       |19 e6 cf f9 43 79 68 61                          |
|AUTN                                       |61 70 23 ab b2 5a 00 00 93 f3 3b b4              |
                                            9a 2e 11 c6
|KASME                                      |dc 86 08 5f a0 ba cf 31 67 c0 c9 b4              |
                                            20 bf 3f...
```

2.4.1.3   Step 3

The Initial Context Setup is requested by the MME. This message is basically the same as in Message Example 2.7, but without any NAS message included. It contains a new security context, but the same S1-U S-GW TEID that was assigned in the GTP create session procedure on S11 during the earlier initial attach (see Figure 2.4).

2.4.1.4   Step 4

The eNodeB confirms the new Initial Context Setup. Again, the S1 user plane TEID from the previous initial attach scenario is reactivated, this time on the eNodeB side. The S1-U TEID is encoded as the transport layer address information element.

2.4.1.5   Step 5

With the GTP modify bearer request message, the previously old S1-U bearer, temporarily deactivated by the scope indicator flag, is resumed.

2.4.1.6   Steps 6 and 7

The Bearer Modification on S11 triggers a Bearer Update on S5 with new charging information and QoS parameters.

2.4.1.7   Step 8

By sending GTP Modify Bearer Response, the S-GW confirms that the user plane tunnel on S1-U can be used again for payload transfer.

2.4.1.8   Step 9

The following Update Bearer Request that is sent from S-GW to MME on the S11 interface is required to start the charging process. A new charging ID together with the true identity of the subscriber (IMSI) are included.

2.4.1.9   Step 10

The E-RAB is going to be modified on S1. Piggybacked on the S1AP E-RAB Modification Request, we find a NAS Modify EPS Bearer Context Request with the same set of QoS parameters as on S1AP.

2.4.1.10   Step 11

The e-NodeB confirms the E-RAB modification and the UE confirms the EPS bearer modification. Now, all involved network elements are ready for transmission using the re-established user plane bearers. Different from the call flow in Figure 2.7, the NAS message EPS Bearer Modification Response may be transported by a separate S1AP UL NAS transport message.

2.5   Dedicated Bearer Setup

Whenever it is required by the parameters of a QoS profile, a dedicated bearer will be established in parallel to the default bearer. The call flow of this procedure with a focus on the protocol layers, messages, and RAB identifier is shown in Figure 2.8.

In the upper part of Figure 2.8, we see the S1AP Initial Context Setup combined with the establishment of the default bearer (RAB-ID = 5). This part was explained in the initial attach scenario (Section 2.2).

After the default bearer is active, a new user plane connection with a QoS profile different from the one of the default bearer requires the setup of a dedicated bearer that will run in parallel with the default bearer.

In the call flow example shown in Figure 2.8, the request to create this dedicated bearer comes from the S-GW. Indeed, it might have been triggered by the Policy and Charging Rule Function (PCRF) via the Gx interface after an interaction with, for example, a web server or the IP Multimedia Subsystem (IMS) raised the demand for a new bearer with a specific QoS. A typical scenario behind such behavior could be the subscriber browsing the Internet using the default bearer and after finding the streaming video source, the real-time download for such a video stream is activated.

This is the reason why the S-GW (triggered by the PDN-GW) sends a new GTP create bearer request message for RAB-ID = 6. The new Traffic Format Template (TFT) in this message contains a list of specific QoS parameters according to the requirements for streaming video.

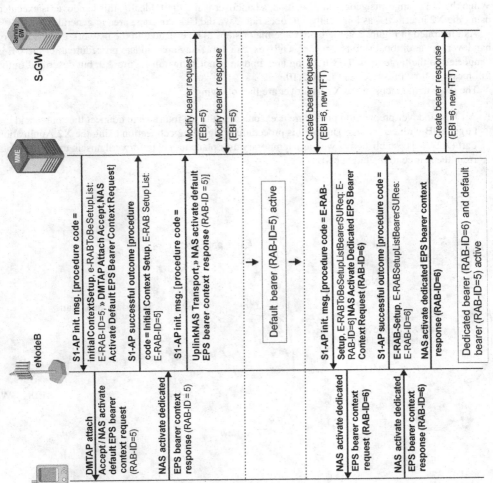

Figure 2.8 Dedicated bearer setup. (*Source:* Tektronix Communications.)

In the end, the setup of the dedicated bearer happens in a similar way as the earlier setup of the default bearer. However, the e-RAB-ID, the QoS profile of the new bearer, and the message names on S1AP and the NAS protocol layer are different, as illustrated in Figure 2.8.

2.6   Inter-eNodeB Handover over X2

When the UE changes its geographic position, a handover is required. Ideally, this handover is executed using the X2 interface – as long as the UE does not leave the LTE coverage area in general.

As illustrated in Figure 2.9, there are three major steps to be executed in the network for this kind of handover. An additional step is a subsequently executed tracking area update procedure after the UE is connected to the target cell. This tracking area update is not shown in Figure 2.9, but described in the full message flow illustrated in Figure 2.10.

The three major steps of the X2 handover are the following:

1. An X2 handover preparation procedure is executed on the X2 interface to connect the source and the target eNB with each other. During this procedure, a signaling connection using the X2 Application Part (X2AP) is established as well as a temporary user plane tunnel to forward unsent user plane data from the source to the target eNB.

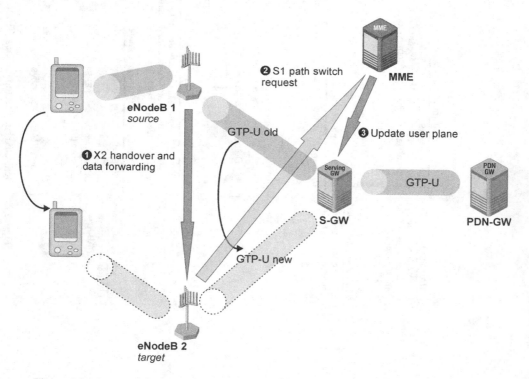

Figure 2.9   Inter-eNodeB handover over X2 – overview. (*Source*: Tektronix Communications.)

2. Using a S1AP path switch request procedure, the target eNB updates the MME with the new geographic position of the UE and requests a new route for the GTP-U tunnel on the S1-U interface.
3. To enable this new route for user plane data on S1-UE, the MME needs to communicate with S-GW, basically to negotiate the new endpoints of the GTP-U tunnel.

Surely, not just the S1-U tunnel must be switched, but also the radio bearer on the air interface. In fact, the UE changing cell on the radio interface performs a RRC reconfiguration procedure and must enter the new cell using the random access procedure described in Section 3.1.1. This mandatory random access during handover is a new functionality of the LTE radio interface signaling. In 2G and 3G radio networks, random access was only used to establish a new connection between the UE and network, but not during handover.

A look at Figure 2.10 reveals all the signaling messages of the procedure that can be monitored in the E-UTRAN including their most important information elements.

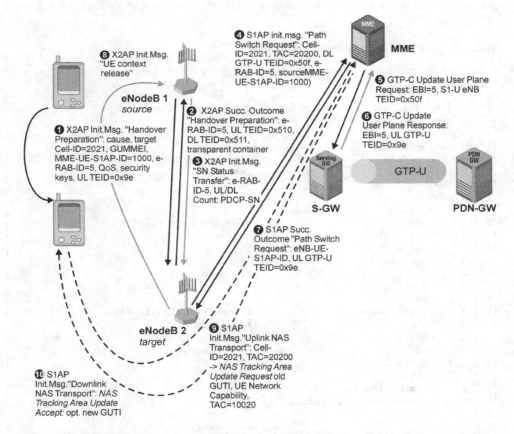

Figure 2.10 Inter-eNodeB handover over X2 – message flow. (*Source*: Tektronix Communications.)

2.6.1 Procedure

2.6.1.1 Step 1

Triggered by an RRC DL quality measurement result or by an internal UL quality measurement result, the source eNB starts the X2 handover procedure by sending an X2AP initiating message for handover preparation (3GPP message name: X2AP Handover Request) to the target eNB. Details of this message are highlighted in Message Example 2.12. The message contains the cause value related to the handover request (a typical one is the radio network cause "handover desirable for radio reasons"), the cell-ID of the source cell, the GUMMEI to identify the MME that manages mobility procedures for this particular call, the MME-UE-S1AP-ID that is the unique identity of this call in the MME S1AP protocol entity, e-RAB-ID, and appropriate QoS parameters of this e-RAB and an UL TEID value. If there is more than just one e-RAB active for this radio connection, multiple e-RAB-IDs and QoS parameter lists will be found in the X2AP message. The UL TEID signaled in the X2AP initiating message for handover preparation is the one currently assigned by S-GW for the IP tunnel on the S1-U reference point. Furthermore, the message contains the RRC context of the radio connection that includes, for example, all security parameters. Also, access stratum security information like the key-eNB is transmitted from the source eNB to target eNB.

Message Example 2.12: X2AP Handover Request

```
+----------------------------------+------------------------------------+
|ID Name                           |Comment or Value                    |
+----------------------------------+------------------------------------+
|X2AP 3GPP TS 36.423 V8.4.0 (2008-12) (X2AP) initiatingMessage          |
|(= initiatingMessage) |                                                 |
|x2apPDU                                                                 |
|1 initiatingMessage                                                     |
|1.1 procedureCode                 |id-handoverPreparation              |
|1.2 criticality                   |reject                              |
|1.3 value                                                               |
|1.3.1 protocolIEs                                                       |
|1.3.1.1 sequence                                                        |
|1.3.1.1.1 id                      |id-Old-eNB-UE-X2AP-ID               |
|1.3.1.1.2 criticality             |reject                              |
|1.3.1.1.3 value                   |1                                   |
|1.3.1.2 sequence                                                        |
|1.3.1.2.1 id                      |id-Cause                            |
|1.3.1.2.2 criticality             |ignore                              |
|1.3.1.2.3 value                                                         |
|1.3.1.2.3.1 radioNetwork          |handover-desirable-for-            |
|                                   radio-reasons                        |
|1.3.1.3 sequence                                                        |
|1.3.1.3.1 id                      |id-TargetCell-ID                    |
|1.3.1.3.2 criticality             |reject                              |
|1.3.1.3.3 value                                                         |
|1.3.1.3.3.1 pLMN-Identity         |'299000'                            |
|1.3.1.3.3.2 eUTRANcellIdentifier  |'2021'                              |
|1.3.1.4 sequence                                                        |
|1.3.1.4.1 id                      |id-GUMMEI-ID                        |
|1.3.1.4.2 criticality             |reject                              |
|1.3.1.4.3 value                                                         |
|1.3.1.4.3.1 gU-Group-ID                                                 |
```

```
|1.3.1.4.3.1.1 pLMN-Identity           |'299000'                         |
|1.3.1.4.3.1.2 mMME-Group-ID           |'0001'                           |
|1.3.1.4.3.2 mMME-Code                 |'01'                             |
|1.3.1.5 sequence                      |                                 |
|1.3.1.5.1 id                          |id-UE-ContextInformation         |
|1.3.1.5.2 criticality                 |reject                           |
|1.3.1.5.3 value                       |                                 |
|1.3.1.5.3.1 mME-UE-S1AP-ID            |1000                             |
|1.3.1.5.3.2 uESecurityCapabilities    |                                 |
|1.3.1.5.3.3 aS-SecurityInformation    |                                 |
|1.3.1.5.3.3.1 key-eNB                 |'01000101000110101011100101     |
                                        1110111'B
|1.3.1.5.3.5 e-RABs-ToBeSetup-List     |                                 |
|1.3.1.5.3.5.1.3.1 e-RAB-ID            |5                                |
|1.3.1.5.3.5.1.3.2 e-RAB-Level-QoS-Parameters                            |
|1.3.1.5.3.5.1.3.2.1 qCI               |9                                |
|1.3.1.5.3.5.1.3.4                     |                                 |
uL-GTPtunnelEndpoint
|1.3.1.5.3.5.1.3.4.1                   |127.112.112.112                  |
transportLayerAddress
|1.3.1.5.3.5.1.3.4.2 gTP-TEID          |'9e'H                            |
|1.3.1.5.3.6 rRC-Context               |0a 10 0c 82 3e 04 00 00 00 08 00... |
```

2.6.1.2   Step 2

The target eNB responds to the source eNB with an X2AP successful outcome message for "Handover Preparation" (3GPP message name: X2AP Handover Request Acknowledge – shown in Message Example 2.13), including the same e-RAB-ID(s) found previously in the initiating message. Furthermore, there is an UL TEID and a DL TEID together with the transport layer address of the appropriate network element enclosed in this message. These TEIDs will be used to forward unsent payload packets from the source eNB to the target eNB across the X2 interface. It must be emphasized that both the UL and DL TEIDs are assigned by the target eNB. This is because the temporary GTP-U tunnel on the X2 reference point during the handover procedure is used as a unidirectional connection on which all remaining packets of the connection still stored in the source eNB's buffer are forwarded to the target eNB without any acknowledgment from the receiver side. In case of packet loss, higher user plane layers like Transmission Control Protocol (TCP) have to order the necessary retransmissions using the end-to-end IP connection between the UE and application server. The reason why separate TEIDs for UL and DL are assigned by the target eNB is to separate UL payload traffic that still needs to be sent to the S-GW from DL payload traffic still needing to be sent to the UE across the radio interface. The transparent container embedded in the X2AP handover request acknowledge message contains the radio resources, especially scheduling information, initially assigned by the target cell to serve this particular radio connection.

Message Example 2.13: X2AP Handover Request Acknowledge

```
+----------------------------------------+----------------------------------------+
|ID Name                                 |Comment or Value                        |
+----------------------------------------+----------------------------------------+
|X2AP 3GPP TS 36.423 V8.4.0 (2008-12) (X2AP) successfulOutcome
(= successfulOutcome) |
|x2apPDU                                                                           |
|1 successfulOutcome                                                               |
```

```
|1.1 procedureCode                    |id-handoverPreparation            |
|1.2 criticality                      |reject                            |
|1.3 value                            |                                  |
|1.3.1 protocolIEs                    |                                  |
|1.3.1.1 sequence                     |                                  |
|1.3.1.1.1 id                         |id-Old-eNB-UE-X2AP-ID             |
|1.3.1.1.2 criticality                |ignore                            |
|1.3.1.1.3 value                      |1                                 |
|1.3.1.2 sequence                     |                                  |
|1.3.1.2.1 id                         |id-New-eNB-UE-X2AP-ID             |
|1.3.1.2.2 criticality                |ignore                            |
|1.3.1.2.3 value                      |1                                 |
|1.3.1.3 sequence                     |                                  |
|1.3.1.3.1 id                         |id-E-RABs-Admitted-List           |
|1.3.1.3.2 criticality                |ignore                            |
|1.3.1.3.3 value                      |                                  |
|1.3.1.3.3.1 sequence                 |                                  |
|1.3.1.3.3.1.1 id                     |id-E-RABs-Admitted-Item           |
|1.3.1.3.3.1.2 criticality            |ignore                            |
|1.3.1.3.3 1.3 value                  |                                  |
|1.3.1.3.3 1.3.1 e-RAB-ID             |5                                 |
|1.3.1.3.3.1.3.2                      |                                  |
uL-GTP-TunnelEndpoint                 |                                  |
|1.3.1.3.3.1.3.2.1                    |'10.254.0.4'                      |
transportLayerAddress                 |                                  |
|1.3.1.3.3.1.3.2.2 gTP-TEID           |'510'H                            |
|1.3.1.3.3.1.3.3                      |                                  |
uL-GTP-TunnelEndpoint                 |                                  |
|1.3.1.3.3.1.3.3.1                    |'10.254.0.4'                      |
transportLayerAddress                 |                                  |
|1.3.1.3.3.1.3.3.2 gTP-TEID           |'511'H                            |
|1.3.1.4 sequence                     |                                  |
|1.3.1.4.1 id                         |id-TargeteNBtoSource-eNBTranspare...|
|1.3.1.4.2 criticality                |ignore                            |
|1.3.1.4.3 value                      |02 49 10 59 a0 54 ca 91 68 23 cb...|
```

2.6.1.3   Step 3

The source eNB sends an X2AP initiating message for "SN Status Transfer," which is shown in Message Example 2.14, including the impacted e-RAB-ID(s) and Packet Data Convergence Protocol (PDCP) SNs for UL and DL data transfer across the radio interface. These PDCP SN count values are mandatory for a seamless continuation of ciphering and integrity protection functions after successful handover.

Message Example 2.14: X2AP SN Status Transfer

```
+-----------------------------------+------------------------------------------+
|ID Name                            |Comment or Value                          |
+-----------------------------------+------------------------------------------+
|X2AP 3GPP TS 36.423 V8.4.0 (2008-12) (X2AP) initiatingMessage
(= initiatingMessage) |
|x2apPDU                                                                        |
```

```
|1 initiatingMessage                                                       |
|1.1 procedureCode                        |id-snStatusTransfer             |
|1.2 criticality                          |ignore                          |
|1.3 value                                |                                |
|1.3.1 protocolIEs                        |                                |
|1.3.1.1 sequence                         |                                |
|1.3.1.1.1 id                             |id-Old-eNB-UE-X2AP-ID           |
|1.3.1.1.2 criticality                    |reject                          |
|1.3.1.1.3 value                          |1                               |
|1.3.1.2 sequence                         |                                |
|1.3.1.2.1 id                             |id-New-eNB-UE-X2AP-ID           |
|1.3.1.2.2 criticality                    |reject                          |
|1.3.1.2.3 value                          |1                               |
|1.3.1.3 sequence                         |                                |
|1.3.1.3.1 id                             |id-E-RABs-SubjectToStatusTransfer... |
|1.3.1.3.2 criticality                    |ignore                          |
|1.3.1.3.3 value                          |                                |
|1.3.1.3.3.1 sequence                     |                                |
|1.3.1.3.3.1.1 id                         |id-E-RABs-SubjectToStatusTransfer... |
|1.3.1.3.3.1.2 criticality                |ignore                          |
|1.3.1.3.3 1.3 value                      |                                |
|1.3.1.3.3 1.3.1 e-RAB-ID                 |5                               |
|1.3.1.3.3.1.3.2 uL-COUNTvalue            |                                |
|1.3.1.3.3.1.3.2.1 pDCP-SN                |0                               |
|1.3.1.3.3.1.3.2.2 hFN                    |0                               |
|1.3.1.3.3.1.3.3 dL-COUNTvalue            |                                |
|1.3.1.3.3.1.3.3.1 pDCP-SN                |0                               |
|1.3.1.3.3.1.3.3.2 hFN                    |0                               |
```

2.6.1.4   Step 4

Now it is time to switch the S1-U GTP tunnel toward the target eNB. To request this change in routing of payload packets, the target eNB sends an S1AP initiating message for Path Switch Request to the MME. The decoder output of that message can be found in Message Example 2.15. It contains the new location of the UE represented by the target cell ID and the TAC of the target cell. The sourceMME-UE-S1AP-ID allows identification of the connection of the particular handset unambiguously within the MME's S1AP entity. The DL GTP-U TEID is the TEID to be used for sending payload packets in the DL direction on the S1-U interface after the path switch procedure has been successfully completed. In addition, the ID(s) of E-RAB that should be transported in the new GTP-U tunnel is(are) listed in the message.

Message Example 2.15: S1AP Path Switch Request

```
+-----------------------------------------+------------------------------------------------+
|ID Name                                  |Comment or Value                                |
+-----------------------------------------+------------------------------------------------+
|S1AP 3GPP TS 36.413 V8.4.0 (2008-12), R3-090477 (S1AP) initiatingMessage
(= initiatingMessage) |
|s1apPDU                                  |                                                |
|1 initiatingMessage                      |                                                |
|1.1 procedureCode                        |id-PathSwitchRequest                            |
```

```
|1.2 criticality                         |reject                              |
|1.3 value                               |                                    |
|1.3.1 protocolIEs                       |                                    |
|1.3.1.1 sequence                        |                                    |
|1.3.1.1.1 id                            |id-eNB-UE-S1AP-ID                   |
|1.3.1.1.2 criticality                   |reject                              |
|1.3.1.1.3 value                         |114                                 |
|1.3.1.2 sequence                        |                                    |
|1.3.1.2.1 id                            |id-E-RABToBeSwitchedDLList          |
|1.3.1.2.2 criticality                   |reject                              |
|1.3.1.2.3 value                         |                                    |
|1.3.1.2.3.1 sequence                    |                                    |
|1.3.1.2.3.1.1 id                        |id-E-RABToBeSwitchedDLItem          |
|1.3.1.2.3.1.2 criticality               |reject                              |
|1.3.1.2.3.1.3 value                     |                                    |
|1.3.1.2.3.1.3.1 e-RAB-ID                |5                                   |
|1.3.1.2.3.1.3.2                         |10.254.0.4                          |
transportLayerAddress                    |                                    |
|1.3.1.2.3.1.3.3 gTP-TEID                |'50f'H                              |
|1.3.1.3 sequence                        |                                    |
|1.3.1.3.1 id                            |id-SourceMME-UE-S1AP-ID             |
|1.3.1.3.2 criticality                   |reject                              |
|1.3.1.3.3 value                         |1000                                |
|1.3.1.4 sequence                        |                                    |
|1.3.1.4.1 id                            |id-EUTRAN-CGI                       |
|1.3.1.4.2 criticality                   |ignore                              |
|1.3.1.4.3 value                         |                                    |
|1.3.1.4.3.1 pLMNidentity                |'299000'                            |
|1.3.1.4.3.2 cell-ID                     |'2021'                              |
|1.3.1.5 sequence                        |                                    |
|1.3.1.5.1 id                            |id-TAI                              |
|1.3.1.5.2 criticality                   |ignore                              |
|1.3.1.5.3 value                         |                                    |
|1.3.1.5.3.1 pLMNidentity                |'299000'                            |
|1.3.1.5.3.2 tAC                         |'20200'                             |
|1.3.1.6 sequence                        |                                    |
|1.3.1.6.1 id                            |id-UESecurityCapabilities           |
|1.3.1.6.2 criticality                   |ignore                              |
|1.3.1.6.3 value                         |                                    |
|1.3.1.6.3.1 encryptionAlgorithms        |'a000'H                             |
|1.3.1.6.3.2                             |'4000'H                             |
integrityProtectionAlgorithms            |                                    |
```

2.6.1.5   Step 5

Since the GTP-U tunnel needs to be switched on the interface between eNB and S-GW, the MME sends a GTP-C Update User Plane Request (Message Example 2.16) to the S-GW that contains the EBIs of the RABs to be switched and the S1-U eNB TEID for DL payload transmission using the new tunnel as was sent by the target eNB to the MME in step 4.

Message Example 2.16: S11 GTP-C Update Bearer Request

```
+------------------------------------------+------------------------------------------+
|ID Name                                   |Comment or Value                          |
+------------------------------------------+------------------------------------------+
|Tunnelling Protocol for Control plane (GTPv2-C) 3GPP TS 29.274 V8.0.0
(2008-12) (GTP_C) UPDUPREQ (= Update User Plane Request) |
|1 Update User Plane Request                                                          |
|Version                                   |GTPv2                                     |
|Spare                                     |0                                         |
|T-Bit                                     |TEID is present in the GTP-C header       |
|Spare                                     |0                                         |
|Message Type                              |Update User Plane Request                 |
|Message Length                            |27                                        |
|TEID                                      |'00000002'H                               |
|Sequence Number                           |2                                         |
|Spare                                     |0                                         |
|1.1 Bearer Context to be updated                                                     |
|1.1.1.1 EPS Bearer ID (EBI)                                                          |
|EPS Bearer ID (EBI)                       |5                                         |
|1.1.1.2 S1-U eNodeB F-TEID                                                           |
|Interface Type                            |S1-U eNB GTP-U interface                  |
|TEID/GRE Key                              |'50f'H                                    |
```

2.6.1.6   Step 6

The S-GW returns a GTP-C Update User Plane Response (Message Example 2.17) containing again the EBI and UL GTP-U TEID of the new tunnel. In this sample scenario, the UL GTP-U TEID remains unchanged. Thus, it has the same value as signaled with the X2AP handover request message in step 1.

Message Example 2.17: S11 GTP-C Update Bearer Response

```
+------------------------------------------+------------------------------------------+
|ID Name                                   |Comment or Value                          |
+------------------------------------------+------------------------------------------+
|Tunnelling Protocol for Control plane (GTPv2-C) 3GPP TS 29.274 V8.0.0
(2008-12) (GTP_C) UPDUPRSP (= Update User Plane Response) |
|1 Update User Plane Response                                                         |
|Version                                   |GTPv2                                     |
|Spare                                     |0                                         |
|T-Bit                                     |TEID is present in the GTP-C header       |
|Spare                                     |0                                         |
|Message Type                              |Update User Plane Response                |
|Message Length                            |36                                        |
|TEID                                      |'00000003'H                               |
|Sequence Number                           |2                                         |
|Spare                                     |0                                         |
|1.1 Cause (without embedded
offending IE)                                                                        |
```

```
|Cause                            |Request accepted                   |
|1.2 Bearer Context updated       |                                   |
|EPS Bearer ID (EBI)              | 5                                 |
|1.2.1.3 S1 SGW F-TEID            |                                   |
|Interface Type                   |S1-U eNB GTP-U interface           |
|TEID/GRE Key                     | '9e'H                             |
```

2.6.1.7   Step 7

After a successful update user plane procedure on S11, the MME now sends the successful outcome message of the S1AP patch switch request procedure to the target eNB – see the details in Message Example 2.18. This message confirms that the UL GTP-U TEID on S1-U will remain the same.

Message Example 2.18: S1AP Path Switch Acknowledge

```
+------------------------------------+------------------------------------+
|ID Name                             |Comment or Value                    |
+------------------------------------+------------------------------------+
|S1AP 3GPP TS 36.413 V8.4.0 (2008-12), R3-090477 (S1AP) successfulOutcome |
|(= successfulOutcome) |                                                  |
|s1apPDU                             |                                    |
|1 successfulOutcome                 |                                    |
|1.1 procedureCode                   |id-PathSwitchRequest                |
|1.2 criticality                     |reject                              |
|1.3 value                           |                                    |
|1.3.1 protocolIEs                   |                                    |
|1.3.1.1 sequence                    |                                    |
|1.3.1.1.1 id                        |id-MME-UE-S1AP-ID                   |
|1.3.1.1.2 criticality               |ignore                              |
|1.3.1.1.3 value                     |1000                                |
|1.3.1.2 sequence                    |                                    |
|1.3.1.2.1 id                        |id-eNB-UE-S1AP-ID                   |
|1.3.1.2.2 criticality               |ignore                              |
|1.3.1.2.3 value                     |114                                 |
|1.3.1.3 sequence                    |                                    |
|1.3.1.3.1 id                        |id-E-RABToBeSwitchedULList          |
|1.3.1.3.2 criticality               |ignore                              |
|1.3.1.3.3 value                     |                                    |
|1.3.1.3.3.1 sequence                |                                    |
|1.3.1.3.3.1.1 id                    |id-E-RABToBeSwitchedULItem          |
|1.3.1.3.3.1.2 criticality           |ignore                              |
|1.3.1.3.3.1.3 value                 |                                    |
|1.3.1.3.3.1.3.1 e-RAB-ID            | 5                                  |
|1.3.1.3.3.1.3.2                     |127.112.112.112                     |
transportLayerAddress                                                     |
|1.3.1.3.3.1.3.3 gTP-TEID            | '9e'H                              |
|1.3.1.4 sequence                    |                                    |
|1.3.1.4.1 id                        |id-SecurityContext                  |
```

In addition, some parameters for the security context are enclosed, such as next hop chaining count, which is used to derive a new K_{eNB}. Thus, the source eNB does not forward its internal security keys to the target eNB.

2.6.1.8 Step 8

An X2AP UE release request message is sent by the target eNB to the source eNB to indicate the successfully completed radio interface handover and path switching. The particular UE connection is unambiguously identified by the pair of Old-eNB-UE-X2AP-ID and New-eNB-UE-X2AP-ID as shown in Message Example 2.19. When receiving this message, the source eNB will release all UE-specific information and radio resources. However, it may continue user plane data forwarding to the target eNB as long as this is necessary.

Message Example 2.19: X2AP UE Context Release

```
+--------------------------------------+----------------------------------------+
|ID Name                               |Comment or Value                        |
+--------------------------------------+----------------------------------------+
|X2AP 3GPP TS 36.423 V8.4.0 (2008-12) (X2AP) initiatingMessage
(= initiatingMessage)  |
|x2apPDU                                                                        |
|1 initiatingMessage                                                            |
|1.1 procedureCode                     |id-uEContextRelease                     |
|1.2 criticality                       |ignore                                  |
|1.3 value                                                                      |
|1.3.1 protocolIEs                                                              |
|1.3.1.1 sequence                                                               |
|1.3.1.1.1 id                          |id-Old-eNB-UE-X2AP-ID                   |
|1.3.1.1.2 criticality                 |reject                                  |
|1.3.1.1.3 value                       |1                                       |
|1.3.1.2 sequence                                                               |
|1.3.1.2.1 id                          |id-New-eNB-UE-X2AP-ID                   |
|1.3.1.2.2 criticality                 |reject                                  |
|1.3.1.2.3 value                       |1                                       |
```

2.6.1.9 Step 9

Although the inter-eNB handover is already successfully completed in the access stratum at this point, and payload data transfer using the new transport resources on the radio interface and S1-U may have already started, it is still necessary to update the UE location on the NAS layer. Surely, this is only necessary if the target cell belongs to a different tracking area than the source cell – as assumed in the sample scenario of Figure 2.10. The TAC of the target cell is detected by the UE when reading the cell's broadcast information. After detecting the new TAC, the UE sends the NAS tracking area update request message (Message Example 2.20) to the MME. This message is transparently forwarded by the target eNB using the S1AP UL NAS transport message. This message contains the new location of the UE (cell ID of the target cell) and new TAC, while in the NAS tracking area request message, the old GUTI to identify the UE on the NAS layer, the UE network capabilities, and TAC of the old cell (source cell) are included.

Message Example 2.20: Tracking Area Update Request

```
+------------------------------------+---------------------------------------+
|ID Name                             |Comment or Value                       |
+------------------------------------+---------------------------------------+
|S1 Application Protocol TS 36.413 V8.4.0 (2008-12) (including Changes for
ERABToBeSetupItemCtxtSUReq:NAS PDU) (S1-AP) initiatingMessage
(= initiatingMessage) |
|s1apPDU                                                                     |
|1 initiatingMessage                                                         |
|1.1 procedureCode                   |id-uplinkNASTransport                   |
|1.2 criticality                     |ignore                                  |
|1.3 value                                                                   |
|1.3.1 protocolIEs                                                           |
|1.3.1.1 sequence                    |                                        |
|1.3.1.1.1 id                        |id-MME-UE-S1AP-ID                       |
|1.3.1.1.2 criticality               |reject                                  |
|1.3.1.1.3 value                     |3                                       |
|1.3.1.2 sequence                                                            |
|1.3.1.2.1 id                        |id-eNB-UE-S1AP-ID                       |
|1.3.1.2.2 criticality               |reject                                  |
|1.3.1.2.3 value                     |2                                       |
|1.3.1.3 sequence                                                            |
|1.3.1.3.1 id                        |id-NAS-PDU                              |
|1.3.1.3.2 criticality               |reject                                  |
|1.3.1.3.3 value                     |27 00 00 00 00 01 07 48 00 0b f6...     |
|1.3.1.4 sequence                    |                                        |
|1.3.1.4.1 id                        |id-EUTRAN-CGI                           |
|1.3.1.4.2 criticality               |ignore                                  |
|1.3.1.4.3 value                                                             |
|1.3.1.4.3.1 pLMNidentity            |'299000'                                |
|1.3.1.4.3.2 cell-ID                 |'2021'                                  |
|1.3.1.5 sequence                                                            |
|1.3.1.5.1 id                        |id-TAI                                  |
|1.3.1.5.2 criticality               |ignore                                  |
|1.3.1.5.3 value                                                             |
|1.3.1.5.3.1 pLMNidentity            |'299000'                                |
|1.3.1.5.3.2 tAC                     |'20200'                                 |
|NAS LTE TS24.301 V8.0.0 (2008-12) (Plain NAS) TAURQ (= Tracking area
update request) |
|1 Tracking area update request                                              |
|Security header type                |No security protection                  |
|Protocol discriminator              |EPS mobility management messages        |
|Type of Management                  |EPS mobility management                 |
|Message type                        |TAURQ                                   |
|1.1 Spare Half Octet                                                        |
|Spare                               |0                                       |
|1.2 EPS update type                                                         |
|Active Flag                         |No bearer establishment requested       |
|EPS update type                     |TA updating                             |
|1.3 Old GUTI                                                                |
|Length                              |11                                      |
|1.3.1 Contents                                                              |
```

```
|Type of identity                        |GUTI                                |
|Mobile Country Code (MCC)               |299                                 |
|Mobile Network Code (MNC)               |000                                 |
|MME Group ID (MMEGI)                    |1                                   |
|MME Group Code (MMEC)                    |1                                   |
|M-TMSI                                  |'c0000003'H                         |
|1.7 Last visited registered TAI                                              |
|Tag                                     |Tracking area identity              |
|Mobile Country Code (MCC)               |299                                 |
|Mobile Network Code (MNC)               |000                                 |
|TAC                                     |10020                               |
|1.8 EPS bearer context status                                               |
|Tag                                     |EPS bearer context status           |
|Length                                  |2                                   |
|EBI(7)                                  |BEARER CONTEXT-INACTIVE              |
|EBI(6)                                  |BEARER CONTEXT-INACTIVE              |
|EBI(5)                                  |BEARER CONTEXT-ACTIVE                |
|EBI(4)                                  |BEARER CONTEXT-INACTIVE              |
```

2.6.1.10   Step 10

Responding to the UE's Tracking Area Update Request, the MME sends a NAS Tracking Area Update Accept (Message Example 2.21) back to the mobile. If a new GUTI is assigned depending on the MME's configuration parameters for subscriber management, the new GUTI will be included in this message and the handset will confirm reception of this new temporary identity with a TAC message.

Message Example 2.21: Tracking Area Update Accept

```
+----------------------------------------+----------------------------------------+
|ID Name                                 |Comment or Value                        |
+----------------------------------------+----------------------------------------+
|NAS LTE TS24.301 V8.0.0 (2008-12) (Plain NAS) TAUAC (= Tracking area             |
|update accept) |                                                                 |
|1 Tracking area update accept                                                    |
|Security header type                    |No security protection                  |
|Protocol discriminator                  |EPS mobility management messages        |
|Type of Management                      |EPS mobility management                 |
|Message type                            |TAUAC                                   |
|1.1 Spare Half Octet                                                             |
|Spare                                   |0                                       |
|1.2 EPS update result                                                            |
|Spare                                   |'0'B                                    |
|EPS update result                       |TA updated                              |
|1.3 T3412 value                                                                  |
|Tag                                     |GPRS timer                              |
|Timer value unit                        |value is incremented in multiples...    |
|Timer value                             |20                                      |
|1.4 T3402 value                                                                  |
|Tag                                     |GPRS timer                              |
|Timer value unit                        |value is incremented in multiples...    |
|Timer value                             |12                                      |
```

2.7 S1 Handover

Whenever two neighbor eNBs are not connected via the X2 interface, the applicable mobility procedure when the UE needs to change the serving cell is the S1 handover procedure. The S1 handover procedure is also used to prepare and execute inter-RAT handovers. Differentiation is possible by looking at the handover type information element and into the various transparent containers embedded in the handover messages.

Figure 2.11 shows the principle of an intra-LTE S1 handover using the three steps of handover preparation, handover resource allocation, and modification of the S1-U bearer. Handover preparation is triggered by the source eNB typically after receiving an RRC measurement report that indicates the need for a serving cell change.

When the MME receives the relocation preparation request of the source eNB, it starts the handover resource allocation procedure to request the necessary radio resource from the target eNB. After the target eNB sends the required radio interface parameters embedded in a handover command message, the MME forwards this handover command message transparently to the UE, which executes the handover.

Now the GTP tunnel on S1-U that is used to send payload packets in the UL/DL direction also needs to be switched and this is realized by performing a bearer modification procedure on S11 that is triggered by the MME and executed by S-GW. In addition, a temporary tunnel for forwarding of unsent downlink payload data is established between the source and target eNodeB. This temporary tunnel is deleted at S1 handover completion.

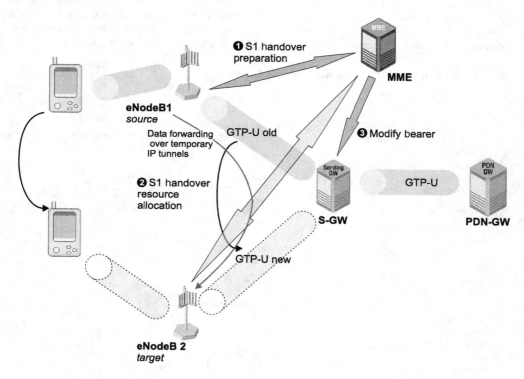

Figure 2.11   Overview of S1 handover. (*Source*: Tektronix Communications.)

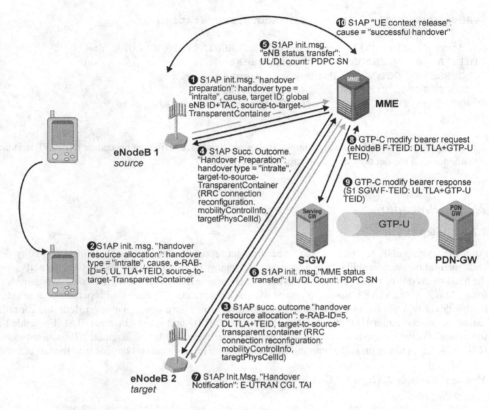

Figure 2.12 S1 handover message flow. (*Source*: Tektronix Communications.)

When looking at the two different S1 interfaces, it is also common (especially in performance measurement counters) to name the handover preparation procedure "outgoing handover" or "outgoing leg," while the relocation resource allocation procedure is referred to as the "incoming handover" or "incoming leg."

In fact, there is another terminology introduced by 3GPP that groups the handover required message of the outgoing leg with the handover request message of the incoming leg together as the "handover preparation phase."

In turn, all S1 messages transporting the handover command to be sent to the UE are grouped as "handover execution phase." These S1 messages are the handover request acknowledge message of the incoming leg and the S1 handover command of the outgoing leg.

Looking at the different signaling messages involved in the S1 handover, the more detailed steps shown in Figure 2.12 can be used to explain the progress of the procedure. The ASN.1 encoding rules of the involved S1AP messages are defined as follows:

```
handoverPreparation S1AP-ELEMENTARY-PROCEDURE ::= {
  INITIATING MESSAGE   HandoverRequired
  SUCCESSFUL OUTCOME   HandoverCommand
  UNSUCCESSFUL OUTCOME   HandoverPreparationFailure
```

```
 PROCEDURE CODE            id-HandoverPreparation
}
handoverResourceAllocation S1AP-ELEMENTARY-PROCEDURE ::= {
 INITIATING MESSAGE   HandoverRequest
 SUCCESSFUL OUTCOME   HandoverRequestAcknowledge
 UNSUCCESSFUL OUTCOME    HandoverFailure
 PROCEDURE CODE            id-HandoverResourceAllocation
}
```

It is important to map the message names as they are used to describe the procedures in 3GPP standard documents with the decoder output of monitoring systems.

2.7.1 Procedure

2.7.1.1 Step 1

After the serving eNB receives an RRC measurement report that indicates a changing DL quality or a node internal measurement indicates a changing UL quality of an active RRC connection, the eNB sends the handover required message as shown in Message Example 2.22 to the MME. This message is encoded using ASN.1 as the S1AP initiating message with procedure code "Relocation Preparation." It contains the handover type (in the example: intra-LTE handover), the cause (e.g., "handover desirable for radio reason"), and the target ID that consists of the target eNB-ID and selected (target) TAI. Embedded in the message is the source-to-target-TransparentContainer that includes all essential UE capabilities and RRC parameters to continue the ongoing RRC connection after successful handover to the target cell.

Message Example 2.22: S1AP Handover Required

```
+---------------------------------------+-------------------------------------------+
|ID Name                                |Comment or Value                           |
+---------------------------------------+-------------------------------------------+
|S1AP 3GPP TS 36.413 V8.5.1 (2009-03), R3-091397, R3-091144 (S1AP)
initiatingMessage (= initiatingMessage) |
|s1apPDU                                                                            |
|1 initiatingMessage                                                                |
|1.1 procedureCode                      |id-HandoverPreparation                     |
|1.2 criticality                        |reject                                     |
|1.3 value                                                                          |
|1.3.1 protocolIEs                                                                  |
|1.3.1.1 sequence                                                                   |
|1.3.1.1.1 id                           |id-MME-UE-S1AP-ID                          |
|1.3.1.1.2 criticality                  |reject                                     |
|1.3.1.1.3 value                        |552658472                                  |
|1.3.1.2 sequence                                                                   |
|1.3.1.2.1 id                           |id-eNB-UE-S1AP-ID                          |
|1.3.1.2.2 criticality                  |reject                                     |
|1.3.1.2.3 value                        |0                                          |
|1.3.1.3 sequence                                                                   |
|1.3.1.3.1 id                           |id-HandoverType                            |
|1.3.1.3.2 criticality                  |reject                                     |
|1.3.1.3.3 value                        |intralte                                   |
```

```
|1.3.1.4 sequence                      |                                    |
|1.3.1.4.1 id                          |id-Cause                            |
|1.3.1.4.2 criticality                 |ignore                              |
|1.3.1.4.3 value                       |                                    |
|1.3.1.4.3.1 radioNetwork              |unspecified                         |
|1.3.1.5 sequence                      |                                    |
|1.3.1.5.1 id                          |id-TargetID                         |
|1.3.1.5.2 criticality                 |reject                              |
|1.3.1.5.3 value                       |                                    |
|1.3.1.5.3.1 targeteNB-ID              |                                    |
|1.3.1.5.3.1.1 global-ENB-ID           |                                    |
|1.3.1.5.3.1.1.1 pLMNidentity          |'299000'H                           |
|1.3.1.5.3.1.1.2 eNB-ID                |                                    |
|1.3.1.5.3.1.1.2.1 macroENB-ID         |'000b2'H                            |
|1.3.1.5.3.1.2 selected-TAI            |                                    |
|1.3.1.5.3.1.2.1 pLMNidentity          |'13f182'H                           |
|1.3.1.5.3.1.2.2 tAC                   |'0001'H                             |
|1.3.1.6 sequence                      |                                    |
|1.3.1.6.1 id                          |id-Source-ToTarget-Transparent      |
|                                      |Container                           |
|1.3.1.6.2 criticality                 |reject                              |
|1.3.1.6.3 value                       |00 78 0a 10 17 80 ba 05 c3 02 14 69 |
|                                      |90 e1 f8...                         |
```

2.7.1.2 Step 2

Based on the target ID, the MME selects the appropriate transport link that connects it with the target eNB. This selection is done on behalf of an MME internal routing table. In the case of configuration errors in this routing table, the target eNB cannot be reached and a handover preparation failure message (ASN.1: S1AP Unsuccessful Outcome with procedure code "Relocation Preparation") is sent back to the source eNB. Otherwise, the MME sends an S1AP handover request message (Message Example 2.23) to the target eNB.

Message Example 2.23: S1AP Handover Request

```
+--------------------------------------+------------------------------------+
|ID Name                               |Comment or Value                    |
+--------------------------------------+------------------------------------+
|S1AP 3GPP TS 36.413 V8.5.1 (2009-03), R3-091397, R3-091144 (S1AP)
initiatingMessage (= initiatingMessage) |
|s1apPDU                               |                                    |
|1 initiatingMessage                   |                                    |
|1.1 procedureCode                     |id-HandoverResourceAllocation       |
|1.2 criticality                       |reject                              |
|1.3 value                             |                                    |
|1.3.1 protocolIEs                     |                                    |
|1.3.1.1 sequence                      |                                    |
|1.3.1.1.1 id                          |id-MME-UE-S1AP-ID                   |
|1.3.1.1.2 criticality                 |reject                              |
|1.3.1.1.3 value                       |542232756                           |
|1.3.1.2 sequence                      |                                    |
```

```
|1.3.1.2.1 id                              |id-HandoverType
|1.3.1.2.2 criticality                     |reject
|1.3.1.2.3 value                           |intralte
|1.3.1.3 sequence
|1.3.1.3.1 id                              |id-Cause
|1.3.1.3.2 criticality                     |ignore
|1.3.1.3.3 value
|1.3.1.3.3.1 radioNetwork                  |unspecified
|1.3.1.4 sequence
|1.3.1.4.1 id                              |id-uEaggregateMaximumBitrate
|1.3.1.4.2 criticality                     |reject
|1.3.1.4.3 value
|1.3.1.4.3.1                               |'027ffff6'H
uEaggregateMaximumBitRateDL
|1.3.1.4.3.2                               |'013ffff6'H
uEaggregateMaximumBitRateUL
|1.3.1.5 sequence
|1.3.1.5.1 id                              |id-E-RABToBeSetupListHOReq
|1.3.1.5.2 criticality                     |reject
|1.3.1.5.3 value
|1.3.1.5.3.1 sequence
|1.3.1.5.3.1.1 id                          |id-E-RABToBeSetupItemHOReq
|1.3.1.5.3.1.2 criticality                 |reject
|1.3.1.5.3.1.3 value
|1.3.1.5.3.1.3.1 e-RAB-ID                  |5
|1.3.1.5.3.1.3.2                           |'0a0a6102'H
transportLayerAddress
|1.3.1.5.3.1.3.3 gTP-TEID                  |'046d1400'H
|1.3.1.5.3.1.3.4
e-RABlevelQosParameters
|1.3.1.5.3.1.3.4.1 qCI                     |5
|1.3.1.5.3.1.3.4.2
allocationRetentionPriority
|1.3.1.5.3.1.3.4.2.1 priorityLevel         |2
|1.3.1.5.3.1.3.4.2.2                       |shall-not-trigger-pre-emption
pre-emptionCapability
|1.3.1.5.3.1.3.4.2.3                       |pre-emptable
pre-emptionVulnerability
|1.3.1.6 sequence
|1.3.1.6.1 id                              |id-Source-ToTarget-Transparent
                                           Container
|1.3.1.6.2 criticality                     |reject
|1.3.1.6.3 value                           |00 78 0a 10 17 80 ba 05 c3 02 14 69
                                           90 e1 f8...
|1.3.1.7 sequence
|1.3.1.7.1 id                              |id-UESecurityCapabilities
|1.3.1.7.2 criticality                     |reject
|1.3.1.7.3 value
|1.3.1.7.3.1 encryptionAlgorithms          |'e000'H
|1.3.1.7.3.2                               |'c000'H
integrityProtectionAlgorithms
|1.3.1.8 sequence
|1.3.1.8.1 id                              |id-SecurityContext
```

```
|1.3.1.8.2 criticality              |reject                                        |
|1.3.1.8.3 value                    |                                              |
|1.3.1.8.3.1 nextHopChainingCount   |1                                             |
|1.3.1.8.3.2 nextHopParameter       |'00001101000001011001011110010000            |
                                     111000110100'B
```

The handover request message (ASN.1: S1AP initiating message with procedure code "Handover Resource Allocation") contains the same handover type, cause value, and source-to-target-Transparent Container as the Handover Required. It further contains the list of e-RABs and their respective QoS parameters for which the S1-U GTP tunnel needs to be switched. Appended are the UE security capabilities and security context information that is required to continue ciphering and integrity protection of the connection after successful handover on the physical radio interface layer.

2.7.1.3   Step 3

After receiving Handover Request, the target eNB allocates the necessary radio resources for taking over the RRC connection of the UE. This information needs to be signaled to the handset. Hence, an RRC handover command (in fact, an RRC connection reconfiguration message that contains, for example, the physical cell identity of the target cell) is constructed by the target eNB and sent back to the source eNB using the target-to-source-TransparentContainer information element.

The target-to-source-TransparentContainer is first sent to the MME together with E-RAB parameters in the S1AP handover request acknowledge message (ASN.1: S1AP successful outcome message with procedure code "Handover Resource Allocation" as shown in Message Example 2.24) on the incoming leg of the handover.

Message Example 2.24: S1AP Handover Request Acknowledge

```
+-------------------------------------------+----------------------------------------+
|ID Name                                    |Comment or Value                        |
+-------------------------------------------+----------------------------------------+
|S1AP 3GPP TS 36.413 V8.5.1 (2009-03), R3-091397, R3-091144 (S1AP) suc-
cessfulOutcome (= successfulOutcome) |
|s1apPDU                                    |                                        |
|1 successfulOutcome                        |                                        |
|1.1 procedureCode                          |id-HandoverResourceAllocation           |
|1.2 criticality                            |reject                                  |
|1.3 value                                  |                                        |
|1.3.1 protocolIEs                          |                                        |
|1.3.1.1 sequence                           |                                        |
|1.3.1.1.1 id                               |id-MME-UE-S1AP-ID                       |
|1.3.1.1.2 criticality                      |ignore                                  |
|1.3.1.1.3 value                            |542232756                               |
|1.3.1.2 sequence                           |                                        |
|1.3.1.2.1 id                               |id-eNB-UE-S1AP-ID                       |
|1.3.1.2.2 criticality                      |ignore                                  |
|1.3.1.2.3 value                            |0                                       |
|1.3.1.3 sequence                           |                                        |
|1.3.1.3.1 id                               |id-E-RABAdmittedList                    |
|1.3.1.3.2 criticality                      |ignore                                  |
|1.3.1.3.3 value                            |                                        |
```

```
|1.3.1.3.3.1 sequence                    |
|1.3.1.3.3.1.1 id                        |id-E-RABAdmittedItem          |
|1.3.1.3.3.1.2 criticality               |ignore                        |
|1.3.1.3.3.1.3 value                     |                              |
|1.3.1.3.3.1.3.1 e-RAB-ID                |5                             |
|1.3.1.3.3.1.3.2                         |'0a0a280a'H                   |
transportLayerAddress
|1.3.1.3.3.1.3.3 gTP-TEID                |'de114432'H                   |
|1.3.1.4 sequence                        |                              |
|1.3.1.4.1 id                            |id-Target-ToSource-Transparent|
                                         Container
|1.3.1.4.2 criticality                   |reject                        |
|1.3.1.4.3 value                         |00 27 01 29 00 58 cb 21 82 80 2b 3d |
                                         87 a4 ca...
```

2.7.1.4   Step 4

The MME forwards the target-to-source-TransparentContainer to the source eNB by using the S1AP handover command message (ASN.1: S1AP successful outcome message with procedure code "Handover Preparation" as shown in Message Example 2.25).

Message Example 2.25: S1AP Handover Command

```
+-------------------------------------+------------------------------------------+
|ID Name                              |Comment or Value                          |
+-------------------------------------+------------------------------------------+
|S1AP 3GPP TS 36.413 V8.5.1 (2009-03), R3-091397, R3-091144 (S1AP)
successfulOutcome (= successfulOutcome) |
|s1apPDU
|1 successfulOutcome                  |                                          |
|1.1 procedureCode                    |id-HandoverPreparation                    |
|1.2 criticality                      |reject                                    |
|1.3 value                            |                                          |
|1.3.1 protocolIEs                    |                                          |
|1.3.1.1 sequence                     |                                          |
|1.3.1.1.1 id                         |id-MME-UE-S1AP-ID                         |
|1.3.1.1.2 criticality                |reject                                    |
|1.3.1.1.3 value                      |552658472                                 |
|1.3.1.2 sequence                     |                                          |
|1.3.1.2.1 id                         |id-eNB-UE-S1AP-ID                         |
|1.3.1.2.2 criticality                |reject                                    |
|1.3.1.2.3 value                      |0                                         |
|1.3.1.3 sequence                     |                                          |
|1.3.1.3.1 id                         |id-HandoverType                           |
|1.3.1.3.2 criticality                |reject                                    |
|1.3.1.3.3 value                      |intralte                                  |
|1.3.1.4 sequence                     |                                          |
|1.3.1.4.1 id                         |id-Target-ToSource-Transparent           |
                                      Container
|1.3.1.4.2 criticality                |reject                                    |
|1.3.1.4.3 value                      |00 27 01 29 00 58 cb 21 82 80 2b 3d       |
                                      87 a4 ca...
```

2.7.1.5   Step 5

Now the source eNB sends the RRC handover command to the UE and the UE performs the handover on the physical radio interface layer. As soon as the UE leaves the source cell, the source eNB sends the S1AP eNB status transfer message to the MME that is shown in Message Example 2.26.

Message Example 2.26: S1AP eNodeB Status Transfer

```
+-------------------------------------+---------------------------------------+
|ID Name                              |Comment or Value                       |
+-------------------------------------+---------------------------------------+
|S1AP 3GPP TS 36.413 V8.5.1 (2009-03), R3-091397, R3-091144 (S1AP)
initiatingMessage (= initiatingMessage) |
|s1apPDU                                                                      |
|1 initiatingMessage                                                          |
|1.1 procedureCode                    |id-eNBStatusTransfer                   |
|1.2 criticality                      |ignore                                 |
|1.3 value                                                                    |
|1.3.1 protocolIEs                                                            |
|1.3.1.1 sequence                                                             |
|1.3.1.1.1 id                         |id-MME-UE-S1AP-ID                      |
|1.3.1.1.2 criticality                |reject                                 |
|1.3.1.1.3 value                      |552658472                              |
|1.3.1.2 sequence                                                             |
|1.3.1.2.1 id                         |id-eNB-UE-S1AP-ID                      |
|1.3.1.2.2 criticality                |reject                                 |
|1.3.1.2.3 value                      |0                                      |
|1.3.1.3 sequence                                                             |
|1.3.1.3.1 id                         |id-eNB-StatusTransfer-Transparent      |
|                                     |Container                              |
|1.3.1.3.2 criticality                |reject                                 |
|1.3.1.3.3 value                                                              |
|1.3.1.3.3.1                                                                  |
bearers-SubjectToStatusTransferList                                           |
|1.3.1.3.3.1.1 sequenceOf                                                     |
|1.3.1.3.3.1.1.1 id                   |id-Bearers-SubjectToStatusTransfer-    |
|                                     |Item                                   |
|1.3.1.3.3.1.1.2 criticality          |ignore                                 |
|1.3.1.3.3.1.1.3 value                                                        |
|1.3.1.3.3.1.1.3.1 e-RAB-ID           |5                                      |
|1.3.1.3.3.1.1.3.2 uL-COUNTvalue                                              |
|1.3.1.3.3.1.1.3.2.1 pDCP-SN          |914                                    |
|1.3.1.3.3.1.1.3.2.2 hFN              |7                                      |
|1.3.1.3.3.1.1.3.3 dL-COUNTvalue                                              |
|1.3.1.3.3.1.1.3.3.1 pDCP-SN          |4                                      |
|1.3.1.3.3.1.1.3.3.2 hFN              |0                                      |
```

This status message contains the last PDCP SNs sent/received by the source eNB in the UL and DL direction. In other words, it is indicated at which point user plane data transfer stopped and needs to be resumed by the target eNB.

Different from the X2 handover procedure, there is no data forwarding from the source eNB to target eNB.

2.7.1.6   Step 6

The PDCP SNs received from the source eNB are forwarded to the target eNB using the S1AP MME status transfer message shown in Message Example 2.27.

Message Example 2.27: S1AP MME Status Transfer

```
+---------------------------------------+-----------------------------------------+
|ID Name                                |Comment or Value                         |
+---------------------------------------+-----------------------------------------+
|S1AP 3GPP TS 36.413 V8.5.1 (2009-03), R3-091397, R3-091144 (S1AP)
initiatingMessage (= initiatingMessage) |
|s1apPDU                                                                          |
|1 initiatingMessage                                                              |
|1.1 procedureCode                      |id-MMEStatusTransfer                     |
|1.2 criticality                        |ignore                                   |
|1.3 value                                                                        |
|1.3.1 protocolIEs                                                                |
|1.3.1.1 sequence                                                                 |
|1.3.1.1.1 id                           |id-MME-UE-S1AP-ID                        |
|1.3.1.1.2 criticality                  |reject                                   |
|1.3.1.1.3 value                        |542232756                                |
|1.3.1.2 sequence                                                                 |
|1.3.1.2.1 id                           |id-eNB-UE-S1AP-ID                        |
|1.3.1.2.2 criticality                  |reject                                   |
|1.3.1.2.3 value                        |0                                        |
|1.3.1.3 sequence                                                                 |
|1.3.1.3.1 id                           |id-eNB-StatusTransfer-Transparent        |
                                        Container                                 |
|1.3.1.3.2 criticality                  |reject                                   |
|1.3.1.3.3 value                                                                  |
|1.3.1.3.3.1                                                                      |
bearers-SubjectToStatusTransferList                                              |
|1.3.1.3.3.1.1 sequenceOf                                                         |
|1.3.1.3.3.1.1.1 id                     |id-Bearers-SubjectToStatusTransfer-      |
                                        Item                                      |
|1.3.1.3.3.1.1.2 criticality            |ignore                                   |
|1.3.1.3.3.1.1.3 value                                                            |
|1.3.1.3.3.1.1.3.1 e-RAB-ID             |5                                        |
|1.3.1.3.3.1.1.3.2 uL-COUNTvalue                                                  |
|1.3.1.3.3.1.1.3.2.1 pDCP-SN            |914                                      |
|1.3.1.3.3.1.1.3.2.2 hFN                |7                                        |
|1.3.1.3.3.1.1.3.3 dL-COUNTvalue                                                  |
|1.3.1.3.3.1.1.3.3.1 pDCP-SN            |4                                        |
|1.3.1.3.3.1.1.3.3.2 hFN                |0                                        |
```

2.7.1.7   Step 7

When the UE arrives in the target cell and the RRC connection is up and running again after performing the random access procedure, the target eNB sends the S1AP handover notification message

(Message Example 2.28) to the MME to signal that the handover was successfully completed and the radio as well as transport resources in the source eNB can be released.

Message Example 2.28: S1AP Handover Notification

```
+--------------------------------------+----------------------------------------+
|ID Name                               |Comment or Value                        |
+--------------------------------------+----------------------------------------+
|S1AP 3GPP TS 36.413 V8.5.1 (2009-03), R3-091397, R3-091144 (S1AP)             |
|initiatingMessage (= initiatingMessage) |                                      |
|s1apPDU                                                                        |
|1 initiatingMessage                                                            |
|1.1 procedureCode                     |id-HandoverNotification                 |
|1.2 criticality                       |ignore                                  |
|1.3 value                                                                      |
|1.3.1 protocolIEs                                                              |
|1.3.1.1 sequence                                                               |
|1.3.1.1.1 id                          |id-MME-UE-S1AP-ID                       |
|1.3.1.1.2 criticality                 |reject                                  |
|1.3.1.1.3 value                       |542232756                               |
|1.3.1.2 sequence                                                               |
|1.3.1.2.1 id                          |id-eNB-UE-S1AP-ID                       |
|1.3.1.2.2 criticality                 |reject                                  |
|1.3.1.2.3 value                       |0                                       |
|1.3.1.3 sequence                                                               |
|1.3.1.3.1 id                          |id-EUTRAN-CGI                           |
|1.3.1.3.2 criticality                 |ignore                                  |
|1.3.1.3.3 value                                                                |
|1.3.1.3.3.1 pLMNidentity              |'299000'H                               |
|1.3.1.3.3.2 cell-ID                   |'000b2b2'H                              |
|1.3.1.4 sequence                                                               |
|1.3.1.4.1 id                          |id-TAI                                  |
|1.3.1.4.2 criticality                 |ignore                                  |
|1.3.1.4.3 value                                                                |
|1.3.1.4.3.1 pLMNidentity              |'299000'H                               |
|1.3.1.4.3.2 tAC                       |'0001'H                                 |
```

The handover notification message also contains the E-UTRAN Cell Global Identity (CGI) and tracking as the identity of the new serving cell.

If the serving tracking area changed as a result of the handover, the UE may perform the tracking area update procedure to align the TAI on the NAS protocol entities of the UE and MME.

2.7.1.8   Step 8

Now it is time for the MME to deal with the necessary switch of the S1-U GTP tunnel. In particular, it is only necessary to change the DL tunnel endpoint from source eNB to target eNB while the UL tunnel endpoint at the S-GW may remain unchanged.

The GTP-C modify bearer request message that is sent by the MME to the S-GW is shown in Message Example 2.29. It contains the EBI and the eNB F-TEID as the new DL tunnel endpoint of the user plane transport.

Message Example 2.29: GTP-C Modify Bearer Request

```
+-----------------------------------------+-----------------------------------------+
| ID Name                                 | Comment or Value                        |
+-----------------------------------------+-----------------------------------------+
| GTPv2-C LTE TS29.274 V8.1.1 (2009-03) + TS29.280 V8.1.0 (2009-03)
| (GTPv2-C) MBREQ (= Modify Bearer Request) |
| 1 Modify Bearer Request                 |                                         |
| Version                                 | GTPv2                                   |
| P-Bit                                   | no piggybacked message shall be         |
|                                         | present                                 |
| T-Bit                                   | TEID is present in the GTP-C header     |
| Spare                                   | 0                                       |
| Message Type                            | Modify Bearer Request                   |
| Message Length                          | 54                                      |
| TEID                                    | '022803ba'H                             |
| Sequence Number                         | 1137                                    |
| IPv4 Address                            | is included                             |
| IPv6 Address                            | is not included                         |
| Spare                                   | 0                                       |
| Interface Type                          | S11 MME GTP-C interface                 |
| TEID/GRE Key                            | '0bf10040'H                             |
| IPv4 Address                            | 10.10.97.3                              |
| 1.3 Bearer Context to be modified       |                                         |
| Type                                    | Bearer Context                          |
| Length                                  | 18                                      |
| Spare                                   | 0                                       |
| Instance                                | 0                                       |
| 1.3.1 EPS Bearer ID (EBI)               |                                         |
| Type                                    | EPS Bearer ID (EBI)                     |
| Length                                  | 1                                       |
| Spare                                   | 0                                       |
| Instance                                | 0                                       |
| Spare                                   | 0                                       |
| EPS Bearer ID (EBI)                     | 5                                       |
| 1.3.2 S1 eNodeB F-TEID                  |                                         |
| Type                                    | Fully Qualified Tunnel Endpoint         |
|                                         | Identifier (F...                        |
| Length                                  | 9                                       |
| Spare                                   | 0                                       |
| Instance                                | 0                                       |
| IPv4 Address                            | is included                             |
| IPv6 Address                            | is not included                         |
| Spare                                   | 0                                       |
| Interface Type                          | S1-U eNodeB GTP-U interface             |
| TEID/GRE Key                            | 'de114432'H                             |
| IPv4 Address                            | 10.10.40.10                             |
| 1.4 Recovery                            |                                         |
| Type                                    | Recovery (Restart Counter)              |
| Length                                  | 1                                       |
| Spare                                   | 0                                       |
| Instance                                | 0                                       |
| Recovery                                | 152                                     |
+-----------------------------------------+-----------------------------------------+
```

2.7.1.9 Step 9

The S-GW responds with a GTP-C modify bearer response message (Message Example 2.30) that signals the S1 SGW F-TEID back to the MME. Now the two tunnel endpoints have been successfully renegotiated and payload transfer is ready to start.

Message Example 2.30: GTP-C Modify Bearer Response

```
+------------------------------------+------------------------------------------+
|ID Name                             |Comment or Value                          |
+------------------------------------+------------------------------------------+
|GTPv2-C LTE TS29.274 V8.1.1 (2009-03) + TS29.280 V8.1.0 (2009-03)
(GTPv2-C) MBRSP (= Modify Bearer Response) |
|1 Modify Bearer Response                                                       |
|Version                             |GTPv2                                     |
|P-Bit                               |no piggybacked message shall be           |
                                      present                                   
|T-Bit                               |TEID is present in the GTP-C header       |
|Spare                               |0                                         |
|Message Type                        |Modify Bearer Response                    |
|Message Length                      |47                                        |
|TEID                                |'0bf10040'H                               |
|Sequence Number                     |1137                                      |
|Spare                               |0                                         |
|1.1 Cause                                                                      |
|Type                                |Cause                                     |
|Length                              |2                                         |
|Spare                               |0                                         |
|Instance                            |0                                         |
|Cause                               |Request accepted                          |
|Spare                               |0                                         |
|OI (Originating Indication)         |originated by the node sending the        |
                                      message                                   
|1.2 Bearer Context to be modified                                              |
|Type                                |Bearer Context                            |
|Length                              |24                                        |
|Spare                               |0                                         |
|Instance                            |0                                         |
|1.2.1 EPS Bearer ID (EBI)                                                      |
|Type                                |EPS Bearer ID (EBI)                       |
|Length                              |1                                         |
|Spare                               |0                                         |
|Instance                            |0                                         |
|Spare                               |0                                         |
|EPS Bearer ID (EBI)                 |5                                         |
|1.2.2 Cause                                                                    |
|Type                                |Cause                                     |
|Length                              |2                                         |
|Spare                               |0                                         |
|Instance                            |0                                         |
|Cause                               |Request accepted                          |
|Spare                               |0                                         |
```

```
|OI (Originating Indication)          |originated by the node sending the   |
                                      |message
|1.2.3 S1 SGW F-TEID                                                        |
|Type                                 |Fully Qualified Tunnel Endpoint      |
                                      |Identifier (F...
|Length                               |9                                    |
|Spare                                |0                                    |
|Instance                             |0                                    |
|IPv4 Address                         |is included                          |
|IPv6 Address                         |is not included                      |
|Spare                                |0                                    |
|Interface Type                       |S1-U SGW GTP-U interface             |
|TEID/GRE Key                         |'046d1400'H                          |
|IPv4 Address                         |10.10.97.2                           |
```

2.7.1.10   Step 10

The last remaining step for the MME is to release the transport and radio resources that have been used on the outgoing leg of handover to serve the connection of the UE that was handed over to the target cell. The MME triggers the release of all such resources by sending the S1AP UE context release command with a cause value "Successful Handover" as shown in Message Example 2.31.

Message Example 2.31: S1AP UE Context Release Command due to "Successful Handover"

```
+-------------------------------------+-------------------------------------------+
|ID Name                              |Comment or Value                           |
+-------------------------------------+-------------------------------------------+
|S1AP 3GPP TS 36.413 V8.5.1 (2009-03), R3-091397, R3-091144 (S1AP)
initiatingMessage (= initiatingMessage) |
|s1apPDU                                                                           |
|1 initiatingMessage                                                               |
|1.1 procedureCode                    |id-UEContextRelease                        |
|1.2 criticality                      |reject                                     |
|1.3 value                                                                         |
|1.3.1 protocolIEs                                                                 |
|1.3.1.1 sequence                                                                  |
|1.3.1.1.1 id                         |id-UE-S1AP-IDs                             |
|1.3.1.1.2 criticality                |reject                                     |
|1.3.1.1.3 value                                                                   |
|1.3.1.1.3.1 uE-S1AP-ID-pair                                                       |
|1.3.1.1.3.1.1 mME-UE-S1AP-ID         |552658472                                  |
|1.3.1.1.3.1.2 eNB-UE-S1AP-ID         |0                                          |
|1.3.1.2 sequence                                                                  |
|1.3.1.2.1 id                         |id-Cause                                   |
|1.3.1.2.2 criticality                |ignore                                     |
|1.3.1.2.3 value                                                                   |
|1.3.1.2.3.1 radioNetwork             |successful-handover                        |
```

The successful release of all resources will finally be confirmed by the old eNB with a successful outcome message (procedure code = "UE Context Release").

2.8 Dedicated Bearer Release

How the dedicated bearer is released when necessary is illustrated in Figure 2.13.

In this example, the bearer release is triggered by the network, for example, by user inactivity timer expiry on the S-GW side. As a result, the S-GW sends a GTP Delete Bearer Request including the EBI of the impacted bearer to the MME via the S11 interface. It should be noted that there is no cause value specified as a mandatory information element in this GTP Delete Bearer Request by 3GPP. Thus, the reason for the network-triggered bearer deletion can only be indicated in the Private Extension sequence of the message following the proprietary encoding rules of the NEMs.

The reception of this message by the MME triggers simultaneously an E-RAB-Release procedure on the S1AP layer between the MME and eNodeB and a deactivate EPS bearer context procedure on the NAS layer between the MME and UE.

Once the NAS and S1AP procedures are completed, the MME sends a GTP delete bearer response message back to the S-GW. This message has a cause value included, but typically it only signals that the previous request was accepted.

After the Dedicated Bearer Release, the default bearer remains active in this example as the only bearer of the connection. Looking at the size of the EBI information element, theoretically 256 different bearers can be addressed for a single UE, but most likely not more than two or three bearers with different QoS parameters will run in parallel in a single UE connection. In the typical live network scenario, the first bearer is the default bearer, all the others are dedicated bearers that can be established and released at any time on demand.

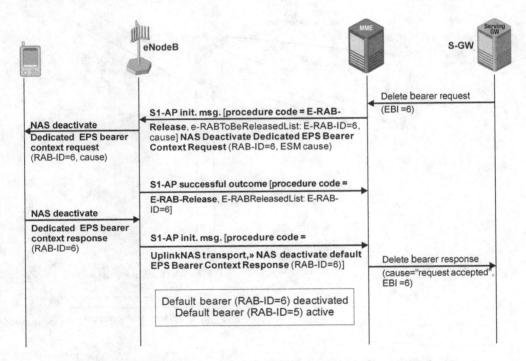

Figure 2.13   Dedicated bearer release. (*Source*: Tektronix Communications.)

2.9   Detach

Figure 2.14 shows the procedure that is used if the network cancels the registration of a particular UE. Surely it is also possible for a UE to request to be detached – that is, the official name for deleting a registered mobile from the MME's database of active subscribers.

The detach procedure shown in Figure 2.14 is a network-triggered detach that starts with a timer expiry. The timer may have expired because the UE has not performed the periodic tracking area updates that have been requested by the network. Hence, the MME considers the UE to be inactive and may be moved to a different network as often happens to roaming subscribers.

The MME performs the following steps.

2.9.1   Procedure

2.9.1.1   Step 1

The MME sends a Delete Bearer Request to the S-GW to terminate the S1-U bearer.

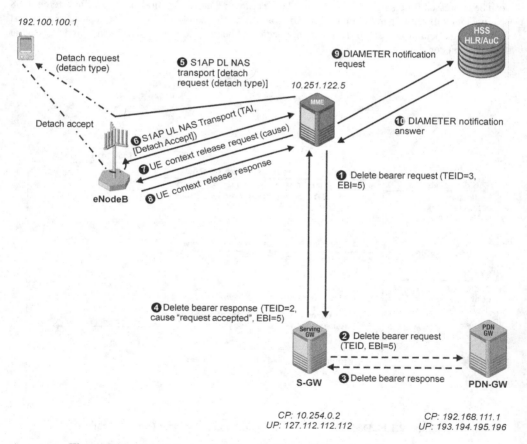

Figure 2.14   Detach (network initiated). (*Source*: Tektronix Communications.)

2.9.1.2   Step 2

Triggered by the Delete Bearer Request that was received from the MME, the S-GW starts the delete bearer procedure on S5 and sends the same GTP signaling message to the PDN-GW.

2.9.1.3   Step 3

The PDN-GW releases the bearer resources and sends a GTP delete bearer response message back to the S-GW.

2.9.1.4   Step 4

Now the S-GW also sends a delete bearer response message back to the MME.

2.9.1.5   Step 5

A NAS detach request message is sent by the MME to the UE. On the S1 interface, this message is transported by the S1AP DL NAS transport message as shown in Message Example 2.32.

Message Example 2.32: Detach Request (Network Initiated) on S1

```
+-------------------------------------+-------------------------------------------+
|ID Name                              |Comment or Value                           |
+-------------------------------------+-------------------------------------------+
|S1 Application Protocol TS 36.413 V8.4.0 (2008-12) (including Changes
for ERABToBeSetupItemCtxtSUReq:NAS PDU) (S1-AP) initiatingMessage
(= initiatingMessage) |
|s1apPDU                                                                          |
|1 initiatingMessage                                                              |
|1.1 procedureCode                    |id-downlinkNASTransport                    |
|1.2 criticality                      |ignore                                     |
|1.3 value                                                                        |
|1.3.1 protocolIEs                                                                |
|1.3.1.1 sequence                                                                 |
|1.3.1.1.1 id                         |id-MME-UE-S1AP-ID                          |
|1.3.1.1.2 criticality                |reject                                     |
|1.3.1.1.3 value                      |2                                          |
|1.3.1.2 sequence                                                                 |
|1.3.1.2.1 id                         |id-eNB-UE-S1AP-ID                          |
|1.3.1.2.2 criticality                |reject                                     |
|1.3.1.2.3 value                      |3                                          |
|1.3.1.3 sequence                                                                 |
|1.3.1.3.1 id                         |id-NAS-PDU                                 |
|1.3.1.3.2 criticality                |reject                                     |
|1.3.1.3.3 value                      |27 00 00 00 00 01 07 45 02                 |
|NAS LTE TS24.301 V8.0.0 (2008-12) (Secured NAS) IPCMSG (= Integrity
protected and ciphered NAS message) |
|1 Integrity protected and ciphered                                               |
NAS message
|Security header type                 |Integrity protected and ciphered           |
```

```
|Protocol discriminator               |EPS mobility management messages     |
|Message authentication code          |'00000000'H                          |
|Sequence number                      |1                                    |
|NAS message(s) FFS                   |07 45 02                             |
|NAS LTE TS24.301 V8.0.0 (2008-12) (Plain NAS) DTRQ (= Detach request)      |
|1 Detach request                                                           |
|Security header type                 |No security protection               |
|Protocol discriminator               |EPS mobility management messages     |
|Type of Management                   |EPS mobility management              |
|Message type                         |DTRQ-TERM                            |
|1.1 Spare Half Octet                                                       |
|Spare                                |0                                    |
|1.2 Detach type                                                            |
|Spare                                |0                                    |
|Detach type                          |IMSI detach / re-attach not required |
```

2.9.1.6 Step 6

The UE answers the NAS Detach Request with a Detach Accept. This certainly applies in scenarios where the UE is still able to respond to the network signaling message.

2.9.1.7 Step 7

Independent of the UE's response (that must be considered to be never received), the MME starts the release procedure of the UE context by sending the S1AP UE context release message that is shown in Message Example 2.33 to the eNodeB.

Message Example 2.33: UE Context Release due to Detach

```
+-------------------------------------------+-------------------------------------------+
|ID Name                                    |Comment or Value                           |
+-------------------------------------------+-------------------------------------------+
|S1 Application Protocol TS 36.413 V8.4.0 (2008-12) (including Changes
for ERABToBeSetupItemCtxtSUReq:NAS PDU) (S1-AP) initiatingMessage
(= initiatingMessage) |
|s1apPDU                                                                                 |
|1 initiatingMessage                                                                     |
|1.1 procedureCode                          |id-UEContextRelease                        |
|1.2 criticality                            |reject                                     |
|1.3 value                                                                               |
|1.3.1 protocolIEs                                                                       |
|1.3.1.1 sequence                                                                        |
|1.3.1.1.1 id                               |id-UE-S1AP-IDs                             |
|1.3.1.1.2 criticality                      |reject                                     |
|1.3.1.1.3 value                                                                         |
|1.3.1.1.3.1 uE-S1AP-ID-pair                                                             |
|1.3.1.1.3.1.1 mME-UE-S1AP-ID               |2                                          |
|1.3.1.1.3.1.2 eNB-UE-S1AP-ID               |3                                          |
|1.3.1.2 sequence                                                                        |
|1.3.1.2.1 id                               |id-Cause                                   |
|1.3.1.2.2 criticality                      |ignore                                     |
|1.3.1.2.3 value                                                                         |
|1.3.1.2.3.1 nas                            |detach                                     |
```

This message contains a cause value. The S1AP cause values are sorted into groups to allow better identification of which entity or part of the network trigged a particular cause.

The group for the "detach" cause is the NAS cause group. It indicates the higher NAS layer requested to release the UE context on the lower S1AP layer.

2.9.1.8   Step 8

The eNodeB confirms the release of the UE context.

2.9.1.9   Step 9

The MME informs the HSS that the UE was detached using the DIAMETER notification procedure.

2.9.1.10   Step 10

The HSS acknowledges the notification sent earlier by the MME.

2.9.1.11   UE-Triggered Detach

In the case of a UE-triggered detach, it is the UE that sends the detach request message. If the power-off flag in this message is set to "True," there will be no Detach Accept sent by the network, because the UE can no longer receive it. The UE context release on S1 and the GTP bearer deletion procedures on S11 and S5 are performed in the same way as described for the network-initiated detach.

2.10   Failure Cases in E-UTRAN and EPC

Since the previously described signaling scenarios contain a mix of messages sent on different interfaces and different layers, a summary of typical failure cases for all the procedures given in this chapter is sorted for the different layers in Tables 2.1–2.4.

2.11   Voice over LTE (SIP) Call – Complete Scenario

In fact, the first edition of "LTE Signaling" already comprised all necessary information to understand and explain the scenario of a SIP call, which is the typical call scenario behind the Voice over LTE (VoLTE) feature.

In particular, a SIP call is a combination of network access including user plane bearer setup as described in previous chapters plus the SIP in-band signaling to establish the connection between the terminals as described in Chapter 1.10.14.

Now, a complete SIP call shall be explained using a different view on the call flow procedure. What is shown in Figure 2.15 is the decoder output of signaling messages as it can be found in a protocol tester like the Tektronix Network and Service Analyzer (NSA) where the screenshot was taken from. The most left column "From" shows the interface on which the signaling messages were recorded. Looking at the SIP messages REGISTER and RESPONSE, one can see that each of these messages occurs three times in the call flow, because we see them on Gi interface (called "Pure_IP-GI") and S5 interface (named "GTP-C-S5" for both GTP control plane and user plane) and S1-U interface

Table 2.1   S1AP failure messages

Message	Description	Impact on call
S1AP layer UE Context Release Request	The UE context release request message is sent by the serving eNodeB if a previously established UE context is terminated. The two most typical cases are: 1. *User inactivity:* The eNodeB detected that for a timer-guarded period no user plane payload packets have been transmitted for a particular connection and, thus, it requests the core network represented by the MME to release the connection so that network resources blocked by the inactive user become available for other connections 2. One of the following typical failure causes indicate an abnormal termination of the UE context and the entire connection between the UE and the network: (a) Unspecified (b) Radio connection with UE lost (c) Failure in the radio interface procedure (d) O&M (Operation and Maintenance) intervention (e) Release due to E-UTRAN generated reason	*User inactivity:* A new connection will be established when necessary on request as described in Section 2.4 *Abnormal termination:* The connection drops and the UE must establish a new RRC connection and UE context from scratch
UE Context Release Command	If the UE context release command is seen with one of the following cause values the call is normally terminated: (a) Normal release (b) Detach (c) User inactivity In the case of any other cause value, the call state either is dropped (especially if the same cause value was seen previously in the UE context release request message) or should be investigated in detail to determine normal or abnormal behavior	In any case (no matter what the cause value is) the call is closed when the UE context release command message is monitored and the UE must request a new RRC connection/new UE context to establish a new connection
Initial Context Setup Failure	The setup of the call already fails at S1AP Initial Context Setup	The call setup is aborted and the UE is expected to request a new RRC connection/new UE context

Table 2.1   *(continued)*

Message	Description	Impact on call
UE Context Modification Failure	The modification of an existing UE context failed. How the network reacts when such an error occurs depends on vendor-specific implementation	Typically it can be expected that the call is continued and the attempt to modify the UE context is repeated
Handover Preparation Failure	A mobility failure that may be caused by incorrect parameter settings in the target network element. Often the target network element is not able to provide the required radio resources for the handover and, hence, does not return the requested handover command message. How the network reacts when such an error occurs depends on vendor-specific implementation	Typically it can be expected that the procedure is repeated until a successful handover preparation or normal/abnormal termination of the call. In the latter case check cause value of the UE Context Release Request/Release Command
Handover Failure	The target network element has provided the requested handover command message and radio resources for the handover, but handover execution failed, for example, because the UE rejected the handover command. Possible root causes can be incorrect parameter settings for radio resources, for example, false values in frequency or scheduling information parameters. How the network reacts when such an error occurs depends on vendor-specific implementation	Typically it can be expected that the network will attempt a new handover using either the same or new parameters
Path Switch Request Failure	The MME cannot switch the path as required for the X2 handover. How the network reacts when such an error occurs depends on vendor-specific implementation	Typically it can be expected that the network will make two or more additional attempts to switch the path. If not successful, the X2 handover will be aborted

Table 2.2 X2AP failure messages

Message	Description	Impact on call
X2AP layer		
Handover Preparation Failure	The target eNodeB cannot provide the necessary resources for the handover or cannot interpret the contents of the handover request message	The call will be continued. Another attempt to prepare the X2 handover will be made
Handover Cancel	If the source eNodeB does not receive a response to its X2AP handover request message from the target eNodeB before timer $T_{RELOCprep}$ expires, the source eNodeB will send the handover cancel message	The call will be continued. When applicable a new Handover Request will be sent to the target eNodeB

(named "USERPLANE-S1-U"). The SIP REGISTER messages goes hop-by-hop the way S1-U -> S5 -> Gi while the SIP RESPONSE comes the other way from Gi via S5 down to S1-U.

In the context of this protocol tester, the terms IP_HIGH and UDP_HIGH indicate the IP/UDP layers on top of GTP-U; this means: the user plane inside the IP tunnels of the EPC or the plain IP/UDP as seen on Gi interface. There are also IP_LOW/UDP_LOW layers (not shown in the figure) that represent the transport network.

Going back to the beginning of the call, it can be seen that first of all the UE attaches to the network. During Attach procedure, the security functions are performed and as described in Section 2.2, the default bearer is established after initial context setup.

The DIAMETER message pairs AIR/AIA (Authentication Information Request/Authentication Information Answer) and ULR/ULA (Update Location Request/Update Location Answer) on S6a interface have been already discussed in Sections 2.3 and 2.4. A new DIAMETER procedure is this call is the message pair CCR/CCA (Credit Control Request/Credit Control Answer). This credit control procedure has a similar function as the MAP Insert Subscriber Data procedure known from 2G/3G core network signaling. With the Credit Control Request message, the MME polls the QoS parameters like QCI and maximum request bit rate for uplink/downlink data transfer from the HSS. These HSS QoS parameters of the subscriber are a set of technical parameters that represent what is agreed in the contract between the subscriber and his/her service provider. Besides the QoS parameters, a charging rule name, flow description and flow status, and a list of event triggers as known from CAMEL procedures in 2G/3G core network are included in the Credit Control Answer message.

Besides the example shown in Figure 2.15, there can be various flavors of the S1AP and NAS signaling procedures for SIP call establishments. As a rule, there are three different E-RABs established before first SIP signaling message is sent. These E-RABs are given as follows:

- The default bearer for any kind of non-real-time IP payload such as e-mail that may arrive while the VoLTE conversation is ongoing.
- An E-RAB with QCI = 5 for the IMS Signaling (SIP signaling).
- An E-RAB with QCI = 1 for transport of the VoLTE packets (user plane).

It is possible that all three bearers are established simultaneously during S1AP Initial Context Setup or that we see a step-by-step establishment procedure that starts with default bearer setup as part of the

Table 2.3 NAS EMM failure messages

Message	Description	Impact on call
NAS EMM layer		
Attach Reject	The network does not allow the UE to attach to the network. It can be distinguished between correct rejections, for example, if there is no roaming agreement between the subscriber's home network operator and the operator of the network that sends the attach reject message. On the other hand, a possible root cause for a rejected attach can be communication problems in the core network between the MME of the visited network and the HSS of the home network	Typically rejected UEs will try to attach repeatedly. There is no limit on the number of subsequently established radio connections to be used to send Attach Requests
Tracking Area Update Reject	This failure only occurs when the subscriber was already successfully attached to the network. The root causes are varied and described in detail in 3GPP 24.301	Normally the UE will try to repeat the tracking area update procedure. In general the detailed actions recommended in 3GPP 24.301, apply
Service Reject	A PS connection that was previously set on hold by "user inactivity" as described in Section 2.3 cannot be resumed. This failure will be perceived by the subscriber as missing PS connectivity or – if a repeated attempt is successful – delay in access to PS user plane contents	The setup of the call/new UE context will be aborted and the UE must start a new call setup attempt from scratch
Authentication Reject	This message is sent by the network to the UE to indicate that the authentication procedure has failed and that the UE should abort all activities	The setup of the call/new UE context will be aborted and the UE must start a new call setup attempt from scratch
Authentication Failure	The UE sends this message to the MME to indicate that a failure during the authentication procedure occurred. The most common cause value is "synchronization failure" to indicate that existing and newly assigned security parameters are in conflict with each other	Typically the call setup is aborted and the UE must start a new call setup attempt from scratch
Security Mode Reject	The UE refused to activate the security functions requested by the network, for example, due to incorrect security parameter values	The network will decide if the security mode function is repeated or if call setup is aborted

Table 2.4 NAS ESM failure messages

Message	Description	Impact on call
NAS ESM layer		
Activate Default EPS Bearer Context Reject	The UE signals to the network that it cannot establish the default EPS bearer with the set of parameters defined by the network	The setup of the initial connection between the UE and network fails and the UE must start a new call setup attempt
Activate Dedicated EPS Bearer Context Reject	The UE signals to the network that it cannot establish a particular dedicated EPS bearer that uses the same IP address and APN as the default bearer with the set of parameters defined by the network	A particular user plane connection cannot be established, but the call and appropriate UE context remain active without limitation
Modify EPS Bearer Context Reject	The UE refuses to modify the parameters (e.g., QoS attributes) of a particular EPS bearer that uses same the IP address and APN as the default bearer	It is expected that the EPS connection is continued using the old parameters
PDN Connectivity Reject	The network rejects the establishment of PDN connectivity between the UE and the PDN gateway	The call setup is aborted and, hence, the default bearer between the UE and network cannot be established. However, the UE may successfully attach to the network
PDN Disconnect Reject	The release of the PDN connection between the UE and PDN gateway is rejected. A possible reason could be that in the EPC there is still user plane payload to be sent to the handset. In any case the subscriber's experience is not adversely affected by this error	The call is normally continued
Bearer Resource Allocation Reject	The UE signals to the network that it cannot establish a particular dedicated EPS bearer that uses a different IP address and APN as the default bearer with the set of parameters defined by the network	A particular user plane connection cannot be established, but the call and appropriate UE context remain active without limitation
Bearer Resource Modification Reject	The UE refuses to modify the parameters (e.g., QoS attributes) of a particular EPS bearer that uses different a IP address and APN as the default bearer	It is expected that the EPS connection is continued using the old parameters

From	Long Time	4. Prot	4. MSG	5. Prot	5. MSG	6. Prot	6. MSG	7. Prot	7. MSG	8. Prot	8. MSG
S1-AP-S1-MME	4:42:30 PM,241,543	S1AP	initiatingMessage	NAS-SECURED	IPCMSG	NAS-EMM	AIRQ	NAS-ESM	PCONRQ	PPP	DTGRS
S1-AP-S1-MME	4:42:30 PM,372,894	S1AP	initiatingMessage	NAS-EMM	IDRQ						
S1-AP-S1-MME	4:42:30 PM,390,392	S1AP	initiatingMessage	NAS-EMM	IDRP						
Diameter-S6a	4:42:30 PM,392,634	DIAMETER	AIR								
Diameter-S6a	4:42:30 PM,433,133	DIAMETER	AIA								
S1-AP-S1-MME	4:42:30 PM,434,742	S1AP	initiatingMessage	NAS-EMM	AUTREQ						
S1-AP-S1-MME	4:42:31 PM,255,408	S1AP	initiatingMessage	NAS-EMM	AUTREP						
S1-AP-S1-MME	4:42:31 PM,320,809	S1AP	initiatingMessage	NAS-SECURED	IPNEMMSG	NAS-EMM	SECCMO				
S1-AP-S1-MME	4:42:31 PM,353,343	S1AP	initiatingMessage	NAS-SECURED	IPCNEMMSG	NAS-EMM	SECCPL				
Diameter-S6a	4:42:31 PM,360,312	DIAMETER	ULR								
Diameter-S6a	4:42:31 PM,369,679	DIAMETER	ULA								
S1-AP-S1-MME	4:42:31 PM,402,778	S1AP	initiatingMessage	NAS-SECURED	IPCMSG	NAS-ESM	ESMINFRQ				
S1-AP-S1-MME	4:42:31 PM,425,187	S1AP	initiatingMessage	NAS-SECURED	IPCMSG	NAS-ESM	ESMINTRP	PPP	DTGRS	IPCP	CREQ
GTP-C-S5	4:42:31 PM,464,173	GTPv2-C	CCREQ	PPP	DTGRS	IPCP	CREQ				
Diameter-Gx	4:42:31 PM,474,467	DIAMETER	CCR								
Diameter-Gx	4:42:31 PM,489,741	DIAMETER	CCA								
GTP-C-S5	4:42:31 PM,485,410	GTPv2-C	CCRSP	PPP	DTGRS	IPCP	CNAK				
S1-AP-S1-MME	4:42:31 PM,527,736	S1AP	initiatingMessage	NAS-SECURED	IPCMSG	NAS-EMM	ATAC	NAS-ESM	ACTDEFRQ	PPP	DTGRS
S1-AP-S1-MME	4:42:31 PM,561,208	S1AP	initiatingMessage	LTE_URACI	uERadioAcc...						
S1-AP-S1-MME	4:42:31 PM,580,962	S1AP	successfulOutcome								
S1-AP-S1-MME	4:42:31 PM,595,913	S1AP	initiatingMessage	NAS-SECURED	IPCMSG	NAS-EMM	ACOM	NAS-ESM	ACTDEFAC		
Diameter-S6a	4:42:31 PM,609,467	DIAMETER	NOR								
Diameter-S6a	4:42:31 PM,621,964	DIAMETER	NOA								
USERPLANE-S1-U	4:42:35 PM,905,918	GTPv1-U	GPDU	IP_HIGH	IPv4	UDP_HIGH	DTGR	SIP	REGISTER		
GTP-C-S5	4:42:35 PM,906,871	GTPv1-U	GPDU	IP_HIGH	IPv4	UDP_HIGH	DTGR	SIP	REGISTER		
Pure_IP-GI	4:42:35 PM,907,617	SIP	REGISTER								
Pure_IP-GI	4:42:35 PM,925,299	SIP	RESPONSE								
GTP-C-S5	4:42:35 PM,925,984	GTPv1-U	GPDU	IP_HIGH	IPv4	UDP_HIGH	DTGR	SIP	RESPONSE		
USERPLANE-S1-U	4:42:35 PM,927,201	GTPv1-U	GPDU	IP_HIGH	IPv4	UDP_HIGH	DTGR	SIP	RESPONSE		
USERPLANE-S1-U	4:42:35 PM,951,912	GTPv1-U	GPDU	IP_HIGH	IPv4	UDP_HIGH	DTGR	SIP	REGISTER		
GTP-C-S5	4:42:35 PM,952,763	GTPv1-U	GPDU	IP_HIGH	IPv4	UDP_HIGH	DTGR	SIP	REGISTER		
Pure_IP-GI	4:42:35 PM,953,439	SIP	REGISTER								
Pure_IP-GI	4:42:35 PM,970,112	SIP	RESPONSE								
GTP-C-S5	4:42:35 PM,970,803	GTPv1-U	GPDU	IP_HIGH	IPv4	UDP_HIGH	DTGR	SIP	RESPONSE		
USERPLANE-S1-U	4:42:35 PM,972,201	GTPv1-U	GPDU	IP_HIGH	IPv4	UDP_HIGH	DTGR	SIP	RESPONSE		
USERPLANE-S1-U	4:42:36 PM,019,838	GTPv1-U	GPDU	IP_HIGH	IPv4						
GTP-C-S5	4:42:36 PM,020,555	GTPv1-U	GPDU	IP_HIGH	IPv4						
USERPLANE-S1-U	4:42:36 PM,844,614	GTPv1-U	GPDU	IP_HIGH	IPv4						
GTP-C-S5	4:42:36 PM,845,491	GTPv1-U	GPDU	IP_HIGH	IPv4	UDP_HIGH	DTGR	SIP	INVITE	SDP	SDP
USERPLANE-S1-U	4:42:48 PM,718,787	GTPv1-U	GPDU	IP_HIGH	IPv4	UDP_HIGH	DTGR	SIP	INVITE	SDP	SDP
GTP-C-S5	4:42:48 PM,719,769	GTPv1-U	GPDU	IP_HIGH	IPv4	UDP_HIGH	DTGR	SIP	INVITE	SDP	SDP
Pure_IP-GI	4:42:48 PM,720,526	SIP	INVITE	SDP	SDP						
Pure_IP-GI	4:42:48 PM,729,109	SIP	RESPONSE								
GTP-C-S5	4:42:48 PM,729,644	GTPv1-U	GPDU	IP_HIGH	IPv4	UDP_HIGH	DTGR	SIP	RESPONSE		
USERPLANE-S1-U	4:42:48 PM,730,692	GTPv1-U	GPDU	IP_HIGH	IPv4	UDP_HIGH	DTGR	SIP	RESPONSE		
Pure_IP-GI	4:42:48 PM,743,531	SIP	RESPONSE								
GTP-C-S5	4:42:48 PM,744,565	GTPv1-U	GPDU	IP_HIGH	IPv4	UDP_HIGH	DTGR	SIP	RESPONSE		
USERPLANE-S1-U	4:42:48 PM,745,269	GTPv1-U	GPDU	IP_HIGH	IPv4	UDP_HIGH	DTGR	SIP	ACK		
USERPLANE-S1-U	4:42:48 PM,774,760	GTPv1-U	GPDU	IP_HIGH	IPv4	UDP_HIGH	DTGR	SIP	ACK		
GTP-C-S5	4:42:48 PM,775,657	GTPv1-U	GPDU	IP_HIGH	IPv4	UDP_HIGH	DTGR	SIP	ACK		
Pure_IP-GI	4:42:48 PM,776,305	SIP	ACK								
USERPLANE-S1-U	4:42:48 PM,776,780	GTPv1-U	GPDU	IP_HIGH	IPv4	UDP_HIGH	DTGR	SIP	INVITE	SDP	SDP
GTP-C-S5	4:42:48 PM,777,786	GTPv1-U	GPDU	IP_HIGH	IPv4	UDP_HIGH	DTGR	SIP	INVITE	SDP	SDP
Pure_IP-GI	4:42:48 PM,778,411	SIP	INVITE	SDP	SDP						
Pure_IP-GI	4:42:48 PM,905,322	SIP	RESPONSE								
GTP-C-S5	4:42:48 PM,905,121	GTPv1-U	GPDU	IP_HIGH	IPv4	UDP_HIGH	DTGR	SIP	RESPONSE		
USERPLANE-S1-U	4:42:48 PM,906,950	GTPv1-U	GPDU	IP_HIGH	IPv4	UDP_HIGH	DTGR	SIP	RESPONSE		
GTP-C-S5	4:42:49 PM,136,560	GTPv1-U	GPDU	IP_HIGH	IPv4	UDP_HIGH	DTGR	RTP	RTP		
USERPLANE-S1-U	4:42:49 PM,138,128	GTPv1-U	GPDU	IP_HIGH	IPv4	UDP_HIGH	DTGR	RTP	RTP		
GTP-C-S5	4:42:49 PM,136,408	GTPv1-U	GPDU	IP_HIGH	IPv4	UDP_HIGH	DTGR	RTP	RTP		

Figure 2.15 Successful SIP call setup including attach of UE

Long Time	4. Prot	4. MSG	Procedure Code	5. Prot	5. MSG	6. Prot	6. MSG	7. Prot	7. MSG
4:43:14 PM,516,125	GTPv1-U	GPDU		IP_HIGH	IPv4	UDP_HIGH	DTGR	RTP	RTP
4:43:14 PM,516,872	GTPv1-U	GPDU		IP_HIGH	IPv4	UDP_HIGH	DTGR	RTP	RTP
4:43:14 PM,523,345	GTPv1-U	GPDU		IP_HIGH	IPv4	UDP_HIGH	DTGR	RTP	RTP
4:43:14 PM,524,345	GTPv1-U	GPDU		IP_HIGH	IPv4	UDP_HIGH	DTGR	RTP	RTP
4:43:14 PM,528,309	GTPv1-U	GPDU		IP_HIGH	IPv4	UDP_HIGH	DTGR	RTCP	SR
4:43:14 PM,529,062	GTPv1-U	GPDU		IP_HIGH	IPv4	UDP_HIGH	DTGR	RTCP	SR
4:43:14 PM,531,150	GTPv1-U	GPDU		IP_HIGH	IPv4	UDP_HIGH	DTGR	SIP	
4:43:14 PM,531,922	GTPv1-U	GPDU		IP_HIGH	IPv4	UDP_HIGH	DTGR	SIP	BYE
4:43:14 PM,548,596	GTPv1-U	GPDU		IP_HIGH	IPv4	UDP_HIGH	DTGR	RTP	RTP
4:43:14 PM,549,404	GTPv1-U	GPDU		IP_HIGH	IPv4	UDP_HIGH	DTGR	RTP	RTP
4:43:14 PM,554,585	GTPv1-U	GPDU		IP_HIGH	IPv4	UDP_HIGH	DTGR	SIP	RESPONSE
4:43:14 PM,555,335	GTPv1-U	GPDU		IP_HIGH	IPv4	UDP_HIGH	DTGR	SIP	RESPONSE
4:43:14 PM,567,975	DIAMETER	RAR							
4:43:14 PM,569,941	GTPv2-C	BRREQ							
4:43:14 PM,570,220	DIAMETER	RAA							
4:43:14 PM,616,361	S1AP	initiatingMessage	id-E-RABRelease	NAS-SECURED	IPCMSG	NAS-ESM	DEACTRQ		
4:43:14 PM,638,297	S1AP	successfulOutcome	id-E-RABRelease						
4:43:14 PM,656,438	S1AP	initiatingMessage	id-uplinkNASTransport	NAS-SECURED	IPCMSG	NAS-ESM	DEACTAC		
4:43:14 PM,662,620	GTPv2-C	DBRSP							
4:51:52 PM,177,749	GTPv1-U	GPDU		IP_HIGH	IPv4	UDP_HIGH	DTGR		
4:51:52 PM,178,737	GTPv1-U	GPDU		IP_HIGH	IPv4	UDP_HIGH	DTGR		
4:54:52 PM,605,401	S1AP	initiatingMessage	id-uplinkNASTransport	NAS-SECURED	IPCMSG	NAS-EMM	DTRQ		
4:54:52 PM,827,045	GTPv2-C	DSREQ							
4:54:52 PM,628,023	GTPv2-C	DSRSP							
4:54:52 PM,633,711	DIAMETER	NOR							
4:54:52 PM,647,186	DIAMETER	NOA							
4:54:52 PM,656,763	S1AP	initiatingMessage	id-downlinkNASTransport	NAS-SECURED	IPCMSG	NAS-EMM	DTAC		
4:54:52 PM,658,765	S1AP	initiatingMessage	id-UEContextRelease						
4:54:52 PM,662,657	S1AP	successfulOutcome	id-UEContextRelease						

Figure 2.16 SIP call release including detach of UE

S1AP initial context and later on, the E-RABs for IMS signaling and VoLTE user plane are established using dedicated S1AP E-RAB Setup procedures.

The IMS signaling/SIP call itself starts when the SIP calling party sends INVITE message, called party answer with RESPONSE and calling party sends ACK to start the call. As soon as RTP packets are send, it is evident that user plane (VoLTE packets) are transmitted using the connection.

During the ongoing SIP call, there will be RTCP Sender Report (SR) messages connection containing the number of packets and data volume of packets sent in each direction. These Sender Report messages also contain measurement results for lost user plane packets and the inter-arrival jitter.

Figure 2.16 shows how the SIP call ends. After the last RTP packet, there is a message from each party that there is a final RTCP Sender Report followed by SIP BYE to release the logical SIP connection. SIP BYE is again confirmed by the receiver with a SIP RESPONSE message.

In this particular call example, there is E-RAB Release on S1 after the SIP call ended and, finally, the UE detaches from the network. However, such detach is not mandatory.

2.12   Inter-RAT Cell Reselection 4G-3G-4G

Figure 2.17 shows the full end-to-end signaling procedures for a UE that is attached to E-UTRAN first, then redirected to 3G, and comes back to 4G radio access after performing an inter-RAT cell reselection.

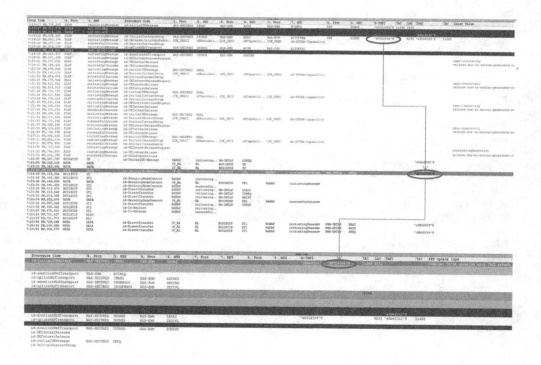

Figure 2.17   Inter-RAT redirection 4G-3G and cell reselection back to 4G

It is especially interesting to follow the changes in the Temporary Mobile Subscriber Identities (TMSIs) during the call flow.

When the UE sends Attach Request, it includes the old GUTI with old m-TMSI (the variable part of GUTI) in the message together with the old (last visited) LAC and TAC. The LAC for sure is related to GERAN/UTRAN environment while the TAC is valid for E-UTRAN.

The LTE Attach Request triggers a DIAMETER Update Location procedure during which a new TMSI for GERAN/UTRAN is provided. The TAC remains the same as it can be seen when comparing the values in Attach Request and Attach Accept. However, there is a new m-TMSI assigned by the MME with the Attach Accept message. This new m-TMSI remains valid as long as the UE is active in the E-UTRAN over multiple RRC connections/S1 UE Contexts. One can see on behalf of S1 release request causes that there are several radio connections of the UE in E-UTRAN. Most of them are terminated due to "user inactivity," except the last, which is due to "interrat-redirection."

The release due to "interrat-redirection" is the trigger for the UE to perform an inter-RAT cell reselection toward 3G.

Arriving on 3G, the UE must perform combined Location/Routing Area Update.

In the 3G Location Area Update Request message, we see the VLR TMSI that was assigned during LTE Attach. As a rule in the following Location Update Accept message, there is no new TMSI assigned.

In the packet domain, things are different.

The 3G Routing Area Update Request messages used the 4G m-TMSI to identify the UE toward the SGSN.

During the following Routing Area Update Accept, a new P-TMSI (valid only in GERAN/UTRAN) is assigned to the UE and when the UE performs Service Request in 3G RAN to request a bearer for user plane transfer, this P-TMSI is used to identify the subscriber.

However, the m-TMSI remains stored in the UE and when the mobile comes back to LTE, we see a combined Tracking/Location Area Updating Request ("with IMSI attach") and this message includes both the P-TMSI assigned in 3G and the last m-TMSI from LTE.

As a result of the Tracking Area Update procedure, a new m-TMSI, as well as a new VLR TMSI (for GERAN/UTRAN CS domain), is assigned and the LAC is updated.

With next Service Request in 4G, the UE uses the new m-TMSI.

For mapping relations between m-TMSI and P-TMSI, see

2.13   Normal/Periodical Tracking Area Update

During normal and periodical Tracking Area Updates, a new m-TMSI does not need to be assigned – see the call flow on S1 in Figure 2.18.

| | | | | | | | | Short View | |
3. Prot	3. MSG	4. Prot	4. MSG	5. Prot	5. MSG	Procedure Code	M-TMSI
S1AP	initiatingMessage	NAS-SECURED	IPMSG	NAS-EMM	TAURQ	id-initialUEMessage	'c3000000'H
S1AP	initiatingMessage	NAS-SECURED	IPCMSG	NAS-EMM	TAUAC	id-downlinkNASTransport	'c3030000'H
S1AP	initiatingMessage	NAS-SECURED	IPCMSG	NAS-EMM	TAUCPL	id-uplinkNASTransport	
S1AP	initiatingMessage					id-UEContextRelease	
S1AP	successfulOutcome					id-UEContextRelease	

Figure 2.18   Normal/Periodical tracking area update on S1

The messages shown in the figure are:

TAURQ – Tracking Area Update Request.
TAUAC – Tracking Area Update Accept.
TAUCPL – Tracking Area Update Complete.

2.14 CS Fallback End-to-End S1/IuCS/IuPS

Figure 2.19 shows a complete CS Fallback scenario including E-UTRAN S1 interface and 3G UTRAN IuCS/IuPS interface procedures. For more information regarding the LTE RRC messages not shown in this scenario, read Chapter 3.2.8.

In case of the call flow shown in Figure 2.19, the UE sends a NAS Service Request message to the MME followed by an Extended Service Request message after successful Initial Context Setup. In this NAS Extended Service Request message, the UE indicates that it needs to establish a voice over LTE (VoLTE) call or CS Fallback. Since there is no VoLTE service in this network available, the MME decides to go for CS Fallback.

The CS Fallback procedure starts with a UE Context Modification on S1 interface where the Initiating Message of the UE Context Modification procedure has information "cs-fallback-required" included.

This leads to a UE Context Release Request sent by eNodeB to MME and an inter-RAT blind redirection of the UE to 3G RAT. On S11 interface between MME and S-GW, the indication of the inter-RAT blind redirection triggers the exchange of GTP-C Deactivate Service Request (DSREQ) and Deactivate Service Response (DSRSP) message. On S6a, the MME informs the HSS that the UE is no longer reachable for paging in the E-UTRAN by sending the DIAMETER Purge UE Request (PUER) message that is answered with DIAMETER Purge UE Answer (PUEA).

When the UE arrives in 3G after performing inter-RAT cell reselection, it sends DMTAP Connection Management Service Request (CMSREQ) on IuCS to the MSC and Routing Area Update Request (RARQ) on IuPS to the SGSN.

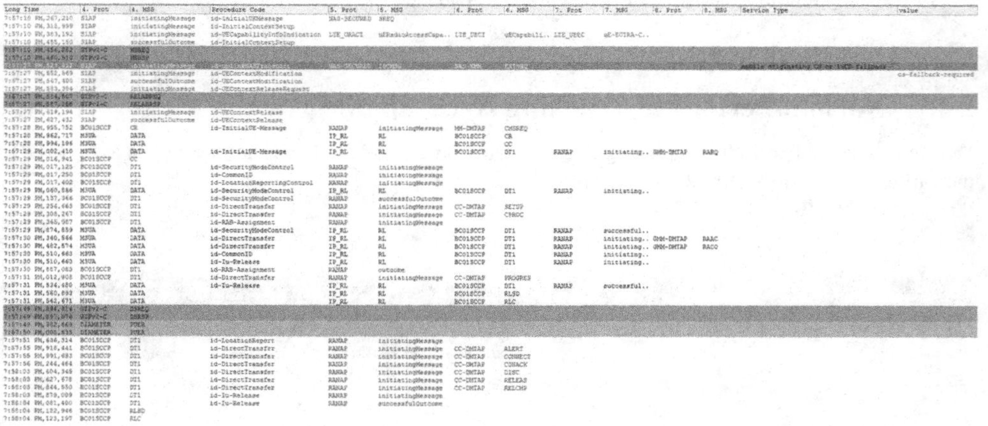

Figure 2.19 CS Fallback S1-IuCS-IuPS

In the following message sequences, the voice call setup and Routing Area Update procedures are successfully executed as described in detail in Kreher/Ruedebusch, *UMTS Signaling, 2nd* edition, Wiley, UK, 2007.

2.15   Paging

The MME initiates the paging procedure by sending the PAGING message to the eNodeB as shown in Figure 2.20.

At the reception of the S1AP Paging message, the eNB shall perform paging of the UE in cells that belong to tracking areas as indicated in the *List of TAIs*. The *CN Domain* IE shall be transferred transparently to the UE.

For each cell that belongs to any of the TA indicated in the *List of TAIs* IE, the eNodeB shall generate at least one page on the radio interface. In typical real-world scenarios, there are three subsequent RRC Paging messages sent per cell for each S1AP paging. Thus, an eNodeB that has three cells will send 9 RRC Paging messages in total on the radio interface after receiving one S1AP Paging.

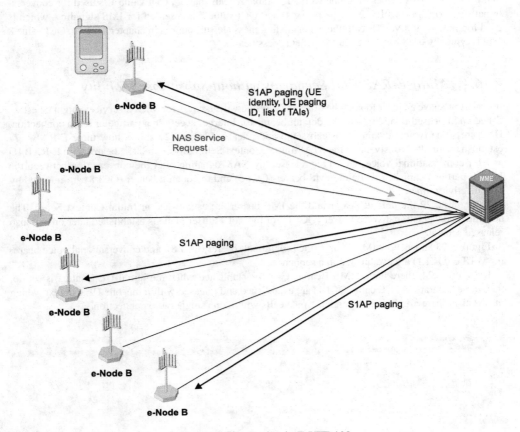

Figure 2.20   Paging in E-UTRAN

In case there is no paging response seen on MME before the paging repetition timer expires, a re-paging will be triggered. The number of re-pagings (paging repetitions), as well as the value of the paging repetition timer, is configurable in the OMC and, thus, depends on the optimization strategy of the network operator.

The paging response is the S1AP Initial UE Message indicating a mobile terminating RRC Establishment Cause, for example, "mt-access" (mobile terminating access). Different from 3G NAS, the NAS Service Request message in E-UTRAN does not indicate if it was sent as a paging response or not.

2.16 Multi-E-RAB Call Scenarios

As mentioned in chapter 2.11, the so-called "VoLTE calls" are always running in parallel with other services due to the always-on-IP concept of the LTE/EPC network. Thus, it would be more accurate to speak about multi-E-RAB call scenarios instead of "VoLTE calls."

The main differentiator for the services running on different E-RABs is the Quality of Service Class Indicator (QCI) that is discussed in Section 1.6. Table 1.6 lists the QCI values that have been defined by 3GPP for standard services of the network. The table reveals that the QCI value 1 is used for conversational voice (means: VoIP user plane packets) and QCI value 5 is reserved for IMS signaling, which is used for, in case of a VoLTE call, the SIP signaling messages introduces in chapter 1.12.14. QCI value 8 or 9 is typically used for the default Internet access bearer.

2.16.1 Multi-E-RAB Call Scenarios without Subscriber Mobility

It is worth to have a closer look at the different ways how the bearers for VoLTE services might be established and released in addition to the default bearer for Internet access in a particular radio connection. Here are some typical examples of call setup and release. Figure 2.21 shows how three E-RABs are established simultaneously using the S1AP Initial Context Setup procedure. The bearer with E-RAB ID = 7 for conversational voice (QCI 1) is released by NAS signaling after the conversation between the VoLTE parties is finished while the IMS bearer (QCI 5) and the default bearer (QCI 8) are released by user inactivity timer.

In Figure 2.22 the default bearer and the IMS bearer are established at Initial Context Setup. The dedicated conversational voice bearer (QCI 1) for the VoLTE user plane is established later on and also released separately.

Figure 2.23 shows how IMS Bearer (QCI 5), default bearer (QCI 8), and conversational voice bearer for VoLTE (QCI 1) are established in a sequence.

In Figure 2.24, there is one IMS Bearer (QCI 5) established when the initial UE Context is set up, then three separate VoLTE calls (QCI 1) are established and released within this one UE Context, which means also: there are three independent voice calls within one single radio connection.

No	Long Time	From	4. Prot	4. MSG	Procedure Code	e-RAB-ID	e-RAB-ID qCI	e-RAB-ID nas	Cause Value
63	13:17:53,657,000	S1-AP-TP-S1-MME			id-InitialUEMessage				
68	13:17:53,693,000	S1-AP-TP-S1-MME	LTE_URACI	uERadioAcc..	id-InitialContextSetup	5, 6, 7	5, 8, 1		
69	13:17:53,699,000	S1-AP-TP-S1-MME			id-InitialContextSetup				
70	13:17:53,700,000	S1-AP-TP-S1-MME			id-CellTrafficTrace				
101	13:18:28,596,000	S1-AP-TP-S1-MME			id-CellTrafficTrace				
102	13:18:28,598,000	S1-AP-TP-S1-MME			id-E-RABRelease			7	normal-release
105	13:18:28,649,000	S1-AP-TP-S1-MME			id-downlinkNASTransport				
108	13:18:28,665,000	S1-AP-TP-S1-MME			id-E-RABRelease				
146	13:19:19,684,000	S1-AP-TP-S1-MME			id-uplinkNASTransport				
147	13:19:19,696,000	S1-AP-TP-S1-MME			id-UEContextReleaseRequest				user-inactivity
149	13:19:19,697,000	S1-AP-TP-S1-MME			id-UEContextRelease				user-inactivity

Figure 2.21   Initial Context Setup with simultaneous establishment of three E-RABs for IMS signaling, VoLTE and default internet traffic

No	Long Time	From	4. Prot	4. MSG	Procedure Code	e-RAB-ID	e-RAB-ID qCI	e-RAB-ID nas	Cause Value
62	10:18:24,765,000	S1-AP-TP-S1-MME			id-initialUEMessage				
63	10:18:24,770,000	S1-AP-TP-S1-MME	LTE_URACI uERadioAcc..		id-InitialContextSetup	5, 6	5, 6		
68	10:18:24,815,000	S1-AP-TP-S1-MME			id-InitialContextSetup				
69	10:18:24,816,000	S1-AP-TP-S1-MME			id-CellTrafficTrace				
70	10:18:24,816,000	S1-AP-TP-S1-MME			id-CellTrafficTrace				
75	10:18:27,313,000	S1-AP-TP-S1-MME			id-E-RABSetup	7	1		
78	10:18:27,355,000	S1-AP-TP-S1-MME			id-E-RABSetup				
80	10:18:27,370,000	S1-AP-TP-S1-MME			id-uplinkNASTransport				
132	10:19:18,593,000	S1-AP-TP-S1-MME			id-E-RABRelease			7	normal-release
133	10:19:18,595,000	S1-AP-TP-S1-MME			id-downlinkNASTransport				
136	10:19:18,612,000	S1-AP-TP-S1-MME			id-E-RABRelease				
139	10:19:18,614,000	S1-AP-TP-S1-MME			id-uplinkNASTransport				
161	10:19:48,794,000	S1-AP-TP-S1-MME			id-UEContextReleaseRequest				user-inactivity
162	10:19:48,806,000	S1-AP-TP-S1-MME			id-UEContextRelease				user-inactivity
164	10:19:48,807,000	S1-AP-TP-S1-MME			id-UEContextRelease				

Figure 2.22 Bearer for VoLTE speech information is established using dedicated E-RAB Setup procedure

No	Long Time	From	4. Prot	4. MSG	Procedure Code	e-RAB-ID	e-RAB-ID qCI	e-RAB-ID nas	Cause Value
8	10:21:42,940,000	S1-AP-TP-S1-MME			id-initialUEMessage				
10	10:21:42,952,000	S1-AP-TP-S1-MME			id-downlinkNASTransport				
12	10:21:42,964,000	S1-AP-TP-S1-MME			id-uplinkNASTransport				
13	10:21:43,015,000	S1-AP-TP-S1-MME			id-downlinkNASTransport				
15	10:21:43,360,000	S1-AP-TP-S1-MME			id-uplinkNASTransport				
16	10:21:43,390,000	S1-AP-TP-S1-MME			id-downlinkNASTransport				
18	10:21:43,405,000	S1-AP-TP-S1-MME			id-uplinkNASTransport				
19	10:21:43,619,000	S1-AP-TP-S1-MME			id-InitialContextSetup	5	5		
22	10:21:43,642,000	S1-AP-TP-S1-MME	LTE_URACI uERadioAcc..		id-UECapabilityInfoIndication				
27	10:21:43,684,000	S1-AP-TP-S1-MME			id-InitialContextSetup				
28	10:21:43,685,000	S1-AP-TP-S1-MME			id-CellTrafficTrace				
29	10:21:43,685,000	S1-AP-TP-S1-MME			id-CellTrafficTrace				
32	10:21:45,694,000	S1-AP-TP-S1-MME			id-uplinkNASTransport				
39	10:21:45,199,000	S1-AP-TP-S1-MME			id-uplinkNASTransport				
40	10:21:45,469,000	S1-AP-TP-S1-MME			id-E-RABSetup	6	6		
43	10:21:45,524,000	S1-AP-TP-S1-MME			id-E-RABSetup				
45	10:21:45,528,000	S1-AP-TP-S1-MME			id-uplinkNASTransport				
57	10:21:59,706,000	S1-AP-TP-S1-MME			id-CellTrafficTrace				
58	10:21:59,706,000	S1-AP-TP-S1-MME			id-CellTrafficTrace				
64	10:22:01,453,000	S1-AP-TP-S1-MME			id-E-RABSetup	7	1		
67	10:22:01,484,000	S1-AP-TP-S1-MME			id-E-RABSetup				
69	10:22:01,488,000	S1-AP-TP-S1-MME			id-uplinkNASTransport				
72	10:22:08,704,000	S1-AP-TP-S1-MME			id-E-RABRelease			7	normal-release
73	10:22:08,706,000	S1-AP-TP-S1-MME			id-downlinkNASTransport				
76	10:22:08,737,000	S1-AP-TP-S1-MME			id-E-RABRelease				
79	10:22:08,741,000	S1-AP-TP-S1-MME			id-uplinkNASTransport				
107	10:22:39,686,000	S1-AP-TP-S1-MME			id-UEContextReleaseRequest				user-inactivity
108	10:22:39,696,000	S1-AP-TP-S1-MME			id-UEContextRelease				user-inactivity
110	10:22:39,699,000	S1-AP-TP-S1-MME			id-UEContextRelease				

Figure 2.23 IMS signaling bearer, default internet traffic bearer and VoLTE bearer are established one by one

No	Long Time	From	4. Prot	4. MSG	Procedure Code	e-RAB-ID	e-RAB-ID qCI	e-RAB-ID nas	Cause Value
60	14:54:49,313,590	S1-AP-S1-MME	S1AP	initiating..	id-initialUEMessage				
61	14:54:49,344,920	S1-AP-S1-MME	S1AP	initiating..	id-InitialContextSetup	5	5		
62	14:54:49,391,213	S1-AP-S1-MME	S1AP	successful..	id-InitialContextSetup				
84	14:55:17,476,600	S1-AP-S1-MME	S1AP	initiating..	id-E-RABSetup	6	1		
85	14:55:17,509,014	S1-AP-S1-MME	S1AP	successful..	id-E-RABSetup				
86	14:55:17,510,597	S1-AP-S1-MME	S1AP	initiating..	id-uplinkNASTransport				
99	14:56:19,910,218	S1-AP-S1-MME	S1AP	initiating..	id-E-RABRelease			6	normal-release
103	14:56:19,939,774	S1-AP-S1-MME	S1AP	successful..	id-E-RABRelease				
104	14:56:19,942,635	S1-AP-S1-MME	S1AP	initiating..	id-uplinkNASTransport				
114	14:56:24,500,809	S1-AP-S1-MME	S1AP	initiating..	id-E-RABSetup	6	1		
115	14:56:24,533,582	S1-AP-S1-MME	S1AP	successful..	id-E-RABSetup				
116	14:56:24,535,167	S1-AP-S1-MME	S1AP	initiating..	id-uplinkNASTransport				
129	15:08:53,066,787	S1-AP-S1-MME	S1AP	initiating..	id-E-RABRelease			6	normal-release
133	15:08:53,098,550	S1-AP-S1-MME	S1AP	successful..	id-E-RABRelease				
134	15:08:53,099,636	S1-AP-S1-MME	S1AP	initiating..	id-uplinkNASTransport				
141	15:08:57,322,728	S1-AP-S1-MME	S1AP	initiating..	id-E-RABSetup	6	1		
145	15:08:57,352,637	S1-AP-S1-MME	S1AP	successful..	id-E-RABSetup				
146	15:08:57,357,284	S1-AP-S1-MME	S1AP	initiating..	id-uplinkNASTransport				
151	15:09:00,368,723	S1-AP-S1-MME	S1AP	initiating..	id-E-RABRelease			6	normal-release
152	15:09:00,404,499	S1-AP-S1-MME	S1AP	successful..	id-E-RABRelease				
153	15:09:00,405,445	S1-AP-S1-MME	S1AP	initiating..	id-uplinkNASTransport				
154	15:09:31,304,720	S1-AP-S1-MME	S1AP	initiating..	id-UEContextReleaseRequest				user-inactivity
155	15:09:31,368,468	S1-AP-S1-MME	S1AP	initiating..	id-UEContextRelease				normal-release
156	15:09:31,371,153	S1-AP-S1-MME	S1AP	successful..	id-UEContextRelease				

Figure 2.24 3 VoLTE bearers ("calls") established and released during same radio connection

2.16.2 Multi-E-RAB Call with Intra-LTE Handover

When it comes to mobility, some additional challenges will occur for those who want to analyze service quality in the RAN.

Figure 2.25   Normal release of a VoLTE call after two failed attempts of handover preparation

Figure 2.26   Successful incoming and outgoing S1 handover of a radio connection with active VoLTE bearer

Figure 2.25 shows a VoLTE call with IMS Bearer, default bearer, and conversational voice bearer is attempting S1 handover, but Handover Preparation Failure is seen. Thus, this outgoing handover is not successful.

It should also be noticed that the QCI is not seen in the S1AP Handover Preparation messages. Thus, it requires tracking of E-RABs and associated QCIs to identify the services involved in an outgoing handover procedure and to peg, for example, performance counters for VoLTE handover attempt, success, and failure.

Figure 2.26 illustrates Incoming and Outgoing successful S1 Handover of a VoLTE call with IMS Bearer (QCI 5), default bearer (QCI 8), and conversational voice bearer (QCI 1). Here the UE enters a new cell and leaves this cell after some seconds. The time the UE remains active (RRC Connected) in a cell is also an important KPI for RAN performance, especially if the UE toggles between some cells overlapping in the same geographical area, an effect that is known as "ping-pong handover." Such "ping-pong handover" deteriorates the service quality perceived by the subscriber as well as generating additional unnecessary signaling load in the RAN.

2.16.3   Inter-RAT Mobility of a Multi-E-RAB Call Using CS Fallback

There are several options for the inter-RAT mobility of radio connections with VoLTE services.

Figure 2.27 shows a VoLTE connection that successfully handed over to 3G UMTS including IMS bearer and conversational voice bearer. Such scenario may happen in case the radio conditions in the LTE network are sub-optimal and 3G RAN is also connected to the IMS. In fact, this is also the typical signaling call flow of a SRVCC (Single Radio Voice Call Continuity) handover. In case of SRVCC handover, the SRVCC flag in the S1AP Handover Preparation Request message is present.

A different case is shown in Figure 2.28. Here the conversational voice bearer is normally terminated before in the next step a CS Fallback is triggered that, in first instance, shall be executed as a 4G-3G inter-RAT handover. However, the inter-RAT handover preparation procedure fails due to a protocol error in the MME, and thus, we can assume that blind inter-RAT redirection is executed afterward. To be sure that it is blind redirection, it would be necessary to see the redirection information in the RRC Connection Release message as described in Chapter 3.3.6 of this book. If only S1AP signaling is available, the

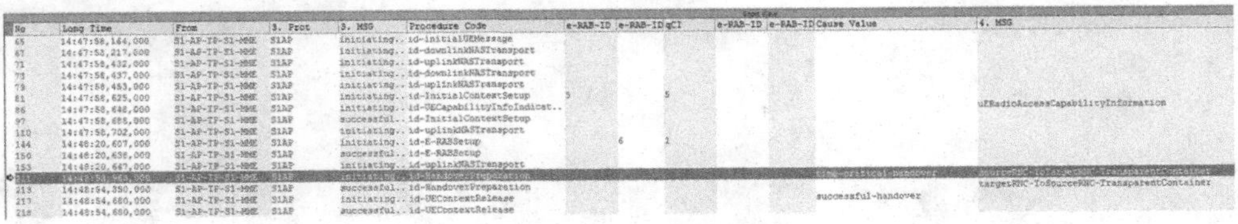

Figure 2.27 Successful outgoing inter-RAT handover of a radio connection with active VoLTE bearer

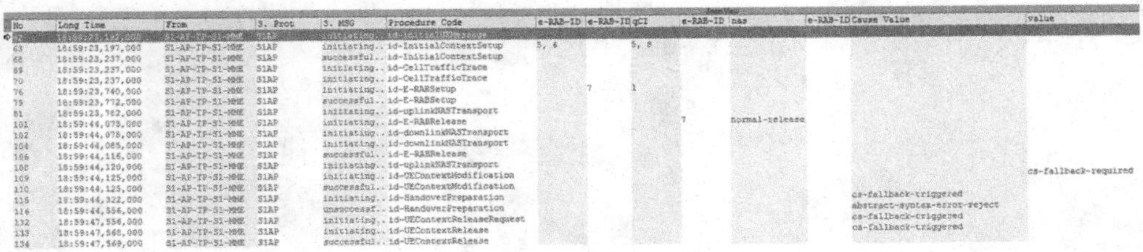

Figure 2.28 Initial Context Setup with simultaneous establishment of three E-RABs for IMS signaling, VoLTE and default internet traffic

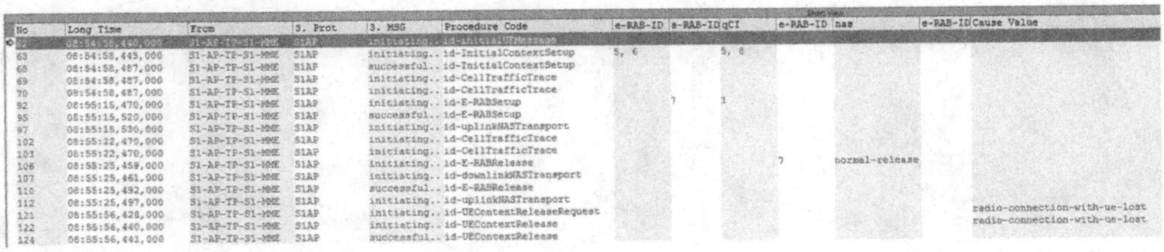

Figure 2.29 Drop of the radio connection after normal release of an previously active VoLTE bearer

appearance of the S1AP UE Context Release Request messages with cause "cs-fallback-triggered" is a strong indication for the blind redirection scenario.

The CS Fallback itself might have been triggered by an incoming voice call, because the first (mobile originating) VoLTE bearer was already released.

2.16.4 Abnormal Releases of Calls with VoLTE Services

When it comes to call release analysis, similar challenges as in case of the handover analysis are seen for the multi-E-RAB call scenarios since the different E-RABs are established and released independently.

Figure 2.29 shows a good example of where a dropping radio connection leads to abnormal release of IMS bearer and default bearer, but conversational voice bearer was already released seconds before. Thus, this abnormal release shall be counted as a call (or better: UE context) drop, but not as a VoLTE drop.

3

Radio Interface Signaling Procedures

As described in Chapter 1, one of the major changes along the evolutionary path from 3G UMTS to LTE is the fact that eNodeBs host the radio resource management functions while in 3G UMTS, all radio resources (frequencies, codes, power thresholds) have been controlled by the RNC. Implementation of the radio resource management in the eNodeB results, in turn, in the fact that also the protocol entity used to negotiate the usage of radio resources between the User Equipment (UE) and network must be part of the eNodeB software. Hence, the Radio Resource Control (RRC) protocol terminates on the network side in the eNodeB. From this it emerges that the eNodeB not only is responsible for setting up and releasing RRC connections and the radio bearer, but also must handle all RRC measurement tasks and resulting handover decisions. Thus, the most interesting and most crucial parameters for radio network optimization in LTE networks are found and can be tuned in the eNodeB.

Looking at the LTE RRC protocol itself as described in 3GPP 36.331, at first sight most of the message names sound familiar or are even identical to what was described in 3GPP 25.331, the RRC protocol standard for 3G UMTS. But significant differences emerge when going into detail and checking the meaning of particular information elements as well as the interaction of the RRC protocol with the underlying Medium Access Control (MAC) transport layer and the upper level Non-Access Stratum (NAS) signaling for which RRC itself acts as the transport layer. It can also be seen that the standardization group working on the protocol definition had the goal to simplify the protocol by introducing a comprehensive set of messages. However, fewer messages compared to 3GPP 25.331 definitions do not necessarily mean a simplification of the analysis of signaling procedures, since 3GPP 25.331 had distinct messages for radio bearer setup and radio bearer release, as well as for physical channel reconfiguration, transport channel reconfiguration, and reconfiguration of the Quality of Service (QoS) parameters of radio bearers. Now, in 3GPP 36.331, a single procedure is used to handle all these tasks and this message is simply called RRC Connection Reconfiguration. One needs to look at every single detail and information element sequence included in such an RRC connection reconfiguration message to find out what is reconfigured or how exactly the RRC connection is changed. With just this one message, radio bearers can be set up and released, measurement tasks can be assigned to UEs, and physical parameters, transport channel parameters, and QoS attributes can be changed – all at once or one after another. Consequently, the new RRC protocol in LTE gives the impression of being more sophisticated than its 3G UMTS predecessor and a distinctive domain of signaling experts and programmers of eNodeB software working for network equipment manufacturers.

LTE Signaling, Troubleshooting and Performance Measurement, Second Edition. Ralf Kreher and Karsten Gaenger.
© 2016 John Wiley & Sons, Ltd. Published 2016 by John Wiley & Sons, Ltd.

The following sections document what is implemented in RRC entities and tested in laboratory and first field trials. In these test scenarios, the setup of an RRC connection, initial attach, and default bearer setup to transmit some user plane data are sufficient. Since this book documents what is implemented in live traffic scenarios, it focuses on describing what was implemented at the time of writing this chapter (spring, 2010). Thus, the reader cannot expect a complete description of all possible RRC procedures, but rather a first insight into this new sophisticated protocol explaining and revealing the principles and most important functions. Further call scenarios including different mobility scenarios will follow in the future.

3.1 RRC Connection Setup, Attach, and Default Bearer Setup

The RRC connection setup procedure is not a stand-alone procedure. It encloses the NAS initial attach and default bearer setup procedures. Thus, it is much more complex than the procedure with the same name in Universal Terrestrial Radio Access Network (UTRAN).

3.1.1 Random Access and RRC Connection Setup Procedure

Also, the random access procedure is different. While in UTRAN, a slotted aloha method is used to access the Physical Random Access Channel (PRACH), E-UTRAN random access preambles are created from a Zadoff–Chu sequence. There are five different formats of random access preambles, each different in length of preamble prefix and sequence part as shown in Table 3.1, which correspond to the generic random access preamble format illustrated in Figure 3.1.

In the case of preamble formats 0, 1, 2, and 3, there is typically one preamble sequence per uplink subframe. The PRACH configuration determines whether an uplink subframe carries a preamble or not. Table 3.2 lists the possible combinations.

The details of random access preamble configuration and generation are not important for understanding the signaling procedure. However, they are essential for troubleshoot and optimizing the random access, which is typically one of the first important tasks in the initial network deployment phase. Besides the already mentioned preamble formats and preamble timing, the power thresholds used by the UE to send the random access preamble are configurable in the network and broadcast to the UE using RRC system information.

Table 3.1 Preamble formats (according to 3GPP 36.211)

Preamble format	T_{CP}	T_{SEQ}
0	$3168 \times T_s$	$24576 \times T_s$
1	$21024 \times T_s$	$24576 \times T_s$
2	$6240 \times T_s$	$2 \times 24576 \times T_s$
3	$21024 \times T_s$	$2 \times 24576 \times T_s$
4 (frame structure type 2 only)	$448 \times T_s$	$4096 \times T_s$

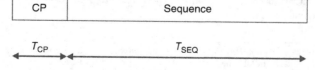

Figure 3.1 Random access preamble format (according to 3GPP 36.211)

Table 3.2 Random access preamble timing for preamble formats 0–3 (according to 3GPP 36.211)

PRACH configuration	System frame number	Subframe number
0	Even	1
1	Even	4
2	Even	7
3	Any	1
4	Any	4
5	Any	7
6	Any	1, 6
7	Any	2, 7
8	Any	3, 8
9	Any	1, 4, 7
10	Any	2, 5, 8
11	Any	3, 6, 9
12	Any	0, 2, 4, 6, 8
13	Any	1, 3, 5, 7, 9
14	Any	0, 1, 2, 3, 4, 5, 6, 7, 8, 9
15	Even	9

As shown in Figure 3.2, the RRC connection setup procedure starts when a MAC random access preamble (Message Example 3.1) with a valid preamble index and Random Access Radio Network Temporary Identity (RA-RNTI) is received by the eNodeB.

Message Example 3.1: MAC Random Access Preamble

```
+-------------------------------------------+----------------------------+
|ID Name                                    |Comment or Value            |
+-------------------------------------------+----------------------------+
|56 05:43:35,511,955 RRC-UU K2AIR-PHY PRACH                              |
|Tektronix K2Air LTE PHY Data Message Header (K2AIR-PHY)                 |
PRACH (= PRACH Message)                                                  |
|1 PRACH Message                                                         |
|1.1 Common Message Header                                               |
|Protocol Version                           |0                           |
|Transport Channel Type                     |RACH                        |
|Physical Channel Type                      |PRACH                       |
|System Frame Number                        |550                         |
|Direction                                  |Uplink                      |
|Radio Mode                                 |FDD                         |
|Internal use                               |0                           |
|Status                                     |Original data               |
|Reserved                                   |0                           |
|Physical Cell ID                           |0                           |
|UE ID/RNTI Type                            |RA-RNTI                     |
|Subframe Number                            |1                           |
```

```
|UE ID/RNTI Value                           |'0002'H
|1.2 PRACH Information
|PRACH Preamble Format                      |0
|Reserved                                   |0
|PRACH Preamble Index                       |1
|Reserved                                   |0
|Number of first PRACH Resource Block       |3
|Logical RSN (Zadoff-Chu)                   |23
|Physical RSN (Zadoff-Chu)                  |838
|Ncs Configuration Index                    |5
|Reserved                                   |0
```

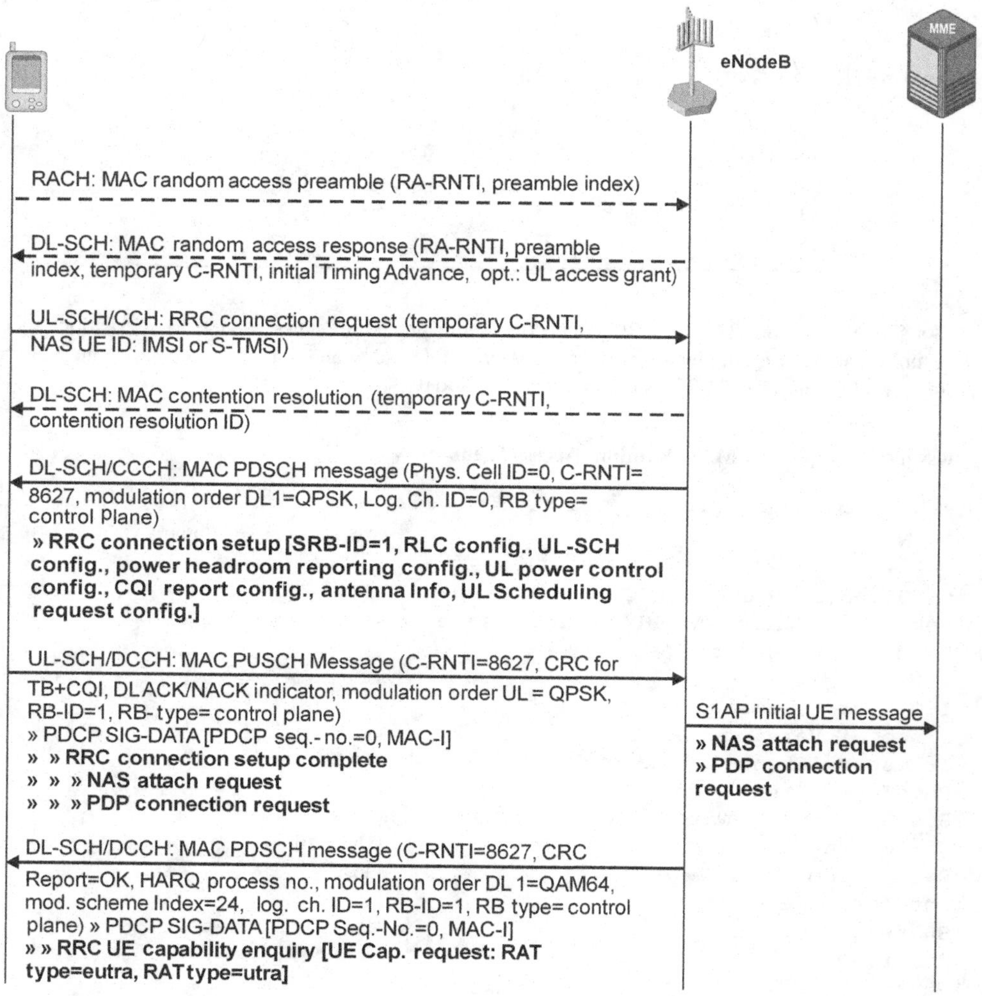

eNodeB

MME

RACH: MAC random access preamble (RA-RNTI, preamble index)

DL-SCH: MAC random access response (RA-RNTI, preamble index, temporary C-RNTI, initial Timing Advance, opt.: UL access grant)

UL-SCH/CCH: RRC connection request (temporary C-RNTI, NAS UE ID: IMSI or S-TMSI)

DL-SCH: MAC contention resolution (temporary C-RNTI, contention resolution ID)

DL-SCH/CCCH: MAC PDSCH message (Phys. Cell ID=0, C-RNTI= 8627, modulation order DL1=QPSK, Log. Ch. ID=0, RB type= control Plane)
» RRC connection setup [SRB-ID=1, RLC config., UL-SCH config., power headroom reporting config., UL power control config., CQI report config., antenna Info, UL Scheduling request config.]

UL-SCH/DCCH: MAC PUSCH Message (C-RNTI=8627, CRC for TB+CQI, DLACK/NACK indicator, modulation order UL = QPSK, RB-ID=1, RB-type= control plane)
» PDCP SIG-DATA [PDCP seq.-no.=0, MAC-I]
» » RRC connection setup complete
» » » NAS attach request
» » » PDP connection request

S1AP initial UE message
» NAS attach request
» PDP connection request

DL-SCH/DCCH: MAC PDSCH message (C-RNTI=8627, CRC Report=OK, HARQ process no., modulation order DL 1=QAM64, mod. scheme Index=24, log. ch. ID=1, RB-ID=1, RB type= control plane) » PDCP SIG-DATA [PDCP Seq.-No.=0, MAC-I]
» » RRC UE capability enquiry [UE Cap. request: RAT type=eutra, RATtype=utra]

Figure 3.2 RRC connection setup procedure 1/3. (*Source:* Tektronix Communications.)

The RA-RNTI is used as a temporary identity of the subscriber during random access and must be replaced by a C-RNTI in the next step when the eNodeB sends a MAC random access response (Message Example 3.2) back to the UE. This random access response is sent like all other downlink signaling on the Downlink Shared Channel (DL-SCH). In E-UTRAN, there is no separate, physical common control channel for signaling on the downlink as in GERAN and UTRAN cells. All information, signaling, and the user plane are mapped onto the DL-SCH.

Message Example 3.2: MAC Random Access Response

```
+-----------------------------------------+-------------------------------------+
|ID Name                                  |Comment or Value                     |
+-----------------------------------------+-------------------------------------+
|57 05:43:35,514,955 RRC-UU K2AIR-PHY PDSCH LTE-RLC/MAC MAC-RAR (DL)            |
|Tektronix K2Air LTE PHY Data Message Header (K2AIR-PHY) PDSCH                   |
|(= PDSCH Message)                                                              |
|1 PDSCH Message                                                                |
|1.1 Common Message Header                                                      |
|Protocol Version                         |0                                    |
|Transport Channel Type                   |DL-SCH                               |
|Physical Channel Type                    |PDSCH                                |
|System Frame Number                      |550                                  |
|Direction                                |Downlink                             |
|Radio Mode                               |FDD                                  |
|Internal use                             |0                                    |
|Status                                   |Original data                        |
|Reserved                                 |0                                    |
|Physical Cell ID                         |0                                    |
|UE ID/RNTI Type                          |RA-RANTI                             |
|Subframe Number                          |5                                    |
|UE ID/RNTI Value                         |'0002'H                              |
|1.2 PDSCH Header                                                               |
|CRC report                               |CRC ok                               |
|HARQ process number                      |0                                    |
|Reserved                                 |0                                    |
|Transport Block Indicator                |single TB info                       |
|Reserved                                 |0                                    |
|1.2.1 Transport Block#1 Information                                            |
|Transport Block#1 Size                   |56                                   |
|Modulation Order DL 1                    |QPSK                                 |
|New Data Indicator DL 1                  |new data                             |
|Redundancy Version DL 1                  |0                                    |
|Reserved                                 |0                                    |
|Modulation Scheme Index DL 1             |1                                    |
|Reserved                                 |0                                    |
|1.2.2 Transport Block Data                                                     |
|TB1 Mac-PDU Data                         |66 00 20 14 8c be b1                 |
|Padding                                  |'14'H                                |
|3GPP LTE-RLC/MAC Rel.8 (MAC TS 36.321 V8.5.0, 2009-03,                         |
|RLC TS 36.322 V8.5.0, 2009-03) (LTE-RLC/MAC) MAC-RAR (DL)                      |
|(= MAC Random Access Response)   |                                           |
|1 MAC Random Access Response                                                   |
|1.1 MAC RAR Header Part                                                        |
|1.1.1 MAC Random Access ID Subheader                                           |
```

```
|Extension field                          |Last MAC PDU subheader           |
|Type Field                               |Random Access Identity           |
|Random Access Identity                   |38                               |
|1.2 MAC RAR Payload Part                 |                                 |
|1.2.1 MAC Random Access Response         |                                 |
|Reserved                                 |0                                |
|Timing Advance Value                     |2                                |
|UpLink Grant Resources                   |'00000001010010001100'B          |
|Temporary C-RNTI Value                   |'beb1'H                          |
```

The UE can identify the MAC random access response sent by the eNodeB on behalf of its RA-RNTI and preamble index. With this message, a temporary C-RNTI is assigned to the handset, the initial timing advance command is signaled, and optionally some uplink resources are granted. For further information about timing advance, see Section 4.3.2.3.

Now, the UE can send the RRC connection request message (Message Example 3.3) containing the previously received temporary C-RNTI and NAS UE identity stored on the USIM card. In case the UE was never previously attached to a network, the International Mobile Subscriber Identity (IMSI) will be used instead of a temporary NAS ID, or else the S-TMSI (the short part of the Globally Unique Temporary UE Identity (GUTI)) will be included.

Message Example 3.3: RRC Connection Request

```
+-------------------------------------------+----------------------------------+
|ID Name                                    |Comment or Value                  |
+-------------------------------------------+----------------------------------+
|59 05:43:35,520,954 RRC-UU K2AIR-PHY PUSCH LTE-RLC/MAC
RLC-REASM-PDU (UL) LTE-RRC_CCCH_UL rrcConnectionRequest                         |
|Tektronix K2Air LTE PHY Data Message Header (K2AIR-PHY) PUSCH
(= PUSCH Message)                                                               |
|1 PUSCH Message                                                                |
|1.1 Common Message Header                                                      |
|Protocol Version                          |0                                 |
|Transport Channel Type                    |UL-SCH                            |
|Physical Channel Type                     |PUSCH                             |
|System Frame Number                       |551                               |
|Direction                                 |Uplink                            |
|Radio Mode                                |FDD                               |
|Internal Use                              |0                                 |
|Status                                    |Reassembled RLC PDU               |
|Reserved                                  |0                                 |
|Physical Cell ID                          |0                                 |
|UE ID/RNTI Type                           |C-RNTI                            |
|Subframe Number                           |1                                 |
|UE ID/RNTI Value                          |'beb1'H                           |
```

```
|RRC (CCCH UL) 3GPP TS 36.331 V8.5.0 (2009-03) (LTE-RRC_CCCH_UL)
rrcConnectionRequest (= rrcConnectionRequest)   |
|uL-CCCH-Message                                                                    |
|1 message                                                                          |
|1.1 Standard                                                                       |
|1.1.1 rrcConnectionRequest                                                         |
|1.1.1.1 criticalExtensions                                                         |
|1.1.1.1.1 rrcConnectionRequest-r8                                                  |
|1.1.1.1.1.1 ue-Identity                                                            |
|1.1.1.1.1.1 s-TMSI                                                                 |
|1.1.1.1.1.1.1 mmec                       |'00'H                                     |
|1.1.1.1.1.1.2 m-TMSI                     |'8009e400'H                               |
|1.1.1.1.1.2 establishmentCause           |mo-Signalling                             |
|1.1.1.1.1.3 spare                        |'0'B                                      |
```

As an option, the MAC layer will send a separate contention resolution message including temporary C-RNTI and contention resolution ID. The purpose of the contention resolution procedure is to turn the temporary C-RNTI into a permanent or semi-permanent C-RNTI that can be used to identify all MAC frames sent to or from a particular UE on the uplink and DL-SCHs of the E-UTRA cell.

In response to the RRC connection request message sent by the UE, the eNodeB will send the RRC connection setup message. Like the MAC random access response before, this message is sent on the DL-SCH but transported in a MAC Physical Downlink Shared Channel (PDSCH) message. The logical channel used to send this RRC message is the Common Control Channel (CCCH). Thus, this message is sent in Radio Link Control (RLC), Unacknowledged Mode (UM), and no dedicated Acknowledgement (ACK) on the RLC layer is expected to be received from the UE.

The MAC PDSCH message that transports the RRC connection setup message contains the physical cell ID. This is the scrambling code used to distinguish the cell's downlink signal unambiguously from those of its neighbor cells that overlap in a geographic region. In fact, the physical cell ID has the same function as the primary scrambling code of a UTRA cell in 3G.

The modulation order for downlink transmission of this message provides important information. From the decoder output, it can be detected that Quadrature Phase Shift Keying (QPSK) is used to transmit this message on the radio interface. The Logical Channel ID (LCID) equal to zero is the one used for RLC UM and the Radio Bearer Type (RB-Type) provides the additional information that control plane information is transported on this logical channel.

The RRC connection setup message (Message Example 3.4) itself contains the identity of the Signaling Radio Bearer (SRB-1) and associated parameters for this bearer. Firstly, there is a sequence of RLC configuration parameters, especially timers and counters for RLC Acknowledged Mode (AM) operation. In Message Example 3.4, the timer for polling (requesting) STATUS reports is set to 45 ms. The number of possible retransmissions of the same STATUS report message is set to "infinity" while the number of retransmissions of an RLC PDU (Packet Data Unit) that contains control plane or user plane data is limited to four by the value set for parameter *maxRetxThreshold*. These parameters are valid for the uplink part of the RLC connection. On the downlink, the value of *t-Reordering* defines the maximum time when the receiving RLC AM or UM entity of the UE should wait for the arrival of a particular RLC Transport Block (TB) – in Message Example 3.4, 35 ms. If the timer expires before the TB is received, a retransmission of this TB will be ordered by sending a STATUS message.

Message Example 3.4: RRC Connection Setup Message

```
+------------------------------------------------------------+--------------------+
|ID Name                                                     |Comment or Value    |
+------------------------------------------------------------+--------------------+
|56 10:41:32,406,966 RRC-UU K2AIR-PHY PDSCH LTE-RLC/MAC
RLC-REASM-PDU (DL) LTE-RRC_CCCH_DL rrcConnectionSetup        |
|Tektronix K2Air LTE PHY Data Message Header (K2AIR-PHY) PDSCH
(= PDSCH Message)    |
|1 PDSCH Message                                                                  |
|1.1 Common Message Header                                                        |
|Protocol Version                                           |0                    |
|Transport Channel Type                                     |DL-SCH               |
|Physical Channel Type                                      |PDSCH                |
|Physical Cell ID                                           |0                    |
|UE ID/RNTI Type                                            |C-RNTI               |
|Subframe Number                                            |0                    |
|UE ID/RNTI Value                                           |'8627'H              |
|Modulation Order DL 1                                      |QPSK                 |
|1.3 Additional Call related Info                                                 |
|Number Of Logical Channel Information                      |1                    |
|1.3.1 Logical Channel Information                                                 |
|LCID                                                       |0                    |
|RLC Mode                                                   |Transparent Mode     |
|Radio Bearer ID                                            |0                    |
|Radio Bearer Type                                          |Control Plane        |
|                                                            (Signa...            |
|Logical Channel Type                                       |CCCH                 |
|RRC (CCCH DL) 3GPP TS 36.331 V8.5.0 (2009-03) (LTE-RRC_CCCH_DL)
rrcConnectionSetup (= rrcConnectionSetup)   |
|dL-CCCH-Message                                                                  |
|1 message                                                                        |
|1.1 Standard                                                                     |
|1.1.1 rrcConnectionSetup                                                         |
|1.1.1.1 rrc-TransactionIdentifier                          |0                    |
|1.1.1.2 criticalExtensions                                                       |
|1.1.1.2.1 c1                                                                     |
|1.1.1.2.1.1 rrcConnectionSetup-r8                                                |
|1.1.1.2.1.1.1 radioResourceConfigDedicated                                       |
|1.1.1.2.1.1.1.1 srb-ToAddModList                                                 |
|1.1.1.2.1.1.1.1.1 sRB-ToAddMod                                                   |
|1.1.1.2.1.1.1.1.1.1 srb-Identity                           |1                    |
|1.1.1.2.1.1.1.1.1.2 rlc-Config                                                   |
|1.1.1.2.1.1.1.1.1.2.1 explicitValue                                              |
|1.1.1.2.1.1.1.1.1.2.1.1 am                                                       |
|1.1.1.2.1.1.1.1.1.2.1.1.1 ul-AM-RLC                                              |
|1.1.1.2.1.1.1.1.1.2.1.1.1.1 t-PollRetransmit               |ms45                 |
|1.1.1.2.1.1.1.1.1.2.1.1.1.2 pollPDU                        |pInfinity            |
|1.1.1.2.1.1.1.1.1.2.1.1.1.3 pollByte                       |kBinfinity           |
|1.1.1.2.1.1.1.1.1.2.1.1.1.4 maxRetxThreshold               |t4                   |
|1.1.1.2.1.1.1.1.1.2.1.1.2 dl-AM-RLC                                              |
|1.1.1.2.1.1.1.1.1.2.1.1.2.1 t-Reordering                   |ms35                 |
|1.1.1.2.1.1.1.1.1.2.1.1.2.2 t-StatusProhibit               |ms0                  |
|1.1.1.2.1.1.1.1.1.3 logicalChannelConfig                                         |
|1.1.1.2.1.1.1.1.1.3.1 explicitValue                                              |
```

```
|1.1.1.2.1.1.1.1.1.3.1.1 ul-SpecificParameters            |
|1.1.1.2.1.1.1.1.1.3.1.1.1 priority                       |1
|1.1.1.2.1.1.1.1.1.3.1.1.2 prioritisedBitRate             |infinity
|1.1.1.2.1.1.1.1.1.3.1.1.3 bucketSizeDuration             |ms100
|1.1.1.2.1.1.1.1.1.3.1.1.4 logicalChannelGroup            |0
|1.1.1.2.1.1.1.2 mac-MainConfig                           |
|1.1.1.2.1.1.1.2.1 explicitValue                          |
|1.1.1.2.1.1.1.2.1.1 ul-SCH-Config                        |
|1.1.1.2.1.1.1.2.1.1.1 maxHARQ-Tx                         |n5
|1.1.1.2.1.1.1.2.1.1.2 periodicBSR-Timer                  |sf20
|1.1.1.2.1.1.1.2.1.1.3 retxBSR-Timer                      |sf320
|1.1.1.2.1.1.1.2.1.1.4 ttiBundling                        |false
|1.1.1.2.1.1.1.2.1.2 timeAlignmentTimerDedicated          |infinity
|1.1.1.2.1.1.1.2.1.3 phr-Config                           |
|1.1.1.2.1.1.1.2.1.3.1 setup                              |
|1.1.1.2.1.1.1.2.1.3.1.1 periodcPHR-Timer                 |sf500
|1.1.1.2.1.1.1.2.1.3.1.2 prohibitPHR-Timer                |sf200
|1.1.1.2.1.1.1.2.1.3.1.3 dl-PathlossChange                |dB3
|1.1.1.2.1.1.1.3 physicalConfigDedicated                  |
|1.1.1.2.1.1.1.3.1 pdsch-ConfigDedicated                  |
|1.1.1.2.1.1.1.3.1.1 p-a                                   |dB-3
|1.1.1.2.1.1.1.3.2 pusch-ConfigDedicated                  |
|1.1.1.2.1.1.1.3.2.1 betaOffset-ACK-Index                 |9
|1.1.1.2.1.1.1.3.2.2 betaOffset-RI-Index                  |6
|1.1.1.2.1.1.1.3.2.3 betaOffset-CQI-Index                 |6
|1.1.1.2.1.1.1.3.3 uplinkPowerControlDedicated            |
|1.1.1.2.1.1.1.3.3.1 p0-UE-PUSCH                          |0
|1.1.1.2.1.1.1.3.3.2 deltaMCS-Enabled                     |en0
|1.1.1.2.1.1.1.3.3.3 accumulationEnabled                  |true
|1.1.1.2.1.1.1.3.3.4 p0-UE-PUCCH                          |0
|1.1.1.2.1.1.1.3.3.5 pSRS-Offset                          |5
|1.1.1.2.1.1.1.3.3.6 filterCoefficient                    |fc4
|1.1.1.2.1.1.1.3.4 cqi-ReportConfig                       |
|1.1.1.2.1.1.1.3.4.1 cqi-ReportModeAperiodic              |rm31
|1.1.1.2.1.1.1.3.4.2 nomPDSCH-RS-EPRE-Offset              |0
|1.1.1.2.1.1.1.3.4.3 cqi-ReportPeriodic                   |
|1.1.1.2.1.1.1.3.4.3.1 setup                              |
|1.1.1.2.1.1.1.3.4.3.1.1 cqi-PUCCH-ResourceIndex          |0
|1.1.1.2.1.1.1.3.4.3.1.2 cqi-pmi-ConfigIndex              |17
|1.1.1.2.1.1.1.3.4.3.1.3 cqi-FormatIndicatorPeriodic      |
|1.1.1.2.1.1.1.3.4.3.1.3.1 widebandCQI                    |0
|1.1.1.2.1.1.1.3.4.3.1.4 ri-ConfigIndex                   |483
|1.1.1.2.1.1.1.3.4.3.1.5 simultaneousAckNackAndCQI        |false
|1.1.1.2.1.1.1.3.5 antennaInfo                            |
|1.1.1.2.1.1.1.3.5.1 explicitValue                        |
|1.1.1.2.1.1.1.3.5.1.1 transmissionMode                   |tm4
|1.1.1.2.1.1.1.3.5.1.2 codebookSubsetRestriction          |
|1.1.1.2.1.1.1.3.5.1.2.1 n2TxAntenna-tm4                  |'111111'B
|1.1.1.2.1.1.1.3.5.1.3 ue-TransmitAntennaSelection        |
|1.1.1.2.1.1.1.3.5.1.3.1 release                          |0
|1.1.1.2.1.1.1.3.6 schedulingRequestConfig                |
|1.1.1.2.1.1.1.3.6.1 setup                                |
|1.1.1.2.1.1.1.3.6.1.1 sr-PUCCH-ResourceIndex             |20
|1.1.1.2.1.1.1.3.6.1.2 sr-ConfigIndex                     |16
|1.1.1.2.1.1.1.3.6.1.3 dsr-TransMax                       |n16
```

In the Uplink Shared Channel (UL-SCH) configuration section of the RRC connection setup message, the maximum number of hybrid Automatic Repeat Request (ARQ) retransmissions is limited to five (*maxHARQ-Tx*) and a periodic Buffer Status Report (BSR) is requested to be sent with a periodicity of 20 subframes =20 ms. Otherwise, a regular BSR is expected to be sent after 320 subframes = 320 ms. Also, power headroom reporting is configured. Here, periodic reports should be sent with a periodicity of 500 subframes =500 ms or such a report should be triggered if the prohibitPHR-Timer of 200 ms expires and the downlink path loss has changed by more than 3 dB.

P0-UE-PUSCH is the UE-specific component in dBm (range: −8 to 7 dBm) that contributes to the formula used to calculate the nominal power for the next uplink resource blocks to be transmitted by the handset on the Physical Uplink Shared Channel (PUSCH) as described in 3GPP 36.213. In the same way, po-UE-PUCCH is used to calculate UE transmission power for information sent on the physical common control channel (i.e., the PUCCH). The power threshold for sending the sounding reference symbols on the uplink is determined by the *pSRS-Offset* value and the value of the flag *deltaMCS-Enabled*. In Message Example 3.1, the deltaMCS-Enabled flag is set to "enabled," which corresponds to a parameter $K_s = 1.25$ as defined in 3GPP 36.213. If $K_s = 1.25$ a semi-static configuration of Sounding Reference Symbol power (pSRS) applies with a 1 dB step size in the range of −3 to 12 dB and the actual parameter value is pSRS-Offset value − 3. Thus, according to settings in Message Example 3.1, the sounding reference symbols will be sent with 2 dB. However, if the deltaMCS-Enabled flag is set to "disabled," which corresponds to a parameter $K_s = 0$, then the pSRS is configured with a 1.5 dB step size in the range −10.5 to 12 dB and the actual parameter value is −10.5 + 1.5 * pSRS-Offset value. Thus, with deltaMCS-Enabled = "disabled," the same value of pSRS-Offset = 5 as seen in Message Example 3.4 would result in a power threshold of −10.5 + 1.5 * 5 = −3 dB. The filter coefficient fc4 is used for RSRP measurements to calculate path loss.

The next section of the RRC connection setup message contains settings for Channel Quality Indicator (CQI) reporting. CQI is the quality feedback sent by the UE to indicate how good or how bad the quality of the downlink channel is perceived on the receiver side. The packet scheduler in the eNodeB can react to this feedback information by assigning new radio resources, for example, the MCS (Modulation Coding Scheme) can be changed or the next data blocks can be transmitted on different subcarriers than those previously sent.

The antenna info section of the message contains the necessary parameters for error-free spatial multiplexing and MIMO.

The last section of the message deals with the configuration of scheduling request parameters. A scheduling request is sent if the uplink data buffer in the UE detects that it is not provided with sufficient uplink radio resources. Thus, all these parameters, antenna info, as well as scheduling request configuration do have a direct impact on the subscriber's perceived QoS.

Now, if the UE successfully received the RRC connection setup message, it responds with an RRC Connection Setup Complete. This message is the first signaling message of a connection sent on the Dedicated Control Channel (DCCH). The RRC connection setup complete message transports the NAS attach request and PDP connection request messages introduced in Section 2.2. It is the first message in this trace transported by the Packet Data Convergence Protocol (PDCP). In the PDCP header, there is the PDCP sequence number for flow control and error detection as well as the message authentication code MAC-I for integrity protection. Since the RRC security mode procedure has not yet been executed, the MAC-I field contains the default value with all bits set to zero.

The PDCP and its upper layers are embedded in a MAC PUCCH message. On the MAC layer, there are the C-RNTI as the UE identity, CRC check for TB, CQI information, and a downlink ACK/NACK (Negative Acknowledgement) indicator to acknowledge successful reception of the previously sent downlink TBs that contained the RRC connection setup message. The modulation order used to send this message in the uplink direction is still QPSK and the radio bearer ID = 1, which is a control plane Signaling Radio Bearer (SRB).

The two NAS messages are piggybacked by the S1 Application Part (S1AP) initial UE message and transparently forwarded to the Mobility Management Entity (MME) across the S1 reference point.

3.1.2   RRC Connection Reconfiguration and Default Bearer Setup

In the next step, a MAC message is sent on the PDSCH that has the same C-RNTI as seen before, CRC Report = OK, a hybrid ARQ process number for radio interface retransmission (if this is required), and new modulation order QAM64. The latter means that the packet scheduler decided to change the modulation scheme due to good reported radio conditions. LCID = 1 is assigned to (signaling) radio bearer = 1, which carries control plane messages. One layer up, we have the PDCP again. As before, on the uplink and also on the downlink path of the connection, the counting of PDCP sequence numbers for a particular logical channel starts with sequence number value 0. On top of PDCP is the RRC UE capability inquiry message that requests the UE to indicate its capability status for E-UTRAN and UTRAN support. The UE responds with the RRC UE capability information message shown in Figure 3.3. This message contains the list of UE capabilities for working in UTRAN and E-UTRAN mode, such as a list of supported frequency bands and handover capabilities.

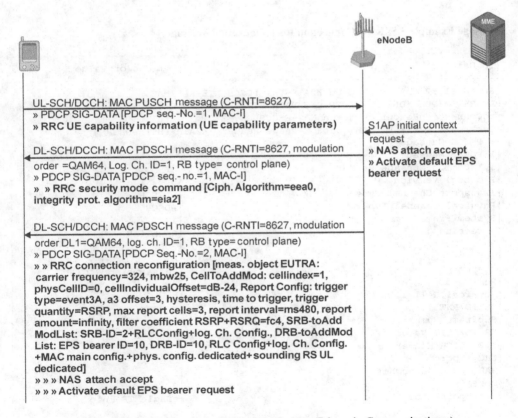

Figure 3.3   RRC connection setup 2/3. (*Source:* Tektronix Communications.)

While the UE capabilities are being inquired by the eNodeB, the NAS messages are processed by the MME and the necessary core network transport and switching resources as described in Section 2.2 are successfully allocated. Thus, the S1AP initial context request message is sent to the eNodeB including NAS messages Attach Accept (may optionally contain a new GUTI) and Activate Default Evolved Packet System (EPS) Bearer Request. On the S1AP layer, security keys are provided as described in Section 1.7. To activate the security functions, the RRC Security Mode Command is sent by the eNodeB to the UE. In the header of the MAC block used to transport this message across the radio interface, we can see the same parameter values as in the previously sent downlink message. On the PDCP layer, the sequence number for downlink transmission on logical channel 1 is increased by one compared to the previous downlink message on the same logical channel. The RRC Security Mode Command itself contains the algorithms to be used for integrity protection and ciphering.

Immediately after the RRC Security Mode Command, another RRC message is sent in the downlink direction. This is an RRC connection reconfiguration message. A look at the contents of this message as listed in Figure 3.3 shows that there are multiple ways to send this message. Actually, the main objective is to activate a dedicated radio bearer for the default EPS bearer on the radio interface. Together with this, the RRC measurements to be executed by the UE are set up and the MAC and physical channel parameters are reconfigured. The full contents of this message including the MAC and PDCP header can be seen when looking at Message Example 3.5.

Message Example 3.5: RRC Connection Reconfiguration Message

```
+----------------------------------------------------+-----------------------------+
| ID Name                                            | Comment or Value            |
+----------------------------------------------------+-----------------------------+
| 60 10:41:32,658,965 RRC-UU K2AIR-PHY PDSCH LTE-RLC/MAC
RLC-REASM-PDU (DL) LTE-PDCP SIG-DATA LTE-RRC_DCCH_DL
rrcConnectionReconfiguration NAS-SECURED ???    |
| Tektronix K2Air LTE PHY Data Message Header (K2AIR-PHY) PDSCH
(= PDSCH Message)    |
| 1 PDSCH Message                                                                   |
| 1.1 Common Message Header                                                         |
| Protocol Version                                    | 0                           |
| Transport Channel Type                              | DL-SCH                      |
| Physical Channel Type                               | PDSCH                       |
| System Frame Number                                 | 172                         |
| Direction                                           | Downlink                    |
| Radio Mode                                          | FDD                         |
| Internal use                                        | 0                           |
| Status                                              | Reassembled RLC PDU         |
| Reserved                                            | 0                           |
| Physical Cell ID                                    | 0                           |
| UE ID/RNTI Type                                     | C-RNTI                      |
| Subframe Number                                     | 2                           |
| UE ID/RNTI Value                                    | '8627'H                     |
| 1.2 PDSCH Header                                                                  |
| CRC report                                          | CRC ok                      |
| HARQ process number                                 | 2                           |
| Reserved                                            | 0                           |
| Transport Block Indicator                           | single TB info              |
```

```
|Reserved                                        |0                    |
|1.2.1 Transport Block#1 Information                                   |
|Transport Block#1 Size                          |624                  |
|Modulation Order DL 1                           |QAM64                |
|New Data Indicator DL 1                          |new data             |
|Redundancy Version DL 1                         |0                    |
|Reserved                                        |0                    |
|Modulation Scheme Index DL 1                    |24                   |
|Reserved                                        |0                    |
|1.2.2 Transport Block Data                                            |
|TB1 Mac-PDU Data                                |02 20 16 15 e8 00 14...|
|1.3 Additional Call related Info                                      |
|Number Of Logical Channel Information           |1                    |
|1.3.1 Logical Channel Information                                     |
|LCID                                            |1                    |
|RLC Mode                                        |Acknowledged Mode    |
|Radio Bearer ID                                 |1                    |
|Radio Bearer Type                               |Control Plane (Signa...|
|Spare                                           |0                    |
|Spare                                           |0                    |
|Logical Channel Type                            |DCCH                 |
|Call ID                                         |'00000001'H          |
|3GPP LTE-RLC/MAC Rel.8 (MAC TS 36.321 V8.5.0, 2009-03, RLC TS         |
|36.322 V8.5.0, 2009-03) (LTE-RLC/MAC) RLC-REASM-PDU (DL)              |
|(= RLC Reassembled PDU (Downlink))                                    |
|1 RLC Reassembled PDU (Downlink)                                      |
|Logical Channel ID                              |1                    |
|Reassembled SDU Data                            |02 20 16 15 e8 00 14...|
|Packet Data Convergence Protocol (PDCP), 3GPP TS 36.323              |
|V8.5.0 (2009-03) (LTE-PDCP) SIG-DATA (= Control plane PDCP            |
|Data PDU SRBs)   |                                                    |
|1 Control plane PDCP Data PDU SRBs                                    |
|reserved                                        |0                    |
|PDCP Sequence Number                            |2                    |
|DATA                                            |20 16 e8 00 14 00... |
|MAC-I (Message Authentication Code)             |'00000000'H          |
|RRC (DCCH DL) 3GPP TS 36.331 V8.5.0 (2009-03) (LTE-RRC_DCCH_DL)       |
|rrcConnectionReconfiguration (= rrcConnectionReconfiguration)   |     |
|dL-DCCH-Message                                                       |
|1 message                                                             |
|1.1 Standard                                                          |
|1.1.1 rrcConnectionReconfiguration                                    |
|1.1.1.1 rrc-TransactionIdentifier               |0                    |
|1.1.1.2 criticalExtensions                                            |
|1.1.1.2.1 c1                                                          |
|1.1.1.2.1.1 rrcConnectionReconfiguration-r8                           |
|1.1.1.2.1.1.1 measConfig                                              |
|1.1.1.2.1.1.1.1 measObjectToAddModList                                |
|1.1.1.2.1.1.1.1.1 measObjectToAddMod                                  |
|1.1.1.2.1.1.1.1.1.1 measObjectId                |1                    |
|1.1.1.2.1.1.1.1.1.2 measObject                                        |
|1.1.1.2.1.1.1.1.1.2.1 measObjectEUTRA                                 |
```

```
|1.1.1.2.1.1.1.1.1.2.1.1 carrierFreq                    |324                       |
|1.1.1.2.1.1.1.1.1.2.1.2 allowedMeasBandwidth           |mbw25                     |
|1.1.1.2.1.1.1.1.1.2.1.3 presenceAntennaPort1           |false                     |
|1.1.1.2.1.1.1.1.1.2.1.4 neighCellConfig                |'01'B                     |
|1.1.1.2.1.1.1.1.1.2.1.5 offsetFreq                     |dB0                       |
|1.1.1.2.1.1.1.1.1.2.1.6 cellsToAddModList              |                          |
|1.1.1.2.1.1.1.1.1.2.1.6.1 cellsToAddMod                |                          |
|1.1.1.2.1.1.1.1.1.2.1.6.1.1 cellIndex                  |1                         |
|1.1.1.2.1.1.1.1.1.2.1.6.1.2 physCellId                 |0                         |
|1.1.1.2.1.1.1.1.1.2.1.6.1.3                            |dB-24                     |
cellIndividualOffset
|1.1.1.2.1.1.1.2 reportConfigToAddModList               |                          |
|1.1.1.2.1.1.1.2.1 reportConfigToAddMod                 |                          |
|1.1.1.2.1.1.1.2.1.1 reportConfigId                     |6                         |
|1.1.1.2.1.1.1.2.1.2 reportConfig                       |                          |
|1.1.1.2.1.1.1.2.1.2.1 reportConfigEUTRA                |                          |
|1.1.1.2.1.1.1.2.1.2.1.1 triggerType                    |                          |
|1.1.1.2.1.1.1.2.1.2.1.1.1 event                        |                          |
|1.1.1.2.1.1.1.2.1.2.1.1.1.1 eventId                    |                          |
|1.1.1.2.1.1.1.2.1.2.1.1.1.1.1 eventA3                  |                          |
|1.1.1.2.1.1.1.2.1.2.1.1.1.1.1.1 a3-Offset              |3                         |
|1.1.1.2.1.1.1.2.1.2.1.1.1.1.1.2 reportOnLeave          |false                     |
|1.1.1.2.1.1.1.2.1.2.1.1.1.2 hysteresis                 |3                         |
|1.1.1.2.1.1.1.2.1.2.1.1.3 timeToTrigger                |ms256                     |
|1.1.1.2.1.1.1.2.1.2.1.2 triggerQuantity                |rsrp                      |
|1.1.1.2.1.1.1.2.1.2.1.3 reportQuantity                 |both                      |
|1.1.1.2.1.1.1.2.1.2.1.4 maxReportCells                 |3                         |
|1.1.1.2.1.1.1.2.1.2.1.5 reportInterval                 |ms480                     |
|1.1.1.2.1.1.1.2.1.2.1.6 reportAmount                   |infinity                  |
|1.1.1.2.1.1.1.3 measIdToAddModList                     |                          |
|1.1.1.2.1.1.1.3.1 measIdToAddMod                       |                          |
|1.1.1.2.1.1.1.3.1.1 measId                             |6                         |
|1.1.1.2.1.1.1.3.1.2 measObjectId                       |1                         |
|1.1.1.2.1.1.1.3.1.3 reportConfigId                     |6                         |
|1.1.1.2.1.1.1.4 quantityConfig                         |                          |
|1.1.1.2.1.1.1.4.1 quantityConfigEUTRA                  |                          |
|1.1.1.2.1.1.1.4.1.1 filterCoefficientRSRP              |fc4                       |
|1.1.1.2.1.1.1.4.1.2 filterCoefficientRSRQ              |fc4                       |
|1.1.1.2.1.1.1.5 measGapConfig                          |                          |
|1.1.1.2.1.1.1.5.1 release                              |0                         |
|1.1.1.2.1.1.1.6 s-Measure                              |0                         |
|1.1.1.2.1.1.1.7 speedStatePars                         |                          |
|1.1.1.2.1.1.1.7.1 release                              |0                         |
|1.1.1.2.1.1.2 dedicatedInfoNASList                     |                          |
|1.1.1.2.1.1.2.1 dedicatedInfoNAS                       |08 02 0f 34 34 30 30...   |
|1.1.1.2.1.1.3 radioResourceConfigDedicated             |                          |
|1.1.1.2.1.1.3.1 srb-ToAddModList                       |                          |
|1.1.1.2.1.1.3.1.1 sRB-ToAddMod                         |                          |
|1.1.1.2.1.1.3.1.1.1 srb-Identity                       |2                         |
|1.1.1.2.1.1.3.1.1.2 rlc-Config                         |                          |
|1.1.1.2.1.1.3.1.1.2.1 explicitValue                    |                          |
|1.1.1.2.1.1.3.1.1.2.1.1 am                             |                          |
|1.1.1.2.1.1.3.1.1.2.1.1.1 ul-AM-RLC                    |                          |
```

```
|1.1.1.2.1.1.3.1.1.2.1.1.1.1 t-PollRetransmit        |ms45                |
|1.1.1.2.1.1.3.1.1.2.1.1.1.2 pollPDU                 |pInfinity           |
|1.1.1.2.1.1.3.1.1.2.1.1.1.3 pollByte                |kBinfinity          |
|1.1.1.2.1.1.3.1.1.2.1.1.1.4 maxRetxThreshold        |t4                  |
|1.1.1.2.1.1.3.1.1.2.1.1.2 dl-AM-RLC                 |                    |
|1.1.1.2.1.1.3.1.1.2.1.1.2.1 t-Reordering            |ms35                |
|1.1.1.2.1.1.3.1.1.2.1.1.2.2 t-StatusProhibit        |ms0                 |
|1.1.1.2.1.1.3.1.1.3 logicalChannelConfig            |                    |
|1.1.1.2.1.1.3.1.1.3.1 explicitValue                 |                    |
|1.1.1.2.1.1.3.1.1.3.1.1 ul-SpecificParameters       |                    |
|1.1.1.2.1.1.3.1.1.3.1.1.1 priority                  |3                   |
|1.1.1.2.1.1.3.1.1.3.1.1.2 prioritisedBitRate        |infinity            |
|1.1.1.2.1.1.3.1.1.3.1.1.3 bucketSizeDuration        |ms100               |
|1.1.1.2.1.1.3.1.1.3.1.1.4 logicalChannelGroup       |0                   |
|1.1.1.2.1.1.3.2 drb-ToAddModList                    |                    |
|1.1.1.2.1.1.3.2.1 dRB-ToAddMod                      |                    |
|1.1.1.2.1.1.3.2.1.1 eps-BearerIdentity              |10                  |
|1.1.1.2.1.1.3.2.1.2 drb-Identity                    |10                  |
|1.1.1.2.1.1.3.2.1.3 pdcp-Config                     |                    |
|1.1.1.2.1.1.3.2.1.3.1 discardTimer                  |infinity            |
|1.1.1.2.1.1.3.2.1.3.2 rlc-AM                        |                    |
|1.1.1.2.1.1.3.2.1.3.2.1 statusReportRequired        |false               |
|1.1.1.2.1.1.3.2.1.3.3 headerCompression             |                    |
|1.1.1.2.1.1.3.2.1.3.3.1 notUsed                     |0                   |
|1.1.1.2.1.1.3.2.1.4 rlc-Config                      |                    |
|1.1.1.2.1.1.3.2.1.4.1 am                            |                    |
|1.1.1.2.1.1.3.2.1.4.1.1 ul-AM-RLC                   |                    |
|1.1.1.2.1.1.3.2.1.4.1.1.1 t-PollRetransmit          |ms80                |
|1.1.1.2.1.1.3.2.1.4.1.1.2 pollPDU                   |p128                |
|1.1.1.2.1.1.3.2.1.4.1.1.3 pollByte                  |kB250               |
|1.1.1.2.1.1.3.2.1.4.1.1.4 maxRetXThreshold          |t4                  |
|1.1.1.2.1.1.3.2.1.4.1.2 dl-AM-RLC                   |                    |
|1.1.1.2.1.1.3.2.1.4.1.2.1 t-Reordering              |ms80                |
|1.1.1.2.1.1.3.2.1.4.1.2.2 t-StatusProhibit          |ms60                |
|1.1.1.2.1.1.3.2.1.5 logicalChannelIdentity          |2                   |
|1.1.1.2.1.1.3.2.1.6 logicalChannelConfig            |                    |
|1.1.1.2.1.1.3.2.1.6.1 ul-SpecificParameters         |                    |
|1.1.1.2.1.1.3.2.1.6.1.1 priority                    |7                   |
|1.1.1.2.1.1.3.2.1.6.1.2 prioritisedBitRate          |infinity            |
|1.1.1.2.1.1.3.2.1.6.1.3 bucketSizeDuration          |ms100               |
|1.1.1.2.1.1.3.2.1.6.1.4 logicalChannelGroup         |2                   |
|1.1.1.2.1.1.3.3 mac-MainConfig                      |                    |
|1.1.1.2.1.1.3.3.1 explicitValue                     |                    |
|1.1.1.2.1.1.3.3.1.1. drx-Config                     |                    |
|1.1.1.2.1.1.3.3.1.1.1 setup                         |                    |
|1.1.1.2.1.1.3.3.1.1.1.1 onDurationTimer             |psf6                |
|1.1.1.2.1.1.3.3.1.1.1.2 drx-InactivityTimer         |psf1920             |
|1.1.1.2.1.1.3.3.1.1.1.3                             |psf16               |
drx-RetransmissionTimer
|1.1.1.2.1.1.3.3.1.1.1.4                             |                    |
longDRX-CycleStartOffset
|1.1.1.2.1.1.3.3.1.1.1.4.1 sf1280                    |14                  |
|1.1.1.2.1.1.3.3.1.2                                 |infinity            |
timeAlignmentTimerDedicated
```

```
|1.1.1.2.1.1.3.4 physicalConfigDedicated                               |
|1.1.1.2.1.1.3.4.1 uplinkPowerControlDedicated                         |
|1.1.1.2.1.1.3.4.1.1 p0-UE-PUSCH                    |0                  |
|1.1.1.2.1.1.3.4.1.2 deltaMCS-Enabled               |en0                |
|1.1.1.2.1.1.3.4.1.3 accumulationEnabled            |true               |
|1.1.1.2.1.1.3.4.1.4 p0-UE-PUCCH                    |0                  |
|1.1.1.2.1.1.3.4.1.5 pSRS-Offset                    |8                  |
|1.1.1.2.1.1.3.4.1.6 filterCoefficient              |fc4                |
|1.1.1.2.1.1.3.4.2                                                      |
soundingRS-UL-ConfigDedicated                                          |
|1.1.1.2.1.1.3.4.2.1 setup                                             |
|1.1.1.2.1.1.3.4.2.1.1 srs-Bandwidth                |bw0                |
|1.1.1.2.1.1.3.4.2.1.2 srs-HoppingBandwidth         |hbw0               |
|1.1.1.2.1.1.3.4.2.1.3 freqDomainPosition           |0                  |
|1.1.1.2.1.1.3.4.2.1.4 duration                     |true               |
|1.1.1.2.1.1.3.4.2.1.5 srs-ConfigIndex              |19                 |
|1.1.1.2.1.1.3.4.2.1.6 transmissionComb             |0                  |
|1.1.1.2.1.1.3.4.2.1.7 cyclicShift                  |cs1                |
|NAS LTE TS24.301 VS.1.0 (2009-03) (NAS-SECURED)  ???                   |
|(= Unknown EPS Message)                                                |
|1 Unknown EPS Message                                                  |
|unknownMsg_Value                          |08 02 0f 34 34 30 30...    |
```

In the MAC header, we have unchanged parameters compared to the other two RRC messages sent previously on the downlink and the message is sent in RLC AM on LCID = 1. The PDCP sequence number was increased to the value 2, because it is the third RRC message sent on the same logical channel in the same direction.

The first part of the RRC connection reconfiguration message deals with the setup of measurement tasks for the UE. In particular, the UE is ordered to measure the Received Signal Reference Power (RSRP) of the cell with physical cell ID = "0." This is the serving cell where the RRC connection was successfully established before.

The parameter *carrierFreq* is the absolute radio frequency channel number for the E-UTRA, commonly known as eARFCN or ARFCN E-UTRA. The downlink carrier frequency in megahertz for the cell to be measured by the UE can be calculated from the following equation where F_{DL_low} and N_{Offs_DL} are given in Table 3.3:

$$F_{DL} = F_{DL\_low} + 0.1(N_{DL} - N_{Offs-DL})$$

While the eARFCN defines the frequency of the cell to be measured, the measurement bandwidth parameter defines the reported measurement sampled from the individual values of a transmission bandwidth of maximum 25 active resource blocks allocated to this particular call. A graphical scheme of the transmission bandwidth is shown in Figure 3.4.

Cell Individual Offset is the cell-specific offset of a neighbor cell to be measured and reported. This parameter is used to define how much a neighbor cell is expected to be better reported than the serving cell or how much the serving cell is expected to be worse than a neighbor cell if the measurement report triggers a handover decision in the eNodeB. The value of −24 dB found in Message Example 3.5 is theoretically possible, but it is not a very realistic value for this parameter. Typically, offset values are expected to be in the range of −5 to 5 dB.

After setting the measurement conditions for the neighbor cell list (that in Message Example 3.5 consists of one cell only), specific event conditions are defined that should trigger UE measurement reports. In the message example, the UE should send A3 Event "Neighbor becomes offset better than serving" if the conditions defined in the section below the event name are fulfilled.

Table 3.3 E-UTRA channel numbers for downlink (according to 3GPP 36.101)

Band	Downlink		
	F_{DL_low}	$N_{Offs-DL}$	Range of N_{DL}
1	2110	0	0–599
2	1930	600	600–1199
3	1805	1200	1200–1949
4	2110	1950	1950–2399
5	869	2400	2400–2649
6	875	2650	2650–2749
7	2620	2750	2750–3449
8	925	3450	3450–3799
9	1844.9	3800	3800–4149
10	2110	4150	4150–4749
11	1475.9	4750	4750–4999
12	–	–	–
13	–	–	–
14	–	–	–
...			
33	1900	26,000	26,000–26,199
34	2010	26,200	26,200–26,349
35	1850	26,350	26,350–26,949
36	1930	26,950	26,950–27,549
37	1910	27,550	27,550–27,749
38	2570	27,750	27,750–28,249
39	1880	28,250	28,250–28,649
40	2300	28,650	28,650–29,649

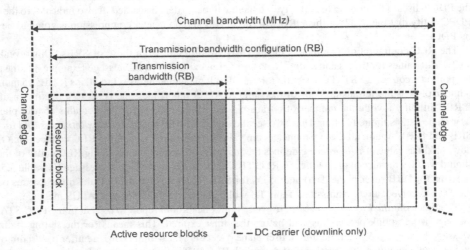

Figure 3.4 Transmission bandwidth (3GPP 36.104)

The information element *Hysteresis* is a parameter used within the entry and leaving condition of an event-triggered reporting condition. The actual value is the integer number seen in Message Example 3.5 multiplied by 0.5 dB. So the hysteresis value in this example is 1.5 dB. With a range of signaled values from 0 to 30, the maximum hysteresis is 15 dB.

Time-to-trigger determines that the trigger condition (e.g., RSRP threshold of the measured neighbor cell) must remain unchanged for 256 ms before the RRC measurement report is sent. This is to prevent the call from making ping-pong handovers due to quickly changing radio conditions, especially at the cell edge.

Trigger quantity defines that only the RSRP is measured. A maximum list of three neighbor cells including the serving call can be reported – if so many cells have been defined in the neighbor list sent previously to the UE.

In case no handover command is received from the eNodeB after sending the measurement report and the measurement conditions are still unchanged, the measurement report should be repeated after 480 ms and repetition should be performed as long as the radio conditions do not change without defining a maximum number of measurement reports to be sent.

The filter coefficient defines how the results measured on different subcarriers during a defined sampling period of, for example, 200 ms are aggregated into a single value that is reported to the eNodeB.

After this measurement setup section in Message Example 3.5, the information field that carries the NAS messages is embedded in the RRC connection reconfiguration message. The NAS messages themselves are not decoded in the message.

Then, we see that a second SRB is added to the active RRC connection. The RLC configuration of this new SRB is the same as before for SRB-1, but its uplink priority value = 3 indicates a lower priority compared to SRB-1. RRC can control the scheduling of uplink data by giving each logical channel a priority, where increasing priority values indicate lower priority levels.

Finally, the DRB that serves the default EPS bearer is requested to be established with the parameter found in the *drb-toAddModList* section. Here, we find the EPS Bearer ID = 10 and DRB Bearer ID = 10. For the PDCP, a discard timer value is set. When the discard timer expires for a user plane packet or fragment of a user plane packet to be transported by the PDCP, the UE should discard the entire PDCP PDU, which means the user plane information including the PDCP header. If the corresponding PDCP PDU has already been submitted to lower layers like RLC/MAC, the discard is indicated to these layers and the data may be discarded in lower layer entities, for example, in the RLC buffer. In Message Example 3.5, the PDCP discard timer is set to "infinity," so there will be no data discarded. It also indicates to the UE's PDPC entity that the RLC for this DRB will run in AM and no header compression is going to be used by PDCP.

The RLC configuration parameters of the DRB are different from those of SRBs. This reveals the analysis of values for timers and counter data correction in RLC AM. In Message Example 3.5, the timer for polling (requesting) STATUS reports for the DRB logical channel is set to 80 ms (SRB: 45 ms). The number of possible retransmissions of the same STATUS report message is set to a maximum of 128 (SRB: "infinity") while the number of retransmissions of an RLC PDU that contains control plane or user plane data is limited to four. Here, we see the same value for parameter *maxRetxThreshold* for both SRB and DRB. Again, all these parameters are valid for the uplink part of the RLC connection. On the downlink, the value of *t-Reordering* defines the maximum time that the receiving RLC entity of the UE should wait for the arrival of a particular RLC TB. According to the values in Message Example 3.5, this is 80 ms for DRB (user plane data) and 35 ms for SRB (control plane data). If the timer expires before the TB is received, a retransmission of this TB will be ordered by sending a STATUS message.

The timer *t-StatusProhibit* is used to trigger a buffer headroom report to be sent in the RLC STATUS message if no uplink packet was sent before the timer expired. This means for the settings found in Message Example 3.5 that if no uplink user plane packet is sent within 60 ms, a buffer headroom report will be sent containing the actual size of the uplink RLC buffer, which is expected to be equal to the size of the next RLC block to be sent.

LCID = 3 is assigned to the DRB and priority for the scheduler is set to seven, which means for this call, first SRB-1, then SRB-2, and lastly DRB RLC data will be scheduled on the uplink. Since the DRB carries the user plane, it is allocated to logical channel group 2.

Further parameters in the RRC connection reconfiguration message are timer values for Discontinuous Reception (DRX) as well as uplink power control parameters and sounding reference symbol configuration parameters that have been already discussed for the RRC connection setup message.

Now, completion of the procedures that have been requested to be performed on the handset side is expected and the first answer sent by the UE is the RRC security mode complete message, which is the message on top of the call flow diagram in Figure 3.5. Looking at the MAC parameters for the UL channel used to transport this message, it becomes evident that the modulation order of the uplink

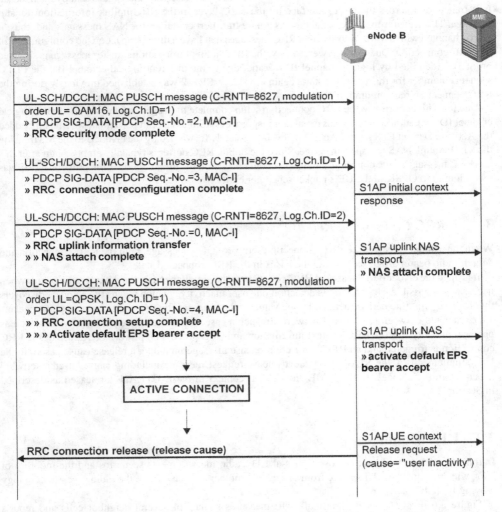

Figure 3.5 RRC connection setup 3/3 and RRC connection release. (*Source:* Tektronix Communications.)

radio transmission was changed from QPSK to 16QAM. On the PDCP layer, the sequence number was increased by 1. The RRC security mode complete message itself does not contain any parameters that need special explanations.

The next message we see is again transmitted on the UL-SCH. It is the RRC Connection Reconfiguration Complete that confirms that all requested parameters of the RRC connection including setup of the new radio bearer have been changed as requested by the eNodeB.

When the RRC Connection Reconfiguration Complete is received by the eNodeB, it is the trigger in the interworking function RRC/S1AP that the S1AP initial context response message should now be sent to the MME. Now, all security functions and bearers are in service, but on the NAS layer, the end-to-end connection setup is not yet complete.

If a new GUTI is assigned to the UE during the attach procedure, the successful reconfiguration of this temporary user identity will be confirmed when the NAS attach complete message is sent by the handset. It is transported across the radio interface by the RRC layer in the RRC uplink information transfer message. This transfer message is handled as a pure transport container for NAS messages and, thus, it can be handled with a lower priority than RRC messages that have a direct impact on the configuration of the RRC connection. Due to this lower priority, the RRC uplink information transfer message is signaled on SRB-2 identified by logical channel ID = 2 and, due to the different logical channel ID, the PDPC sequence number for this transport starts again at value 0. SRB-2 was established previously during the RRC connection reconfiguration procedure.

Due to its higher priority, the RRC connection setup complete message is sent again on SRB-1 (logical channel ID = 1), although this message also transports a NAS message. In particular, we find embedded the Activate Default EPS Bearer Accept that is immediately forwarded by the eNodeB to the MME using the S1AP uplink NAS transport message. Since the uplink RLC buffer is typically empty again after this last RRC message of the connection setup procedure was sent, the uplink modulation order on MAC was reconfigured to QPSK in the UE's packet scheduler entity.

3.1.3   RRC Connection Release

When the eNodeB receives the RRC connection setup complete message, both the RRC connection and default radio bearer are in service and the UE is in the RRC connected state.

It is up to the eNodeB to release the RRC connection and simultaneously the default radio bearer when necessary. A possible trigger point is detection on "user inactivity," which means the eNodeB discovers that for a certain time period (defined by the value of the "user inactivity" timer), no user plane packets are exchanged between the UE and network. In such a case, to save resources in the eNodeB, such as physical memory to store RRC context information and CPU processing power, the release of the RRC connection is triggered and an RRC connection release message including a release cause is sent to the UE. Simultaneously, the S1AP UE context release request message including cause "user inactivity" is sent from the eNodeB to the MME and the UE context in the E-UTRAN is released as described in Section 2.3.

3.2   LTE Mobility

In this chapter, we will have a closer looks at LTE radio measurements scenarios and the mobility of UEs, which means: how UEs move from one cell to another or even change the radio access technology moving from 4G to 3G or 2G.

Figure 3.6 illustrates how a particular UE measures signals of several neighbor cells and reports their signal strength to the eNodeB that hosts the serving cell. This eNodeB is responsible for setting

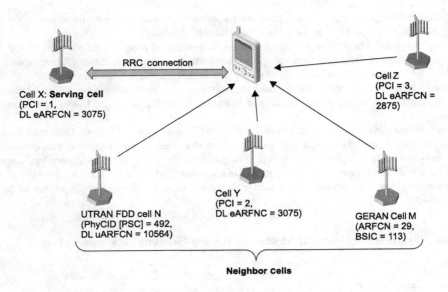

Figure 3.6 LTE neighbor cell measurements

up all measurement jobs, receives all measurement reports, and decides about handovers execution. In E-UTRAN, there is no centralized radio resource management in a network element like RNC (3G UTRAN) or BSC (2G GERAN) and even the Mobility Management Entity (MME) only assists the eNodeB in handover preparation and execution in case of inter-eNodeB and inter-RAT handovers.

The different types of measurement jobs can be characterized according to the type of the reported neighbor cells:

Cell Y is called an intra-frequency neighbor that has a different Physical Cell ID (PCI = 2) but sends its signals on the same carrier frequency (identified by eARFCN = 3075). Measurement Reports indicating the signal strength of cell Y are called *intra-frequency measurement* report.

Cell Z has a different Physical Cell ID (PCI = 3) and works in addition on a different carrier frequency (eARFCN = 2875). Thus, measurement reports for this cell are called *inter-frequency measurement* reports.

Measurement Reports containing measurement results for cell N or M are called *inter-RAT measurement* reports, because the neighbors are working on a different radio access technology.

Cell N is a 3G UMTS neighbor cell working in FDD mode. It sends its signals on the downlink frequency identified by uARFCN = 10,564 with Primary Scrambling Code (PSC) = 492. In 3GPP LTE standard documents, the primary scrambling code is also named "physical call ID" in analogy to the PCI of LTE cells. In fact, also the LTE physical cell ID is a scrambling code.

Cell M is a 2G GERAN neighbor. Within a geographical cluster of GSM cells, it has a unique frequency identified by ARFCN = 29. Additional parameters for unique identification of GSM cells are Base Station Color Code (BCC) and Network Color Code (NCC) that form together the BSIC (Base Station Identity Code).

When talking about design options for inter-frequency and inter-RAT measurements, the entire design of a UE needs to be considered. Measurement of the radio signal strength of a different frequency requires a second radio receiver chain in the UE or – if it has just one receiver – the ongoing radio connection must be interrupted for a short time so that the only radio transceiver unit can be used to measure on different frequency or RAT. Sure, these interruptions of the ongoing radio transmission need to be precisely synchronized between the UE and base station (eNodeB).

In the 3G specification, these interruptions for measurement purposes are called "compressed mode." Now in 4G standards, the same concept applies, but instead of "compressed mode," the standard documents use the terms "idle periods" and "Tx/Rx gaps" to describe the mechanism.

The disadvantage of "compressed mode" measurements is already known from 3G: the continuous switch on/off of the ongoing radio connection bears the risk of call drops and has an impact on service quality.

On the other hand, a second radio receiver makes the UE more "power-hungry." Thus, the UE's battery is running out of power faster. It also makes the UE more expensive due to additional hardware.

Knowing this, the idle period/gap concept is the preferred choice for most handset vendors – especially because simultaneous operations in two RATs are not considered in the 4G standard documents.

An example of idle period/gap parameter definition from RRC signaling is shown in Message Example 3.6.

Message Example 3.6: Inter-RAT Measurement Setup Including Measurement Gap

```
+-------------------------------------------------------+---------------------+
|ID Name                                                |Comment or Value     |
+-------------------------------------------------------+---------------------+
|RRC (DCCH DL) LTERRC 3GPP TS 36.331 V9.5.0 (2010-12) (LTE-RRC_DCCH_DL)
rrcConnectionReconfiguration (= rrcConnectionReconfiguration)                  |
|dL-DCCH-Message                                        |                     |
|1 message                                              |                     |
|1.1 Standard                                           |                     |
|1.1.1 rrcConnectionReconfiguration                     |                     |
|1.1.1.1 rrc-TransactionIdentifier                      |0                    |
|1.1.1.2 criticalExtensions                             |                     |
|1.1.1.2.1 c1                                           |                     |
|1.1.1.2.1.1 rrcConnectionReconfiguration-r8            |                     |
|1.1.1.2.1.1.1 measConfig                               |                     |
|1.1.1.2.1.1.1.1 measObjectToAddModList                 |                     |
|1.1.1.2.1.1.1.1.1 measObjectToAddMod                   |                     |
|1.1.1.2.1.1.1.1.1.1 measObjectId                       |2                    |
|1.1.1.2.1.1.1.1.1.2 measObject                         |                     |
|1.1.1.2.1.1.1.1.1.2.1 measObjectUTRA                   |                     |
|1.1.1.2.1.1.1.1.1.2.1.1 carrierFreq                    |10462                |
|1.1.1.2.1.1.1.1.1.2.1.2 offsetFreq                     |0                    |
|1.1.1.2.1.1.1.1.1.2.1.3 cellsToAddModList              |                     |
|1.1.1.2.1.1.1.1.1.2.1.3.1 cellsToAddModListUTRA-FDD    |                     |
|1.1.1.2.1.1.1.1.1.2.1.3.1.1 cellsToAddModUTRA-FDD      |                     |
|1.1.1.2.1.1.1.1.1.2.1.3.1.1.1 cellIndex                |1                    |
|1.1.1.2.1.1.1.1.1.2.1.3.1.1.2 physCellId               |354                  |
|1.1.1.2.1.1.1.1.1.2.1.3.1.2 cellsToAddModUTRA-FDD      |                     |
|1.1.1.2.1.1.1.1.1.2.1.3.1.2.1 cellIndex                |2                    |
|1.1.1.2.1.1.1.1.1.2.1.3.1.2.2 physCellId               |17                   |
|1.1.1.2.1.1.1.1.1.2.1.3.1.3 cellsToAddModUTRA-FDD      |                     |
|1.1.1.2.1.1.1.1.1.2.1.3.1.3.1 cellIndex                |3                    |
|1.1.1.2.1.1.1.1.1.2.1.3.1.3.2 physCellId               |75                   |
|1.1.1.2.1.1.1.1.1.2.1.3.1.4 cellsToAddModUTRA-FDD      |                     |
|1.1.1.2.1.1.1.1.1.2.1.3.1.4.1 cellIndex                |4                    |
|1.1.1.2.1.1.1.1.1.2.1.3.1.4.2 physCellId               |214                  |
|1.1.1.2.1.1.1.1.1.2.1.3.1.5 cellsToAddModUTRA-FDD      |                     |
```

```
|1.1.1.2.1.1.1.1.1.2.1.3.1.5.1 cellIndex                    |5                       |
|1.1.1.2.1.1.1.1.1.2.1.3.1.5.2 physCellId                   |115                     |
|1.1.1.2.1.1.1.1.1.2.1.3.1.6 cellsToAddModUTRA-FDD          |                        |
|1.1.1.2.1.1.1.1.1.2.1.3.1.6.1 cellIndex                    |6                       |
|1.1.1.2.1.1.1.1.1.2.1.3.1.6.2 physCellId                   |136                     |
|1.1.1.2.1.1.1.1.1.2.1.3.1.7 cellsToAddModUTRA-FDD          |                        |
|1.1.1.2.1.1.1.1.1.2.1.3.1.7.1 cellIndex                    |7                       |
|1.1.1.2.1.1.1.1.1.2.1.3.1.7.2 physCellId                   |251                     |
|1.1.1.2.1.1.1.1.1.2.1.3.1.8 cellsToAddModUTRA-FDD          |                        |
|1.1.1.2.1.1.1.1.1.2.1.3.1.8.1 cellIndex                    |8                       |
|1.1.1.2.1.1.1.1.1.2.1.3.1.8.2 physCellId                   |41                      |
|1.1.1.2.1.1.1.1.1.2.1.3.1.9 cellsToAddModUTRA-FDD          |                        |
|1.1.1.2.1.1.1.1.1.2.1.3.1.9.1 cellIndex                    |9                       |
|1.1.1.2.1.1.1.1.1.2.1.3.1.9.2 physCellId                   |498                     |
|1.1.1.2.1.1.1.1.1.2.1.3.1.10 cellsToAddModUTRA-FDD         |                        |
|1.1.1.2.1.1.1.1.1.2.1.3.1.10.1 cellIndex                   |10                      |
|1.1.1.2.1.1.1.1.1.2.1.3.1.10.2 physCellId                  |130                     |
|1.1.1.2.1.1.1.1.1.2.1.3.1.11 cellsToAddModUTRA-FDD         |                        |
|1.1.1.2.1.1.1.1.1.2.1.3.1.11.1 cellIndex                   |11                      |
|1.1.1.2.1.1.1.1.1.2.1.3.1.11.2 physCellId                  |379                     |
|1.1.1.2.1.1.1.1.1.2.1.3.1.12 cellsToAddModUTRA-FDD         |                        |
|1.1.1.2.1.1.1.1.1.2.1.3.1.12.1 cellIndex                   |12                      |
|1.1.1.2.1.1.1.1.1.2.1.3.1.12.2 physCellId                  |462                     |
|1.1.1.2.1.1.1.2 reportConfigToAddModList                   |                        |
|1.1.1.2.1.1.1.2.1 reportConfigToAddMod                     |                        |
|1.1.1.2.1.1.1.2.1.1 reportConfigId                         |5                       |
|1.1.1.2.1.1.1.2.1.2 reportConfig                           |                        |
|1.1.1.2.1.1.1.2.1.2.1 reportConfigInterRAT                 |                        |
|1.1.1.2.1.1.1.2.1.2.1.1 triggerType                        |                        |
|1.1.1.2.1.1.1.2.1.2.1.1.1 event                            |                        |
|1.1.1.2.1.1.1.2.1.2.1.1.1.1 eventId                        |                        |
|1.1.1.2.1.1.1.2.1.2.1.1.1.1.1 eventB1                      |                        |
|1.1.1.2.1.1.1.2.1.2.1.1.1.1.1.1 b1-Threshold               |                        |
|1.1.1.2.1.1.1.2.1.2.1.1.1.1.1.1.1 b1-ThresholdUTRA         |                        |
|1.1.1.2.1.1.1.2.1.2.1.1.1.1.1.1.1.1 utra-RSCP              |-101 dBm <= CPICH       |
|                                                           RSCP < -100 dBm         |
|                                                           |                        |
|1.1.1.2.1.1.1.2.1.2.1.1.1.2 hysteresis                     |0                       |
|1.1.1.2.1.1.1.2.1.2.1.1.1.3 timeToTrigger                  |ms640                   |
|1.1.1.2.1.1.1.2.1.2.1.2 maxReportCells                     |3                       |
|1.1.1.2.1.1.1.2.1.2.1.3 reportInterval                     |ms480                   |
|1.1.1.2.1.1.1.2.1.2.1.4 reportAmount                       |infinity                |
|1.1.1.2.1.1.1.3 measIdToAddModList                         |                        |
|1.1.1.2.1.1.1.3.1 measIdToAddMod                           |                        |
|1.1.1.2.1.1.1.3.1.1 measId                                 |5                       |
|1.1.1.2.1.1.1.3.1.2 measObjectId                           |2                       |
|1.1.1.2.1.1.1.3.1.3 reportConfigId                         |5                       |
|1.1.1.2.1.1.1.4 quantityConfig                             |                        |
|1.1.1.2.1.1.1.4.1 quantityConfigUTRA                       |                        |
|1.1.1.2.1.1.1.4.1.1 measQuantityUTRA-FDD                   |cpich-RSCP              |
|1.1.1.2.1.1.1.4.1.2 measQuantityUTRA-TDD                   |pccpch-RSCP             |
|1.1.1.2.1.1.1.4.1.3 filterCoefficient                      |fc6                     |
|1.1.1.2.1.1.1.5 measGapConfig                              |                        |
|1.1.1.2.1.1.1.5.1 setup                                    |                        |
|1.1.1.2.1.1.1.5.1.1 gapOffset                              |                        |
|1.1.1.2.1.1.1.5.1.1.1 gp0|                                 |11                      |
```

Here, the gap offset value is used to identify the subframe during which the 4G radio transmission shall be suspended and inter-RAT measurement shall be performed.

Other important parameters in this measurement setup are the UMTS carrier frequency identified by its uARFCN, the list of 3G neighbors to be measured (physical cell ID here means: primary scrambling code of the cell), and the section where event B1 reporting is specified.

The B1 event will be triggered if the RSCP of a 3G cell is measured better than −100 dBm. Then, the RSCP of maximum 3 UMTS neighbors will be reported with a periodicity of 480 milliseconds until eNodeB makes a handover or blind redirect decision.

3.2.1 Intra-eNB Intra-Frequency HO

Figure 3.7 shows which Uu signaling messages are involved in any intra-LTE handover procedures (4G-4G handover).

It is the RRC Connection Reconfiguration message that is used as the "handover command" containing the target cell's physical cell ID and target carrier frequency (eARFCN). Also a new c-RNTI, the unique temporary identity of the UE within the target cell after successful handover is included.

After the UE receives the RRC Connection Reconfiguration messages with the handover parameters, it leaves its serving cell and performs random access procedure in the target cell. This is something never seen before in mobile access networks. In GERAN and UTRAN, the random access procedure is only performed when UE does cell (re-)selection, but not during handover. Now, thanks to the benefits of Zadoff–Chu random access algorithm that allows very quick radio access, the first signal sent by the UE in the target cell is the random access preamble with the new c-RNTI.

The eNodeB will answer with a MAC Random Access Response message including the same new c-RNTI and finally the UE sends RRC Connection Reconfiguration Complete message to confirm the successful handover on access stratum signaling level.

Figure 3.8 shows the full call flow diagram of the intra-eNodeB handover including the measurement configuration (a3-offset: 3 dB + hysteresis: 2 dB means the neighbor cell must be measured 5 dB better than serving cell to trigger the start of the handover procedure in the eNodeB).

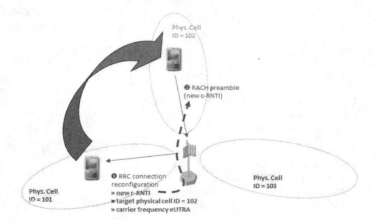

Figure 3.7 Intra-frequency intra-eNodeB handover

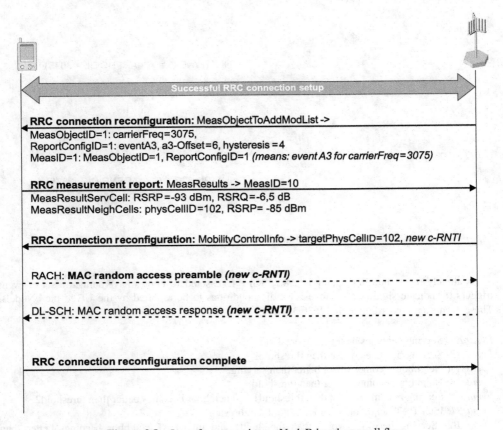

Figure 3.8 Intra-frequency intra-eNodeB handover call flow

3.2.2 Intra-eNodeB Inter-Frequency Handover

In general, 4G-4G handover scenarios are classified as intra- and inter-eNodeB handover. In case of intra-eNodeB handover, the target cell is located in the same eNodeB as the serving cell. In case of inter-eNodeB handover, the target cell is located in a different eNodeB. If the two eNodeBs involved in an inter-eNodeB handover are connected via X2 interface, the inter-eNodeB handover procedure can be classified as X2 Handover. If no X2 connection is available, the S1 handover procedure will apply.

In Figure 3.9, the overview of an intra-eNodeB handover is shown. The UE changes its serving cell from Cell X to Cell Y. Due to the fact that both cells operate on different frequencies indicated by different downlink eARFCN values, the handover can be also described as an inter-frequency handover. Thus, the combined name intra-eNodeB inter-frequency handover applies.

After successful RRC Connection Setup, the measurement reporting of the UE will be configured by the eNodeB using the RRC signaling connection.

In fact, any RRC message sent in downlink direction can be used to transmit measurement settings. In live network environment, typically the RRC Connection Reconfiguration message is used as shown in Figure 3.8.

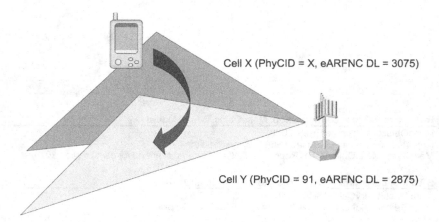

Figure 3.9   Intra-LTE inter-frequency handover

As already mentioned in Chapter 1.12.8.3, 3GPP LTE standard bodies have defined a set of event triggers to indicate significant changes of radio conditions to be reported by the UE to the eNodeB. These events are (according to 3GPP 36.331):

Event A1: Serving becomes better than threshold.
Event A2: Serving becomes worse than threshold.
Event A3: Neighbor becomes offset better than serving.
Event A4: Neighbor becomes better than threshold.
Event A5: Serving becomes worse than threshold1 and neighbor becomes better than threshold2.
Event B1: Inter-RAT neighbor becomes better than threshold.
Event B2: Serving becomes worse than threshold1 and inter-RAT neighbor becomes better than threshold2.

The serving cell is the one used for the ongoing RRC connection. In case a measurement report triggers a handover, the serving cell will be the source cell of the handover procedure, while the neighbor cell will be the target cell of the handover procedure.

Different from the measurement events used in 3G, the LTE event ID (A1, A2, etc.) itself will not be present in the RRC Measurement Reports, but only a Measurement ID (MeasID). This Measurement ID refers to a combination of measurement object (e.g., a particular carrier frequency or cell) and a report configuration that is typically determined by the measurement event. How a valid set of Measurement IDs for a particular UE are configured can be seen in first RRC Connection Reconfiguration message of Figure 3.a. The Measurement-Object-to-Add-or-Modify-List in this measurement defines two different carrier frequencies as measurement objects, six different report configurations, and three different Measurement IDs for the UE.

The carrier frequencies are identified on behalf of their downlink eARFCN value. How the real frequency value can be derived from the eARFCN is explained in Section 3.1.1.

In the report configuration, identical events with different parameter values can be configured. In the example, we see events A3 in ReportConfigID = 1 and 4 with different a3-offset. The a3-Offset value is a factor to be multiplied with 0.5 dB. Thus, a3-offset = 6 means an offset value of 3 dB while a3-offset = 12 equals an offset value of 6 dB. The a3-offset can be seen as a kind of "handover margin" – how stronger the neighbor cell shall be than the serving cell to trigger the measurement report.

If the hysteresis value is not zero (as in the example), this hysteresis offset needs to be added to the a3-offset to get the total handover margin value. Example: If the a3-offset = 3 dB and hysteresis value = 4 * 0.5 dB = 2 dB, then the total handover margin is 5 dB and the A3 reporting will be triggered if a neighbor cell is measured 5 dB better than the serving cell.

There are also two different configurations for measurement event A5. In ReportConfigID = 2, the serving cell's RSRP must fall below –96 dBm while simultaneously the neighbor cell's RSRP must exceed –86 dBm to trigger the measurement report. In ReportConfigID = 3, the RSRP measured for the neighbor cell must only be better than –91 dBm.

All dBm values in these report configurations refer to the Reference Signal Received Power (RSRP). For the definition of RSRP, see Section 4.3.1.

Report configuration number 5 defines an A2 event, which means: the eNodeB shall receive a measurement report in case the serving cell's RSRP falls below –91 dBm no matter if a better neighbor cell is around or not.

Finally, report configuration number 6 will order the UE to send a measurement report in case the serving cell's RSRP exceeds –86 dBm. This will be evidence that the criteria for the A1 event are met and the serving cell's RSRP is excellent.

However, not all the measurement report configurations will be activated immediately. Looking at the list of Measurement IDs sent to the UE, it can be seen that there are only three combinations of measurement object and report configuration activated in the first step.

Measurement ID = 1 will indicate that event A3 was triggered for carrier frequency with downlink eARFCN = 3075, Measurement ID = 2 will indicate event A5, and Measurement ID = 7 will be used to report the event A2, all for the same downlink eARFCN.

In the next step, the RRC Connection Reconfiguration Complete message confirms that the UE received and activated the measurement configuration as requested by the eNodeB.

The first RRC Measurement Report contains RSRP and RSRQ measured for the serving cell as well as the Measurement ID = 7. Looking back into the measurement configuration, it becomes evident that Measurement ID = 7 refers to an event A2 triggered on carrier frequency with downlink ARFCN = 3075. This event indicates that the radio quality of the serving cell is insufficient and a handover candidate needs to be found.

What comes next is another RRC Connection Reconfiguration procedure to modify the measurement settings in the UE. The measurement with Measurement ID = 7 is removed from the UE's measurement task list and new measurement tasks are added:

Measurement ID = 5 defines the event A1 for the carrier frequency with downlink eARFNC = 2875.
Measurement ID = 12 defines the event A5 for the carrier frequency with downlink eARFNC = 2875.
Measurement ID = 13 defines the event A3 with an a3-offset of +10 dB for the carrier frequency with downlink eARFNC = 2875.

Again, RRC Connection Reconfiguration Complete message confirms that the UE received and activated the new measurement settings.

Shortly after sending the RRC Connection Reconfiguration Complete message, the UE sends the next RRC Measurement Report, because the general radio conditions have not changed. Now it is Measurement ID = 13 that was triggered, which means: a neighbor cell on carrier frequency with downlink eARFCN = 2875 was measured with +10 dB more signal strength than the serving cell of the radio connection. Besides the Measurement ID and RSPR/RSRQ values measured for the serving cell, this Measurement Report now also contains the measured RSRP of the neighbor cell and this neighbor cell's physical cell ID (PCI = 91). Comparing the RSRP of the neighbor cell that was measured at −75 dBm and the RSRP of the serving cell measured at −93 dBm, it can be confirmed that the A3 criteria was indeed met: the difference in RSRP between serving and neighbor cells is in fact 18 dB. This is a clear handover trigger.

Hence, the eNodeB decides to execute a handover to the target Physical Cell ID = 91, which is the previously reported neighbor cell. It is once again the RRC Connection Reconfiguration message that is sent to the UE, this time including a "MobilityControlInfo" protocol information section. Whenever this "MobilityControlInfo" is found in an RRC Connection Reconfiguration message, this means: the message is sent to the UE as a handover command. Looking at the eARFCN values in the measurement settings, it can be further analyzed that the type of handover in this call flow is "inter-frequency handover."

The UE will now perform the requested change in the radio connection, and in case of successful handover, the RRC Connection Reconfiguration Complete message will be seen as first RRC message of this new radio connection in the handover target (neighbor) cell.

It should be noticed here that there is no explicit "handover command" message in the LTE RRC. All modifications of the radio connection including intra-frequency and inter-frequency handover are signaled to the UE using the RRC Connection Reconfiguration message (Figure 3.10).

Figure 3.11 shows another RRC connection with a different measurement configuration. It should be noticed that RRC measurement configuration parameters are set by the network operator's engineers and default parameters used immediately after initial deployment often strictly follow the eNodeB vendor's recommendations.

The first significant difference in the RRC Connection Reconfiguration message after successful RRC connection setup is that the measurement object with MeasObjectID = 1 refers not only to the carrier frequency, but also to a particular physical cell ID. This PCI = 72 is the PCI of the serving cell.

In the next section of the message, three report configurations are defined with parameters that trigger the event A3, A1, or A2. In the Measurement IDs, these events are all linked to the physical cell ID = 72 working on carrier frequency with downlink eARFCN = 1275.

These settings are sent to the UE and the UE confirms reception and activation of these measurement tasks by sending the RRC Connection Reconfiguration Complete message.

Shortly after that, the UE sends its first RRC Measurement Report reporting that the A1 criterion for the serving cell was met. In other words, excellent radio conditions for the serving cell have been measured and no further action is required on eNodeB side.

This changes when event A2 is triggered after a while and also reported to eNodeB. The RSRP of the serving cell is now at −108 dBm, which is below the A2 threshold of −105 dBm.

Triggered by the reported A2 event, the eNodeB sets up new additional measurement tasks. In particular, the UE is ordered to measure possible handover candidates on carrier frequencies identified by downlink eARFNC values 6300 and 3075. The event A4 shall be reported if the RSRP of one of these inter-frequency neighbors becomes better than −105 dBm.

In Figure 3.12, the UE confirms the acceptance of the new measurement configuration by sending the already known RRC Connection Reconfiguration Complete message.

Then, the next RRC Measurement Report is seen in the message flow of figure 3.12 containing the Measurement ID = 5 that refers to event A4 for carrier frequency 3075. The measured serving cell's RSRP is −109 dBm now, serving cell's RSRQ is −6.5 dB, and the measured RSRP of the inter-frequency neighbor is −79 dBm. In addition to the neighbor's RSRP, its physical cell ID "330" is reported.

Now, another new reporting function of the LTE RRC comes in: the CGI reporting. CGI is the acronym of Cell Global Identity and as explained in Chapter 1, the CGI is a worldwide unique identifier.

Using another RRC Connection Reconfiguration procedure, the eNodeB orders the UE to continue with periodical (not event-triggered) measurements on carrier frequency with downlink eARFCN 3075 and report the CGI that can be read from the broadcast information of the cells.

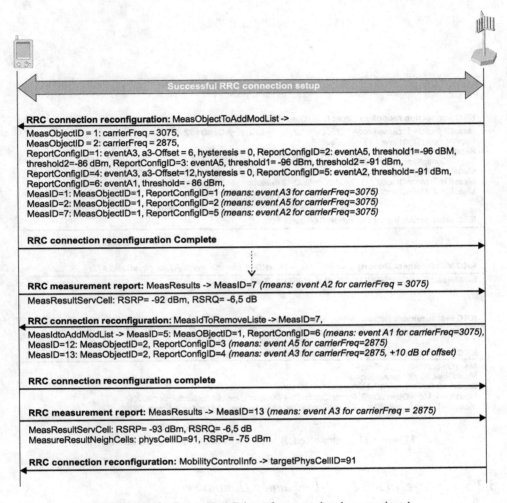

RRC connection reconfiguration: MeasObjectToAddModList ->

MeasObjectID = 1: carrierFreq = 3075,
MeasObjectID = 2: carrierFreq = 2875,
ReportConfigID=1: eventA3, a3-Offset = 6, hysteresis = 0, ReportConfigID=2: eventA5, threshold1=-96 dBM,
threshold2=-86 dBm, ReportConfigID=3: eventA5, threshold1= -96 dBm, threshold2= -91 dBm,
ReportConfigID=4: eventA3, a3-Offset=12,hysteresis = 0, ReportConfigID=5: eventA2, threshold=-91 dBm,
ReportConfigID=6: eventA1, threshold= - 86 dBm,
MeasID=1: MeasObjectID=1, ReportConfigID=1 *(means: event A3 for carrierFreq=3075)*
MeasID=2: MeasObjectID=1, ReportConfigID=2 *(means: event A5 for carrierFreq=3075)*
MeasID=7: MeasObjectID=1, ReportConfigID=5 *(means: event A2 for carrierFreq=3075)*

RRC connection reconfiguration Complete

RRC measurement report: MeasResults -> MeasID=7 *(means: event A2 for carrierFreq = 3075)*

MeasResultServCell: RSRP= -92 dBm, RSRQ= -6,5 dB

RRC connection reconfiguration: MeasIdToRemoveListe -> MeasID=7,

MeasIdtoAddModList -> MeasID=5: MeasOBjectID=1, ReportConfigID=6 *(means: event A1 for carrierFreq=3075),*
MeasID=12: MeasObjectID=2, ReportConfigID=3 *(means: event A5 for carrierFreq=2875)*
MeasID=13: MeasObjectID=2, ReportConfigID=4 *(means: event A3 for carrierFreq=2875, +10 dB of offset)*

RRC connection reconfiguration complete

RRC measurement report: MeasResults -> MeasID=13 *(means: event A3 for carrierFreq = 2875)*

MeasResultServCell: RSRP= -93 dBm, RSRQ= -6,5 dB
MeasureResultNeighCells: physCellID=91, RSRP= -75 dBm

RRC connection reconfiguration: MobilityControlInfo -> targetPhysCellID=91

Figure 3.10   Intra-eNodeB inter-frequency handover, variant 1

The next RRC Measurement Report in the call flow chart is such a periodical Measurement Report and from its details, it can be seen that the cell with physical cell ID = 330 belongs to eNode-BID = 909091/Cell-ID = 0 in the operator's network that is identified by MCC/MNC combination. The tracking area this cell belongs to is identified by the Tracking Area Code (TAC) = 3.

Having now all necessary information at hand, the eNodeB decides that the UE shall perform inter-frequency handover to this previously reported target cell. The Mobility Control Info in the appropriate RRC Connection Reconfiguration message contains the target physical cell ID = 330 and the downlink carrier eARFNC = 3075.

When the UE's RRC Connection Reconfiguration Complete message is received in the target cell (no matter if target cell is in same or different eNodeB), the inter-frequency handover is successfully completed.

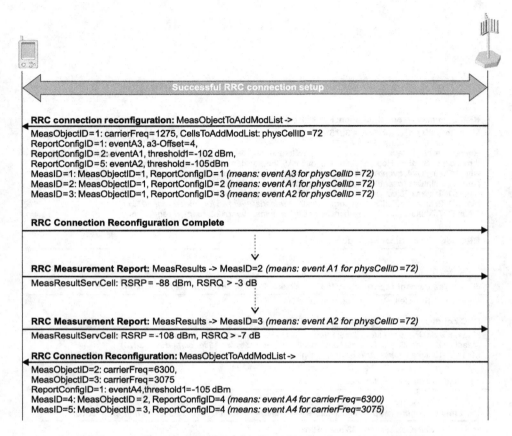

Figure 3.11   Intra-eNodeB inter-frequency handover, variant 2 (1/2)

3.2.3   Inter-eNodeB Intra-Frequency Handover

As already described earlier, there are two options for executing inter-eNodeB handover: the X2 handover and the S1 handover procedure. S1 handover is used if no X2 connectivity is provided.

In fact, X2 handover is a Self-Optimizing Network (SON) feature, because the X2 interface is a virtual connection running over the IP transport network that in a properly designed SON environment can be established on-demand by the serving eNodeB. However, it may also happen that the IP-connectivity between different geographical clusters of eNodeB is limited, for example, for security reasons (firewalls, etc.). In such cases, no X2 interface is available, and thus, S1 handover must be performed.

The overview picture in Figure 3.13 shows a UE that is moving quickly in RRC Connected Mode from one cell X to cell Y and then to cell Z. While there is X2 connectivity between the base stations serving cell X and Y, there is no X2 between eNodeBs of cell Y and cell Z.

So, when the first handover trigger measurement report is received in step 1 after successful establishment of the RRC connection and Measurement Configuration, a X2 Handover Preparation procedure (step 2) is triggered. The purpose of this X2 handover preparation is twofold: at one hand, the neighbor eNodeB assigns radio resources for the UE and creates a handover command message to be sent to the

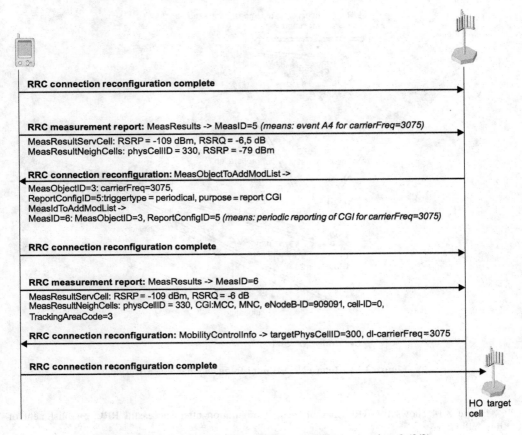

Figure 3.12   Intra-eNodeB inter-frequency handover, variant 2 (2/2)

UE (step 3). On the other hand, a temporary IP-tunnel for forwarding unsent downlink data packets is established between the serving and neighbor eNodeB and the unsent IP packets are transmitted using this temporary tunnel to the target eNodeB.

After the UE arrived successfully in the target cell, the target eNodeB performs the S1 Path Switch procedure (for details, see chapter 2.6) to allocate a new downlink IP address and tunnel endpoint identifier to the S1-U IP-tunnel for user plane transport.

In step 4, another RRC Measurement Report is sent by the UE that indicates that a better cell was found for the connection. This time the new source eNodeB starts the S1 Handover procedure that has the same functionality as X2 handover preparation: handover command is created by target eNodeB and sent to UE via S1 control plane and radio connection still active in serving cell. Temporary IP-tunnels are established for the purpose of data forwarding. Only difference: due to the nature of the S1 interface, two temporary IP-tunnels (one from source eNodeB to S-GW and one from S-GW to target eNodeB) need to be established while in case of X2 handover, one temporary tunnel was sufficient.

In step 6, the new S1-U tunnel between S-GW and target eNodeB is established using the S1 Handover Resource Allocation procedure and simultaneously the handover command is sent over S1 interface to the source eNodeB that will transmit it over radio interface of the UE.

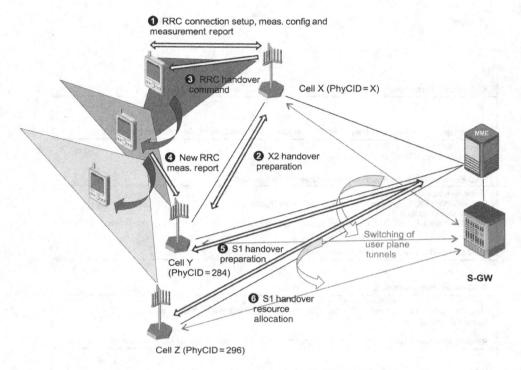

Figure 3.13   Inter-eNodeB intra-frequency handover overview

Figure 3.14 shows the RRC measurement configuration after successful RRC establishment of this call.

Compared to the measurement configurations that have been discussed in the previous sub-chapters, a reporting configuration is activated in this call: ReportConfigID = 3 that is linked to MeasID = 9 defines a periodical reporting of the strongest cells where the maximum number of reported cells is six and the measured signal level is RSRP. Report interval defines that such a periodical measurement report for strongest cells is sent every 10,240 milliseconds and the report amount limits the number consecutive periodical measurement reports to 16. This means: after 16 periodical RRC Measurement Reports have been sent, the UE stops periodical reporting unless a new measurement configuration for periodical reporting is requested to be activated by the eNodeB.

After RRC Connection Reconfiguration Complete that is sent by UE to confirm activation of the measurement, the UE sends the first periodical RRC Measurement Report. It contains measurement results (RSRP and RSRQ) for the serving cell and two neighbor cells identified by their physical cell ID.

After approximately 10 seconds, the next periodical report is received. This time, measurement results for three neighbor cells are reported.

All periodical measurement reports can be identified on behalf of MeasID = 9.

The other measurement tasks identified by other MeasID values are running simultaneously in the UE software. There is no correlation between them and the periodical reporting.

The MeasID = 1 that is seen in first RRC Measurement Report in Figure 3.15 is tied to the A3 event for carrier frequency with eARFCN = 3075. The A3 event is triggered because the neighbor cell with

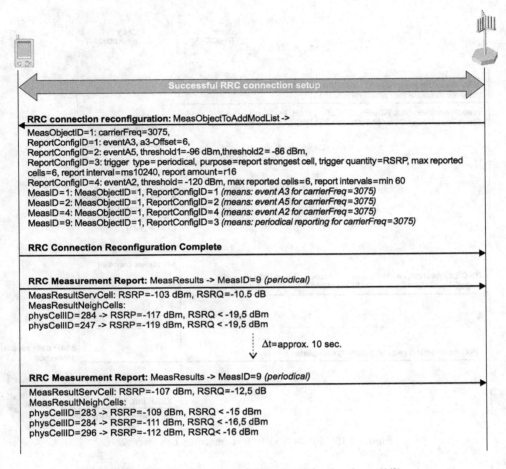

RRC connection reconfiguration: MeasObjectToAddModList ->

MeasObjectID=1: carrierFreq=3075,
ReportConfigID=1: eventA3, a3-Offset=6,
ReportConfigID=2: eventA5, threshold1=-96 dBm,threshold2 = -86 dBm,
ReportConfigID=3: trigger type= periodical, purpose=report strongest cell, trigger quantity=RSRP, max reported
cells=6, report interval=ms10240, report amount=r16
ReportConfigID=4: eventA2, threshold= -120 dBm, max reported cells=6, report intervals=min 60
MeasID=1: MeasObjectID=1, ReportConfigID=1 *(means: event A3 for carrierFreq=3075)*
MeasID=2: MeasObjectID=1, ReportConfigID=2 *(means: event A5 for carrierFreq=3075)*
MeasID=4: MeasObjectID=1, ReportConfigID=4 *(means: event A2 for carrierFreq=3075)*
MeasID=9: MeasObjectID=1, ReportConfigID=3 *(means: periodical reporting for carrierFreq=3075)*

RRC Connection Reconfiguration Complete

RRC Measurement Report: MeasResults -> MeasID=9 *(periodical)*

MeasResultServCell: RSRP=-103 dBm, RSRQ=-10.5 dB
MeasResultNeighCells:
physCellID=284 -> RSRP=-117 dBm, RSRQ < -19,5 dBm
physCellID=247 -> RSRP=-119 dBm, RSRQ < -19,5 dBm

Δt=approx. 10 sec.

RRC Measurement Report: MeasResults -> MeasID=9 *(periodical)*

MeasResultServCell: RSRP=-107 dBm, RSRQ=-12,5 dB
MeasResultNeighCells:
physCellID=283 -> RSRP=-109 dBm, RSRQ < -15 dBm
physCellID=284 -> RSRP=-111 dBm, RSRQ < -16,5 dBm
physCellID=296 -> RSRP=-112 dBm, RSRQ< -16 dBm

Figure 3.14   Inter-eNodeB intra-frequency handover (1/3)

physical cell ID = 284 is reported to be 7 dB stronger than the serving cell. This triggers the handover toward this neighbor cell.

Since source and target eNodeBs are connected via X2 interface, the X2AP Handover Preparation procedure is started. As a result, the target eNodeB assigns radio resources for the UE and sends the handover command message back to the source eNodeB. This handover command messages is the RRC Connection Reconfiguration message piggybacked in X2AP Successful Outcome message of the Handover Preparation procedure.

The X2AP SN Status Transfer message is used to synchronize the sequence numbers for user plane data forwarding. For more details, see Section 2.6.1.3.

The RRC Connection Reconfiguration messages including the target cell's physical cell identity and the new c-RNTI for the UE to be used in the target cell are sent on radio interface toward the mobile.

The UE changes the cell and performs random access procedure in the target cell. After successful random access, it sends RRC Connection Reconfiguration Complete to indicate handover success.

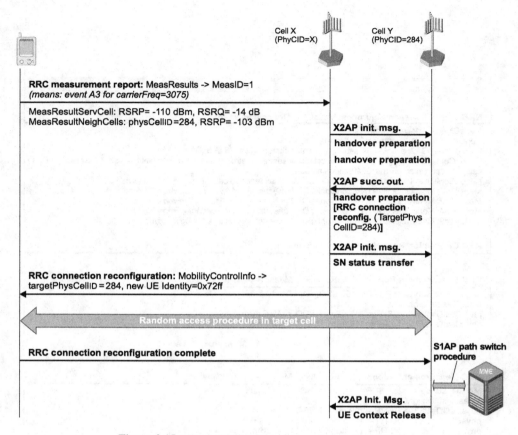

Figure 3.15   Inter-eNodeB intra-frequency handover (2/3)

The reception of RRC Connection Reconfiguration Complete triggers the start of the S1 Path Switch procedure in the target eNodeB as described in detail in Sections 2.6.1.4–2.6.1.7.

The X2AP UE Context Release deletes the UE-specific data stored in the source eNodeB and this is the end of the X2 handover scenario.

The next RRC Measurement Report (Figure 3.16) indicates event A3 for physical cell ID 296. However, since there is no X2 interface between the new target eNodeB and the source eNodeB, the S1 handover procedure is triggered.

It starts with S1 Handover Preparation procedure. The purpose of the handover preparation phase is to get radio resources in the target cell assigned and send a handover command (RRC Connection Reconfiguration message) to the source eNodeB. On the interface between the MME and the target eNodeB, the appropriate procedure is the S1 Handover Resource Allocation.

The S1AP eNodeB Status Transfer and MME Status Transfer messages are once again used to synchronize the PDCP sequence numbers for data forwarding.

Then, RRC Connection Reconfiguration message is sent to the UE and this starts the handover execution phase.

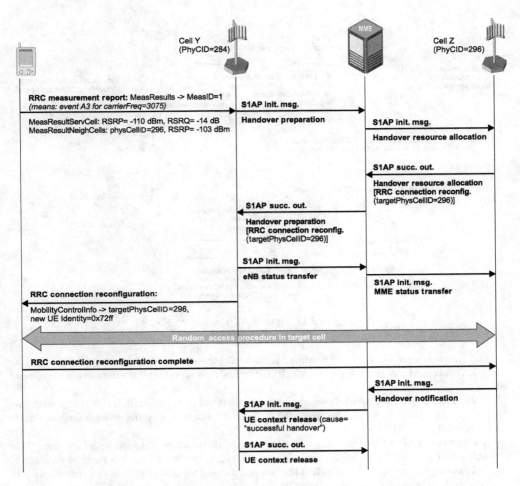

Figure 3.16   Inter-eNodeB intra-frequency handover (3/3)

Again, UE performs random access procedure and sends RRC Connection Reconfiguration Complete message to the target cell. This triggers the S1AP Handover Notification by target eNodeB to MME. Then, after the MME was informed about the successful handover, it releases the UE context in the old eNodeB including the self-explaining S1AP release cause.

3.2.4   Inter-RAT Handover to 3G

Figure 3.17 provides the overview of an inter-RAT handover from E-UTRAN to UTRAN (4G-3G handover).

Different from the blind redirect procedures described later in Section 3.2.6, the handover procedure also requires communication between the involved core network elements.

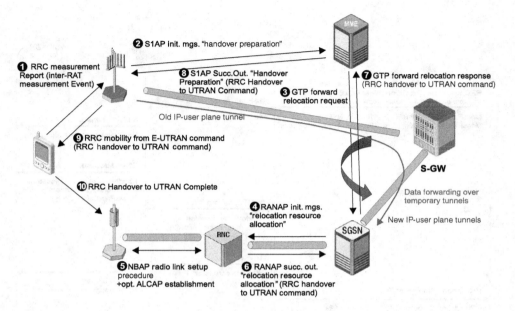

Figure 3.17   Inter-RAT handover to 3G

The typical trigger for a 4G-3G handover is an RRC Measurement Report (step 1) containing an inter-RAT measurement result for a 3G cell (RSCP or Ec/N0 is reported) that refers to reporting event B1 or B2.

In step 2, the S1AP Handover Preparation procedure is started by the eNodeB. It is the same procedure as used in case of 4G-4G S1 Handover, but the embedded containers are different as described already in Section 2.7.

After receiving the S1AP Handover Preparation Request (Initiating Message with Procedure Code "Handover Preparation"), the MME sends in step 3 a GTP Forward Relocation Request message to the 3G SGSN. This message contains IP addresses and tunnel endpoint identifiers for the temporary IP tunnels required for user plane data forwarding.

In step 4, the RANAP Relocation Resource Allocation procedure is started on IuPS interface. Also, here tunnel endpoint identifiers are negotiated for the Iu (user plane) transport bearer.

The RNC performs NBAP Radio Link Setup procedure on Iub in step 5. Optionally, if the Iub transport layer is not IP yet, the ALCAP establishment procedure for setup of ATM transport channels is executed.

After radio resources have been allocated, the RNC constructs the RRC Handover to UTRAN Command that is in the chain of steps 6 and 7 and sent to the eNodeB where it is embedded in the LTE RRC Mobility from E-UTRAN Command message that is in step 9 received by the UE.

After the UE changed the radio access successfully, it sends 3G RRC Handover to UTRAN Complete message to the RNC that will trigger in turn the release of all 4G radio resources and temporary tunnels (not shown in figure).

The regular user plane transport after the 4G-3G handover may involve the S-GW or not. This depends on additional core network signaling procedures, for example, GTP Context Request procedure was executed in addition.

During data forwarding and handover procedures, direct or indirect tunneling of user plane packets can be used. Figure 3.18 explains the differences behind this terminology.

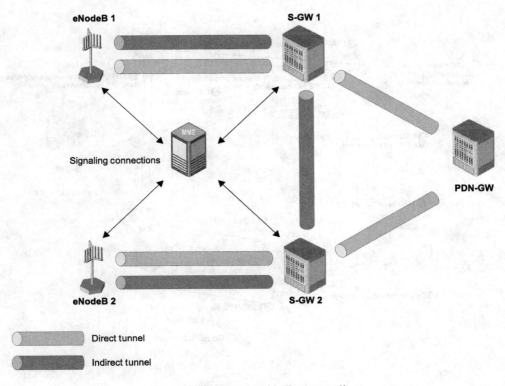

Figure 3.18 Direct and indirect tunneling

As one can see, the indirect tunnel is a user plane connection between two S-GWs while the direct tunnel always involves the PDN-GW.

The data forwarding using indirect tunnel involves the following steps:

– User plane packets buffered in the source eNodeB are sent to source S-GW using the UL TEID that the source S-GW provided before.
– Then, the source S-GW sends these user plane packets to the target S-GW using the DL TEID that target S-GW has provided to the MME before. The MME has forwarded this DL TEID to the source S-GW.
– Now, the target S-GW sends the user plane packets to the target eNodeB using this DL TEID that target eNB has been sent to MME, which was sent to target SGW.

3.2.5   Inter-RAT Handover to 2G

Figure 3.19 shows the overview of an inter-RAT handover from E-UTRAN to GERAN (4G-2G handover).

The trigger for a 4G-2G handover is also an RRC Measurement Report (step 1) containing inter-RAT measurement results, but signal strength for a GERAN cell is measured as GSM RSSI. In addition, the event B1 or B2 must have been triggered to start the 4G-2G handover procedure.

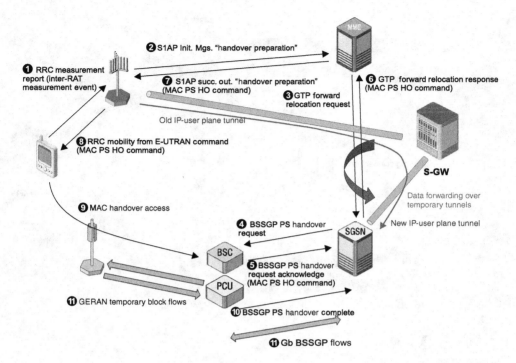

Figure 3.19   Inter-RAT handover to 2G

In step 2, the S1AP Handover Preparation procedure is started by the eNodeB. Again, it is the same procedure as used in the case of 4G-4G S1 Handover, but the embedded containers are different as described in Section 2.7.

After receiving the S1AP Handover Preparation Request (Initiating Message with Procedure Code "Handover Preparation"), the MME sends in step 3 a GTP Forward Relocation Request message to the 2G SGSN. This message contains IP addresses and tunnel endpoint identifiers for the temporary IP tunnels required for user plane data forwarding.

In step 4, a BSSGP PS Handover Request message is sent from SGSN to the BSC/PCU (assuming that the Base Station Controller and Packet Control Unit are combined in the same physical entity).

The BSC/PCU unit builds a MAC PS Handover Command to be sent to the UE in step 5. This MAC PS Handover Command is forwarded in steps 6 and 7 to the eNodeB where it is embedded in the LTE RRC Mobility from E-UTRAN Command message that is in step 8 received by the UE.

After the UE changed the radio access successfully, it sends MAC Handover Access to the GERAN base station in step 9.

The reception of MAC Handover Access triggers sending of BSSGP PS Handover Complete from BSC/PCU to SGSN. Now, SGSN will send a feedback for the completed handover to the MEE that will release all 4G resources subsequently.

Step 11 shows the temporary block flows that are used to transmit both signaling and user plane payload across radio interface and Abis interface. On Gb interface, BSSGP flows are in service for the same purpose.

The regular user plane transport after the 4G-2G handover may involve the S-GW or not. This depends on additional core network signaling procedures, for example, a GTP Context Request procedure is executed in addition.

3.2.6 Inter-RAT Blind Redirection to 3G

In contrast to the real handover, the inter-RAT blind redirect is a rather simple signaling procedure. Here, the network does not need to allocate resources in the target RAN and no data forwarding is required.

The trigger for an inter-RAT blind redirection could be an RRC Measurement Report that indicates serving 4G cell's quality below a certain threshold, for example, RSRP <= –120 dBm and no suitable 4G neighbor for handover around.

In such cases, the eNodeB makes the decision to request UE context release by MME over S1 (Figure 3.20, step 1). The S1AP UE Context Release Request message already contains the cause value "Inter-RAT Redirection," but without specification of the target RAT.

The true redirect carrier info is found in the RRC Connection Release message that is sent by eNodeB to the UE right after UE Context Release Request in step 2. Typically, it contains, in addition to the target technology (here: "utran-fdd"), target frequency in the form of a uARFCN value.

In step 3, the UE is "in between" the radio access technologies and performs a cell reselection in IDLE mode. During this cell reselection, the UE cannot make any connection attempt and cannot be paged.

In step 4, the UE performs a combined Location Area/Routing Area Update after successful establishment of an RRC connection. The combined LAU/RAU results in updated UE location info in the

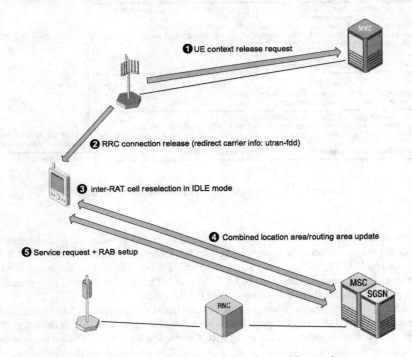

Figure 3.20 Inter-RAT blind redirection to 3G overview

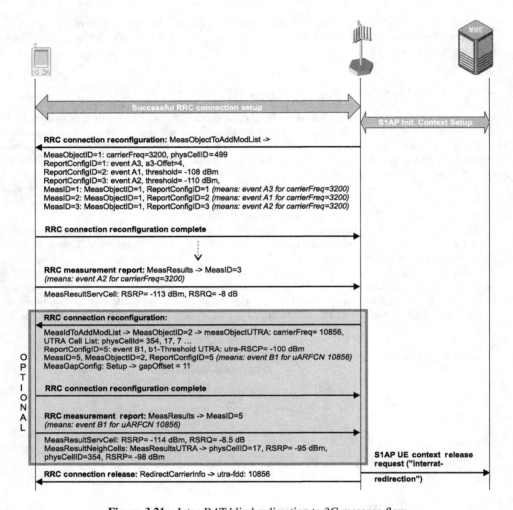

Figure 3.21   Inter-RAT blind redirection to 3G message flow

VLR/SGSN/HSS databases. In turn, the 3G core network elements will receive authentication tokens from HSS to perform 3G security procedures (ciphering, integrity protection) on Uu and Iub interface.

After the combined LAU/RAU, the UE will send a Service Request message and RAB Setup is performed whenever payload data need to be transmitted. The Service Request may contain a paging response indicator in case the UE was paged by SGSN before.

Figure 3.21 shows the detailed message flow of a call that is released due to blind redirect to UTRAN.

After RRC Connection and UE Context have been established successfully in the eNodeB, the measurement configuration is sent to the UE using RRC Connection Reconfiguration message and confirmed with RRC Reconfiguration Complete.

In the next step, the UE sends an RRC Measurement Report triggered by event A2, which means: the RSRP of the serving cell is below the reference threshold of −110 dBm.

Now, an optional configuration for inter-RAT measurement is possible. The availability of such inter-RAT measurement configuration depends highly on the eNodeB vendor, but also if the vendor supports this scenario, it is only rarely triggered, because most blind redirect procedures are executed literally "blind" without checking the quality of the target RAT before.

Anyway, the optional part in Figure 3.21 gives an example of parameters used for UTRAN inter-RAT measurement reporting.

The event defined in the report configuration is event B1 – here the RSCP of a 3G cell must become better than –100 dBm to trigger a report of this cell. The cells that shall be monitored are included in the measurement object configuration as a list of physical cell IDs (3G terminology: list of primary scrambling codes).

The inter-RAT measurement report that is finally received by the eNodeB contains a B1 event for uARFCN 10856 and RSCP of –95 dBm for the 3G cell with primary scrambling code 17 and –98 dBm for cell with primary scrambling code 354. At the same time, the RSRP of the 4G serving cell is down to –114 dBm.

This constellation triggers the inter-RAT redirection that is executed using S1AP UE Context Release Request and RRC Connection Release.

3.2.7   Inter-RAT Blind Redirection to 2G

For the inter-RAT blind redirect to GERAN (2G), the same basic statements apply as for inter-RAT redirect to 3G: the network does not need to allocate resources in the target RAN and no data forwarding is required.

Also, the trigger event is identical, for example, serving 4G cell RSRP <= –120 dBm and no suitable 4G neighbor for handover around.

Also in this case, the eNodeB makes the decision to request UE context release by MME over S1 (Figure 3.22, step 1). The S1AP UE Context Release Request message already contains the cause value "Inter-RAT Redirection," but without specification of the target RAT.

The redirect carrier info is found in the RRC Connection Release message that is sent by eNodeB to the UE right after UE Context Release Request in step 2. Typically, it contains, in addition to the target technology (here: "geran"), target frequency in the form of a BCCH ARFCN value.

In step 3, the UE is "in between" the radio access technologies and performs a cell reselection in IDLE mode. During this cell reselection, the UE cannot make any connection attempt and cannot be paged.

In step 4, the UE performs a combined Location Area/Routing Area Update after successful establishment of an RRC connection. The combined LAU/RAU results in updated UE location info in the VLR/SGSN/HSS databases. In turn, the 2G core network elements will receive authentication tokens from HSS to perform ciphering across Um, Abis, and Gb interface.

Step 5 shows the unidirectional GERAN temporary block flows that transport signaling messages as well as payload packets for PS services.

Figure 3.23 depicts the message flow of an inter-RAT blind redirection to GERAN.

The measurement settings are the same as for the 4G-3G inter-RAT blind redirect and also the trigger event A2 for carrier frequency with eARFCN = 3200 is the same as in case of the redirection to UTRAN.

It is a matter of internal eNodeB configuration (parameters to be set in OMC) if the bind redirection to 3G or 2G RAN is preferred.

That the UE is indeed redirected to GERAN becomes evident when looking at the redirect carrier info of RRC Connection Release messages. This information element indicates "GERAN" and starting ARFCN that represents the most preferred target cell in 2G. The following list of ARFCNs must be seen as a kind of alternative targets in case the preferred target cannot be accessed.

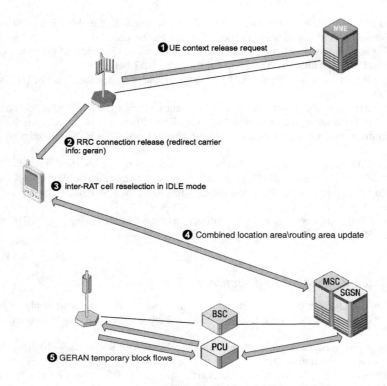

Figure 3.22   Inter-RAT blind redirection to 2G overview

3.2.8   CS Fallback

When designing the LTE standards, it was the intention of the standard bodies that all voice traffic in LTE shall be realized as Voice over IP. However, disadvantage of this idea is that the mandatory prerequisite is an IMS and/or expensive codec converters in media gateways.

Thus, as a more cost–efficient way to handle voice calls for LTE subscribers, the CS Fallback procedure was defined. The principle of the CS Fallback is simple to explain: if the UE signals the network that a voice call shall be made or if it was paged to perform a voice call, then E-UTRAN shall redirect the UE to a different RAT (UTRAN or GERAN) that suits better to voice services.

Figure 3.24 shows the steps of a typical CS Fallback scenario:

In step 1, the UE sends a NAS Service Request after successful RRC Connection establishment.
Then, the UE context is set up in step 2.
Step 3: The UE sends Extended Service Request message that contains a service type indicator set to "CS MO or CSFB". Now, the MME "knows" that the subscriber is attempting to make a voice call.
In step 4, the MME reacts and performs a UE Context Modification procedure with the eNodeB. The Initiating Message of this procedure contains a parameter that indicates "CS Fallback required."

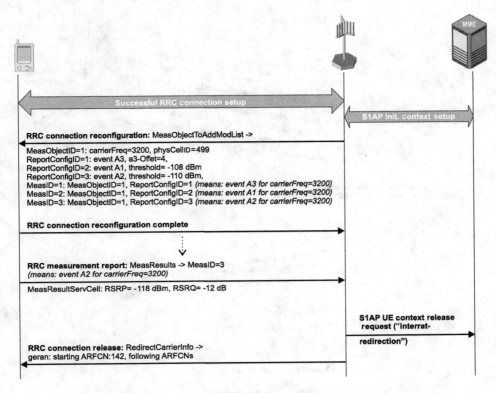

Figure 3.23 Inter-RAT blind redirection to 2G message flow

Now, the eNodeB sends RRC Connection Release message including redirect carrier info for GERAN or UTRAN to the UE in step 5. Almost simultaneously, the UE context is requested to be released due to "CS Fallback triggered" or "UE not available for PS service" (step 6). The cause value to be used depends on the eNodeB vendor. In one case, the authors have observed that these two different causes have been used to indicate the target RAT: "CS Fallback triggered" was used in case of CSFB to 3G and "UE not available for PS service" cause was used in case of CSFB to GSM.

In step 7, the MME orders release of IP tunnels in the core network and on S1-U.

Step 8 sees the UE already connected to UTRAN or GERAN sending a Connection Management Service Request message for mobile originating call (MOC) that is followed by other NAS messages for call setup.

In addition, the UE must perform a Routing Area Update toward the PS domain (SGSN).

Note: There are a couple of implementation options for CS Fallback. A variant that is also often seen has the NAS Extended Service Request as initial UE message so that the "CS Fallback required" indicator is already included in the Initiating Message of S1AP "UE Context Setup" procedure. In this case, the S1AP UE Context Modification procedure does not need to be executed so that step 3 and

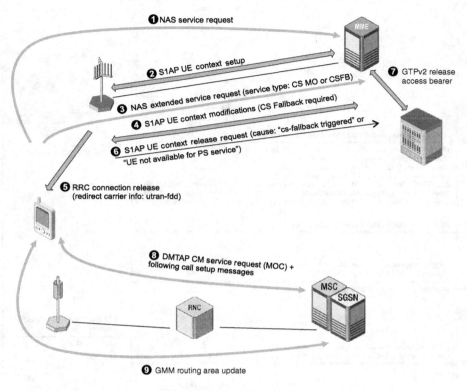

Figure 3.24 CS fallback

step 4 of Figure 3.24 will not be found in the signaling call flow. Also, it is possible that CS Fallback uses the 4G-3G inter-RAT Handover procedure instead of inter-RAT blind redirection. In this case, the NAS signaling described in this chapter is combined with the RAN signaling that was described in Section 3.2.4.

3.3 Failure Cases

During the establishment of the RRC connection and the default radio bearer, the errors shown in Figure 3.25 are the most common ones that are expected to be monitored on the Uu interface.

In the case of Error 1, the eNodeB does not send a RAR on the MAC layer or the RAR is lost on the radio interface, so the UE does not get permission to send the RRC connection request message. It is important to distinguish if the RAR was in fact not sent by the eNodeB – a possible root cause could be CPU overload of the base station – or if the message was sent but not received on the UE side due to a transmission error on the radio interface. Such transmission errors can occur due to interference on the physical resources allocated to the DL-SCH or due to coverage issues in cell edge regions. A coverage issue due to high path loss is likely if the initial timing advance indicates a long distance between the UE and the cell's antenna. A typical counter-strategy to overcome such long-distance problems is the partial frequency reuse approach described in Section 4.1 and illustrated in Figure 4.6.

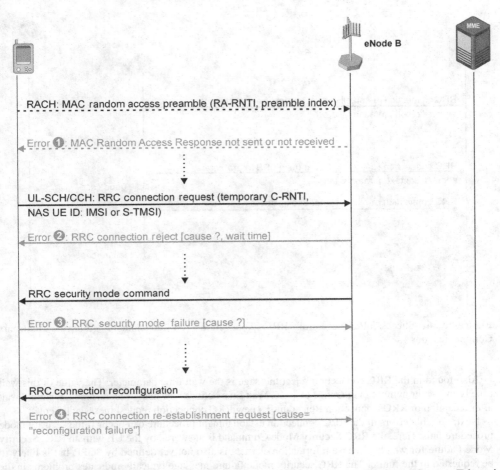

Figure 3.25 Errors during RRC connection setup and default radio bearer setup. (*Source:* Tektronix Communications.)

Error 2 occurs when the eNodeB sends RRC Connection Reject in response to the UE's RRC Connection Request. This typically happens if the cell is congested and the necessary radio resources for the connection setup cannot be provided and/or the eNodeB is overloaded. According to 3GPP 36.331 v.8.9.0 (2010-03), there is no reject cause value specified for the RRC connection reject message in the LTE environment. However, looking at the complementary message with the same name in the 3G UTRAN, it is expected that appropriate cause values will be defined as soon as network operators working in 3GPP specification groups define the need to get more specific information about failures in the RRC protocol. In fact, there are only three RRC causes defined in the 2010-03 version of the standard (the latest available at the time of writing this chapter):

- Reconfiguration failure.
- Handover failure.
- Other failure.

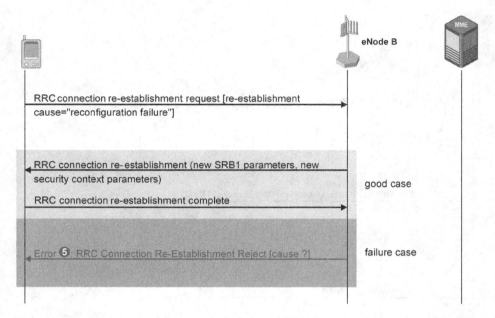

Figure 3.26 Successful/unsuccessful RRC connection re-establishment. (*Source:* Tektronix Communications.)

Also found in the RRC connection reject message is the wait time parameter. The value of this wait time is an integer number in the range of 1–16. Wait time defines how many seconds the UE should wait after reception of RRC Connection Reject until a new RRC connection request message is sent.

Next, possible errors in the RRC connection establishment procedure can occur if the security mode procedure fails. Here, the RRC Security Mode Command is answered by the UE with an RRC Security Mode Failure for which no cause information element is also not yet defined by 3GPP, but is likely to be requested in the future. The RRC security mode failure message typically indicates problems in the handset in the case during the activation of ciphering and/or integrity protection functions.

Last but not least, the RRC connection reconfiguration procedure can also fail. If such a failure occurs in the handset software, the RRC connection re-establishment request message with failure cause "re-configuration failure" will be sent instead of RRC Connection Reconfiguration Complete.

Now, it is up to the eNodeB to decide if the RRC connection should be re-established with new parameters or not. Both options are illustrated in Figure 3.26. If the response to the UE's re-establishment request is positive, the eNodeB will send an RRC connection re-establishment message with new parameters for SRB-1 and new security mode context parameters. If the UE accepts these new parameters, it will return an RRC connection re-establishment complete message. Otherwise, the eNodeB will send an RRC re-establishment reject message (there is also no cause value yet defined for this message in the 2010-03 version of the RRC standard specification) and subsequently the RRC connection will be released.

A special failure case that impacts the RRC signaling connection is retransmission errors in the RLC AM that is used to transport RRC signaling messages. Such errors can only be detected by looking deep into the bit and byte levels of the radio interface transport layer. An example of such an analysis is shown in Figure 3.27. This example was recorded on the 3G Iub interface, but the functionality and parameters are found on 4G Uu.

No	Long Time	2. Prot	RLC: Last Sequence Number	Sequence number	RLC/MAC: C/T Field	FP: Direction	RLC: Polling Bit
1067	15:03:16,173,340	RLC/MAC	45		Logical Channel 2	Uplink	
1068	15:03:16,213,265	RLC/MAC		44	Logical Channel 2	Uplink	Status report not requested
1069	15:03:16,253,272	RLC/MAC		45	Logical Channel 2	Uplink	Request a status report
1076	15:03:16,292,195	RLC/MAC	46		Logical Channel 2	Downlink	
1077	15:03:16,332,417	RLC/MAC		45	Logical Channel 2	Downlink	Request a status report
1078	15:03:16,371,309	RLC/MAC		46	Logical Channel 2	Downlink	Status report not requested
1079	15:03:16,412,418	RLC/MAC		47	Logical Channel 2	Downlink	Request a status report
1082	15:03:16,453,294	RLC/MAC			Logical Channel 2	Uplink	
1083	15:03:16,493,299	RLC/MAC	46		Logical Channel 2	Uplink	
1085	15:03:16,573,231	RLC/MAC		46	Logical Channel 2	Uplink	Request a status report
1091	15:03:16,612,417	RLC/MAC	47		Logical Channel 2	Downlink	
1092	18:03:16,613,235	RLC/MAC		47	Logical Channel 2	Uplink	Request a status report
1093	15:03:16,652,416	RLC/MAC		47	Logical Channel 2	Downlink	Request a status report
1094	15:03:16,653,398	RLC/MAC	48		Logical Channel 2	Uplink	
1095	15:03:16,691,973	RLC/MAC		48	Logical Channel 2	Downlink	Status report not requested
1096	15:03:16,693,400	RLC/MAC		48	Logical Channel 2	Uplink	Status report not requested
1097	15:03:16,731,083	RLC/MAC		49	Logical Channel 2	Downlink	Status report not requested
1098	15:03:16,733,328	RLC/MAC		49	Logical Channel 2	Uplink	Request a status report
1102	15:03:16,772,195	RLC/MAC	50		Logical Channel 2	Downlink	
1104	15:03:16,773,332	RLC/MAC		50	Logical Channel 2	Uplink	Status report not requested
1106	15:03:16,811,972	RLC/MAC		50	Logical Channel 2	Downlink	Status report not requested
1108	15:03:16,813,257	RLC/MAC	48		Logical Channel 2	Uplink	
1109	15:03:16,852,418	RLC/MAC		51	Logical Channel 2	Downlink	Request a status report
1111	15:03:16,853,264	RLC/MAC		51	Logical Channel 2	Uplink	Status report not requested
1112	15:03:16,893,187	RLC/MAC		52	Logical Channel 2	Uplink	Request a status report
1114	15:03:16,931,974	RLC/MAC	53		Logical Channel 2	Downlink	
1115	15:03:16,973,351	RLC/MAC		53	Logical Channel 2	Uplink	Request a status report
1118	15:03:17,051,749	RLC/MAC		51	Logical Channel 2	Downlink	Request a status report
1119	15:03:17,092,195	RLC/MAC	54		Logical Channel 2	Downlink	
1121	15:03:17,251,972	RLC/MAC		51	Logical Channel 2	Downlink	Request a status report
1122	15:03:17,451,973	RLC/MAC		51	Logical Channel 2	Downlink	Request a status report
1123	15:03:17,651,195	RLC/MAC		51	Logical Channel 2	Downlink	Request a status report
1124	15:03:17,851,750	RLC/MAC		51	Logical Channel 2	Downlink	Request a status report
1125	15:03:18,051,084	RLC/MAC		51	Logical Channel 2	Downlink	Request a status report
1127	15:03:18,252,637	RLC/MAC		51	Logical Channel 2	Downlink	Request a status report
1128	15:03:18,451,971	RLC/MAC		51	Logical Channel 2	Downlink	Request a status report
1129	15:03:18,652,194	RLC/MAC		51	Logical Channel 2	Downlink	Request a status report
1130	15:03:18,852,193	RLC/MAC		51	Logical Channel 2	Downlink	Request a status report
1131	15:03:19,052,195	RLC/MAC		51	Logical Channel 2	Downlink	Request a status report
1132	15:03:19,251,091	RLC/MAC		51	Logical Channel 2	Downlink	Request a status report
1133	15:03:19,452,198	RLC/MAC		51	Logical Channel 2	Downlink	Request a status report
1134	15:03:19,651,977	RLC/MAC		51	Logical Channel 2	Downlink	Request a status report
1135	15:03:19,852,422	RLC/MAC		51	Logical Channel 2	Downlink	Request a status report
1136	15:03:20,051,977	RLC/MAC		51	Logical Channel 2	Downlink	Request a status report
1137	15:03:20,251,976	RLC/MAC		51	Logical Channel 2	Downlink	Request a status report
1138	15:03:20,451,978	RLC/MAC		51	Logical Channel 2	Downlink	Request a status report
1139	15:03:20,652,197	RLC/MAC		51	Logical Channel 2	Downlink	Request a status report
1140	15:03:20,852,865	RLC/MAC		51	Logical Channel 2	Downlink	Request a status report
1141	15:03:21,051,309	RLC/MAC		51	Logical Channel 2	Downlink	Request a status report
1142	15:03:21,252,421	RLC/MAC		51	Logical Channel 2	Downlink	Request a status report

Figure 3.27   RLC AM retransmission errors

Looking at the RLC AM frames sent on logic channel 2, it emerges that they all have a unique RLC sequence number. In some of them, the RLC polling bit indicates that a status report is requested. For instance, the frames numbered 1068 and 1069 are sent on the uplink with sequence numbers 44 and 45. For sequence number 45, a status report is requested and this status report is sent in the downlink(!) direction with RLC Last Sequence Number = 45 to indicate that all uplink frames up to this last sequence number have been successfully received by the base station.

In turn, an ACK for the reception of downlink RLC sequence number = 51 is requested multiple times by the base station, but the ACK is missed. Either it is not sent by the UE at all or the received uplink signal of the connection was so badly affected by noise that the NodeB was not able to decode a valid uplink frame.

If repeated retransmissions or repeated status report requests cannot successfully be recovered error-free, the RRC connection is considered as dropped by the eNodeB and the UE context is requested to be released on the S1 interface using S1AP cause values like "failure in the radio interface procedure" or "release due to E-UTRAN generated reason." Which S1AP cause is implemented depends on eNodeB vendor-specific design.

Possible root causes for this kind of RLC AM retransmission error are interference on the uplink or downlink channels or problems with uplink power control algorithms.

4

Key Performance Indicators and Measurements for LTE Radio Network Optimization

This chapter describes how the different interfaces of the Evolved Universal Terrestrial Radio Access Network (E-UTRAN) can be monitored and which measurement equipment can be used. It also outlines which important performance parameters are required to troubleshoot and optimize the network and a set of counter-based Key Performance Indicators (KPIs) is proposed to check the performance of the most essential network functions.

4.1 Monitoring Solutions for LTE Interfaces

The following sections describe approaches of monitoring the LTE (Long Term Evolution) interfaces from the air interface through core network (Enhanced Packet Core - EPC) interfaces. All fixed line interfaces are IP-based protocol stacks; thus, interfaces widely use a physical and link layer predominantly used for IP, like Ethernet.

Although legacy Asynchronous Transfer Mode (ATM) networks are technically able to transport IP-type traffic, capacity constraints should be kept in mind (see Section 4.1.5).

4.1.1 Monitoring the Air Interface (Uu)

The main factor influencing the performance increase is the higher complexity of the air interface. This results in a need to monitor the protocols and procedures on the LTE air interface, because on fixed line interfaces, nothing of these required factors is seen any more as interesting protocols and procedures are only "seen" on the air interface (Section 1.8). Instead of UMTS, the Iub interface between NodeBs and RNCs carries each frame transmitted between the User Equipment (UE) and the base stations. Additionally, common radio cell measurements are available within the NBAP protocol on the Iub interface in UMTS. As a result, this leads to the following dilemma: on the one hand, there is a big decrease in accessibility of needed information for troubleshooting and optimizing network performance and, on the other hand, there is a higher demand on monitoring the protocols and control procedure as the increasing complexity demands it.

LTE Signaling, Troubleshooting and Performance Measurement, Second Edition. Ralf Kreher and Karsten Gaenger.
© 2016 John Wiley & Sons, Ltd. Published 2016 by John Wiley & Sons, Ltd.

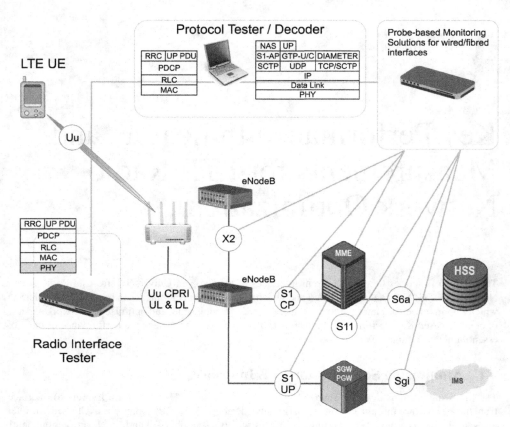

Figure 4.1   Monitoring architecture of an LTE network with fixed line interfaces (subset) and air interface

As illustrated in Figure 4.1, major control protocols of interest such as Radio Resource Controls (RRCs) are terminated within the eNodeB (eNB). Taking just fixed line interfaces into account for troubleshooting and optimization will enable only a surface-based analysis. In order to seek most root causes, a deeper analysis of the LTE air interface is required.

There are different possible approaches to monitor the air interface. Some of them are more useful than others, as the level of insight or the level of interference introduced into the system varies from approach to approach.

Figure 4.2 compares the three basic strategies to access air interface information.

The following sections discuss the three objective options (as eNB vendors control what information is provided at a mirror/monitoring port) to monitor the radio interface: antenna-based monitoring, coax-based monitoring, and digital RF (CPRI)-based monitoring.

Note that eNB vendor-specific monitoring ports are not discussed in detail because those implementations are proprietary and not objective monitoring points, since eNB vendors can decide which information to show and which to hide in specific situations. In other words, in situations like eNB overload or congestion, it is possible that those monitoring ports will suffer from such a system overload as well, which might lead to missing information in conditions where it is needed the most.

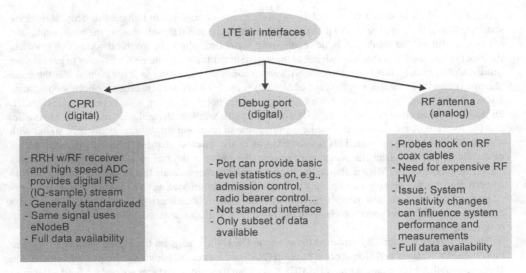

Figure 4.2   Strategies to monitor the air interface

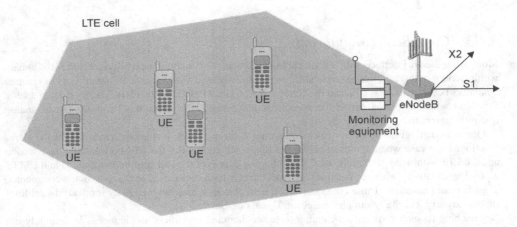

Figure 4.3   Antenna-based monitoring

4.1.2   *Antenna-Based Monitoring*

Antenna-based monitoring means that the monitoring equipment uses its own antenna passively to fetch the RF between base station antennas and the UE within a cell. Figure 4.3 illustrates an antenna-based monitoring point within a cell.

This method seems to be a favorable possibility to monitor the LTE air interface, as it is easy to install and moving from site to site is comfortable. But after deeper analysis, this seems not to be such a good way to monitor, for various reasons. An orthogonal and correct reception of the signal is only possible at the base station site, or at least close to it, as the eNB controls the Uplink (UL) alignment with the timing advance procedure in such a way that UL signals are only orthogonal and aligned directly at the base station antenna. Nevertheless, it should be possible to monitor the RF signals close to the base station site, but how useful are the results? A basic requirement for optimization and troubleshooting is to monitor the exact same receive and transmit signal as the device under test (here the eNB), in order to get relevant results. Unfortunately, this is not the case with antenna-based monitoring as it is essential that the same antenna (or multiple antennas in the case of MIMO) is used to guarantee that the monitoring equipment fetches the same signal, like the eNB does. UL alignment is not the deciding issue here, as a sufficient UL alignment is given close enough to the eNB site. No, it is the wireless channel, the antenna characteristic, and the location of the antenna that make it impossible to receive a meaningful equivalent signal with distinct RF antenna equipment. This is due to fading of the wireless channel, which changes by degree of wavelength of the carrier frequency. For example, the wavelength of an electromagnetic signal with a carrier frequency of 900 MHz is about 1 ft (30 cm). Thus, another antenna that is located just 30 cm from an eNB antenna will receive a completely different fading pattern leading to an incomparable result.

Furthermore, an essential benefit for optimization and troubleshooting is the correlation of data between multiple interfaces, which is often not given with an antenna-based decoupled network monitoring solution.

The air interface is directly ciphered between the UE and the eNB on the Packet Data Convergence Protocol (PDCP) layer and Non-Access Stratum (NAS) messages are additionally ciphered between the UEs and Mobility Management Entity (MME). Keys from fixed line interfaces are needed for deciphering from the S1 and S6a interfaces. Reception of this key information needs to be ensured to the monitoring system in real time.

4.1.3   Coax-Based Monitoring

Monitoring the air interface with a coax cable-based scheme is mainly a way to access the RF domain within vendor or test labs where coax cables are commonly used with variable RF attenuators or channel emulators between the test UEs and the Remote Radio Head (RRH) of the eNB. The RRH is an analog front-end, which amplifies, carrier (de)modulates, and converts the digital baseband signal from analog to digital (respectively, digital to analog).

One must distinguish between two common coax cable use cases: on the one hand, there is the aforementioned use case where the coax cables are implemented to carry the RFs between the base station and the UEs within the "cell in the lab." The other use case is in the still commonly used 2G and UMTS diploid base stations, where coax cables run from the core base station rack to the roof- or tower-mounted amplifiers and antennas. Those coax cables will be replaced, at least with LTE, by (optical) digital lines (fibers) to carry a digitally sampled baseband signal (see Section 4.1.4).

Attaching to such coax cables running to tower-mounted amplifiers and antennas is definitely not recommended (Figure 4.4). This is due to changing such essential system parameters as sensitivity caused by the tapping and splitting of those coax cables.

Tapping coax cables carrying the RF between the eNB and the UEs within the lab use case is a feasible option to access the air interface information. Nevertheless, one should keep in mind that this tapping still introduces attenuation (e.g., 3 dB) and at least some interference.

4.1.4   CPRI-Based Monitoring

A common interface between the core eNB (baseband processing unit) and the RRH is defined by a group of several LTE vendors. The letters CPRI stand for Common Public Radio Interface. The specification definitions can be downloaded from www.CPRI.info. In terms of CPRI, the core base station processing

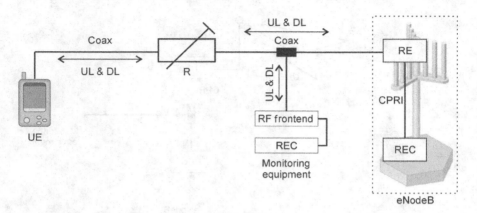

Figure 4.4 Coax-based monitoring of the air interface

unit is called the Radio Equipment Controller (REC) and the RRH (close to the antenna mounted on a tower or roof) is called Radio Equipment (RE). The distance was usually bridged by coax cables carrying an analog signal, which was received or to be transmitted.

CPRI is a digital interface conveying the digital baseband signal of the Downlink (DL) and UL. The digital baseband signal is the digitally sampled RF spectrum already independent of the carrier frequency (digital baseband). Thus, CPRI is independent of the band and carrier frequency conducted in the LTE cell, which is a great advantage for monitoring probes as there is a "one for all" unit. Digital baseband samples are IQ samples as described in Section 1.8, which describes phase and magnitude information as seen on the RF within the LTE cell.

An RE converts the digital IQ sample signal into analog signals and vice versa, depending on the direction. The resulting signal is either modulated to the carry frequency and amplified for transmission in the DL direction or received and down-modulated from the carrier frequency for UL signals.

The REC does all radio baseband processing and protocol decodings and executes the eNB procedures.

Usually, CPRI is transmitted via optical fibers, but an electrical PHY layer is also defined. It is recommended that PHY transceivers with the following specifications are used:

- IEEE 802.3-2005 (line bit rate option 1: 1 GbE, else 10 GbE).
- Fiber channel (FC-PI); ISO/IEC 14165-115.
- Fiber channel (FC-PI-4); INCITS (ANSI) Revision 8, T11/08-138v1.
- Infiniband Volume 2 Release 1.1 (November 2002).

Logically, a Control and Management Plane (C&M Plane), a User Plane (U-Plane, which carries the IQ samples), and Synchronization (Synch) layers are specified. The CPRI physical layer uses line rate options as multiples of a base rate of 614.4 Mbps:

- Line bit rate option 1: 614.4 Mbps.
- Line bit rate option 2: 1228.8 Mbps (2×614.4 Mbps).
- Line bit rate option 3: 2457.6 Mbps (4×614.4 Mbps).
- Line bit rate option 4: 3072.0 Mbps (5×614.4 Mbps).
- Line bit rate option 5: 4915.2 Mbps (8×614.4 Mbps).
- Line bit rate option 6: 6144.0 Mbps (10×614.4 Mbps).

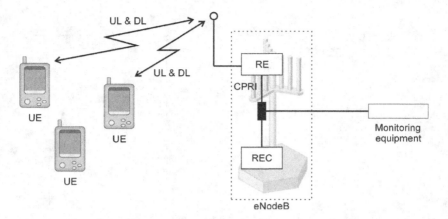

Figure 4.5 CPRI-based monitoring

CPRI defines basic frames with control words and a payload part. A hyper frame aggregates 256 basic frames. The size of a basic frame depends on the conducted CPRI line speed. The control words multiplex C&M Plane and Synch information. The larger U-Plane part conveys the IQ samples of multiple antennas. The position and exact format of the IQ samples within the U-Plane are not defined by CPRI. A common format is a bitwise interleaving of 15-bit words of I and Q per antenna.

Control words convey a fast and a slow C&M Plane. The fast and slow C&M Planes are transmitted asynchronously. The slow C&M Plane is defined with HDLC (High-Level Data Link Control) frames. A pointer indicates the control word area where fast C&M Planes can occur, which are Ethernet frames.

Figure 4.5 shows the basic eNB architecture with a CPRI interface between the REC and the RE. The CPRI interface enables a monitoring probe to access all cell air interface signals without any loss or without introducing any interference. It also processes the same signal as the eNB receives or transmits, which is especially important for troubleshooting and optimization, as already mentioned.

The CPRI signal is accessed either with optical splitters in the case of optical fibers or with repeaters (monitoring points) on electrical CPRI interfaces. Two basic optical architectures are used:

- Two-fiber transmissions, one for the UL and one for the DL direction. See Figure 4.6.
- One-fiber transmission (Wavelength Division Multiplex, WDM), where the UL and DL are transmitted at different wavelengths. See Figure 4.7.

One optical splitter is used for each fiber with the two-fiber architecture. The split UL and DL signals are used as input signals for the Uu monitoring probe (see Figure 4.6).

The one-fiber CPRI architecture with monitoring points is depicted in Figure 4.7. As the UL and DL use different wavelengths, this scheme is called wavelength division multiplex. A WDM splitter (seen in the center of Figure 4.7) is used in order to split the UL and DL signals from one fiber. When a designated WDM splitter is not available, two normal bidirectional splitters can be used as well. Figure 4.7 illustrates a common configuration with 1310 and 1490 nm for the UL and DL, respectively.

4.1.5 Monitoring the E-UTRAN Line Interface

LTE uses protocol stacks based on common IP interfaces on its fixed line interfaces. Usually, the IP frames are transmitted on an Ethernet infrastructure. Line speeds vary from 1 to 10 GbE. Both optical and electrical Ethernet PHY layers are used.

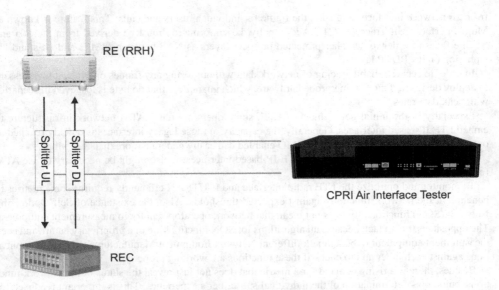

Figure 4.6 Regular two-fiber CPRI architecture

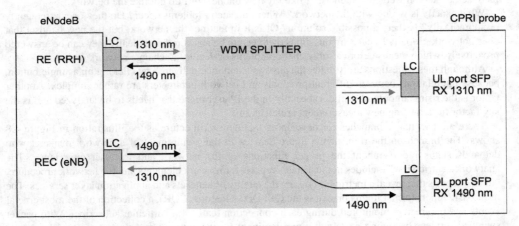

Figure 4.7 WDM CPRI architecture with one fiber for uplink and downlink. LC = Lucent or Local Connector; SFP = Small Form-Factor Pluggable (fiber optic module)

Optical interfaces are monitored in two ways: either by conducting optical splitters to the fibers, as described similarly in Section 4.1.4, or by using mirror ports of manageable network switches. Electrical interfaces use mirror ports only. Tapping electrical lines with T-pieces is no longer done, though it was common with coax cable-based 10BaseT Ethernet interfaces.

When deploying optical splitters in Ethernet fiber, it is especially important that the probe hardware works in a completely passive mode without Ethernet auto negotiation.

Figure 4.1 shows a common use case of monitoring a fixed line interface with high-performance probe hardware together with an air interface probe. The central processing software aggregates the information from all network interfaces and sorts the frame for individual users and calls. This feature is known as Multi-Interface Call Trace (MICT). Additionally, key information must be derived from the S6a and S1 interfaces in order to decipher protected protocol layers as NAS between MMEs and UEs and Uu ciphering on the PDCP layer.

In order to secure reliable tracing of network data without losing any frames or bytes, it is necessary to deploy designated monitoring probe hardware, which guarantees that no byte is lost even in scenarios with peak data rates.

Especially in the initial deployment of LTE, some operators reuse ATM network infrastructure to embed LTE IP-based interfaces. Generally, the capacity of those legacy interfaces is quickly reached as only one eNB with three sectors and MIMO enabled can easily generate more than 300 Mbps.

Hence, ATM can be used to convey LTE IP-based interfaces; it also might be necessary to use ATM (e.g., STM1 or E1/T1) monitoring probes.

To monitor and optimize the LTE radio interface and E-UTRAN efficiently requires some metrics to benchmark the network functions against expected thresholds. Also, the existence of Self-Optimizing Network (SON) functionality does not mean that network operators can forgo measurement equipment. The opposite is true in fact, because all algorithms for SON functions are of a proprietary nature and need measurement equipment to benchmark different Network Equipment Manufacturer's (NEM's) applications against each other and to check if these functions are working properly.

Besides, there is a rising demand for a metric that does not just reveal the status of the network functions, but allows determination of the individual subscriber's experience. This is important feedback for customer care and marketing departments, which have to fight against rising churn rates and individual feedback posted on websites, such as: "I like my new phone, but I do not like the network."

What exactly is wrong with the network? Where do such problems occur? Do they occur frequently or sporadically? Is there a possible solution? Or is it in fact not the network, but the new handset that does not work properly? These are questions that need to be answered and ideally they can be answered proactively, which means before a subscriber complains or decides to change network operator.

Although these questions are simple, the answers cannot be delivered by clicking on a single button, because the relations between different protocols and network parameters are rather complex. Another factor is the rising amount of traffic, especially in the IP user plane that needs to be analyzed and is the key factor for requiring new measurement architectures.

There are two major branches for developing this new architecture as the illustration in Figure 4.8 shows. The branch on the right of the figure deals with the requirement of a network manager who demands status reports about the proper functions of the network and the user experience. The "network-centric view" includes not just status reports for network elements and the network procedure, but also reports that reveal a metric for ranking the quality of handsets and application layer services. The most basic report format in this branch is the Call Detail Record (CDR), a collection of the most crucial events and parameters monitored during each connection (call) of a particular subscriber. With packet switched services becoming more and more dominant for the network operator's business, the details and amount of data to be stored in such CDRs are rising exponentially. There is the definite requirement to capture and analyze all this data 24/7, which means analyze and store in real time. The limit for the CDR-based measurement architecture is actually not the technical feasibility of measurement functions, it is the price factor. If continuous monitoring is a fixed requirement (and it is), then the rising amount of measurement and growing size of databases that need to be handled, transferred, stored, postprocessed, and enriched with additional data (e.g., subscriber-specific information not detectable from monitoring the network) can reach a point where the measurement equipment might become more expensive than the monitored network infrastructure. This is a little bit too expensive.

So in the end, although there is a new architecture for network monitoring on the way, this new architecture will not be able to provide all the desired metrics in real time. There will be compromises adjusting the number of possible KPIs and especially the depth of possible analysis to a level that is bearable for an

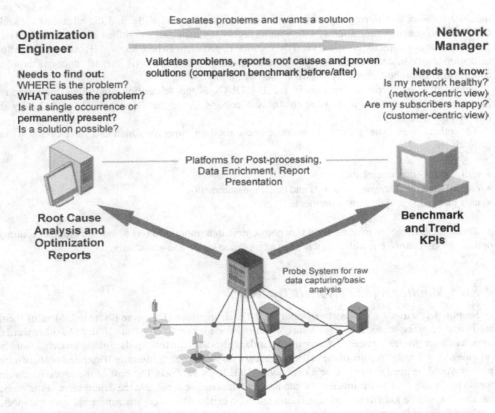

Figure 4.8 Split workflow and measurement architecture for network benchmarking and analysis/optimization

architecture and that should not exceed a certain price, as well as being tied to the 24/7 operating mode and realizing the rising performance of new computer hardware.

Such compromises open the path that is shown on the left of Figure 4.8. Here we see what is required for the network engineer to identify root causes of possible problems and come up with solutions for network troubleshooting and optimization. For this work, the available level of detail is crucial. The demand for in-depth analysis on the bit and byte levels, paired with requirements for highly sophisticated analysis functions like rule-based expert systems that allow the detection of root causes of problems automatically, creates the need for a different kind of measurement equipment that has a different architecture compared to those network monitoring systems that are stuffed with a rising number of high-level KPIs to detect symptoms of malfunctions in the network and in user-perceived quality of experience on an average level. However, it does not help to know that the average quality of a 5 minute call was good, while the insufficient quality during the last 5 seconds led to a call drop.

Root cause analysis requires a depth of detail that cannot be stored in CDRs, but this level of detail is only required for the analysis of a tiny amount of the overall network traffic.

It is a major challenge for all measurement equipment manufacturers to build a probe system for the wired interfaces of the E-UTRAN, UTRAN, and GERAN that can serve both measurement architectures, the 24/7 network monitoring system with its benchmark and trend KPIs, and the advanced network troubleshooting and optimization system with its highly sophisticated algorithms and in-depth analysis functions that can point to root causes and the location of problems (where applicable, even on the geolocalization level). In addition, the analysis of radio interface traffic and eNB internal functions requires another kind of probe, commonly known as the "air interface tester."

Indeed, the most critical function in the E-UTRAN is the scheduling algorithms implemented in the eNB. This function is decisive for the subscriber's Quality of Service (QoS) and Quality of Experience (QoE).

The other three main groups of measurements used to supervise, maintain, and optimize the network are:

- radio quality measurements;
- control plane performance counters and delay measurements;
- user plane QoS and QoE measurements.

These measurements may be found in both continuous monitoring systems and troubleshooting/ optimization systems, but with differences in aggregation levels and granularity.

4.1.6 Monitoring the eNodeB Trace Port

In Section 4.1, various solutions to monitor the LTE air interface (Uu) were discussed. Most of them are hardware probe–based solutions, which cover only one sector (cell) of an eNodeB. In other words, three hardware probes are needed to monitor a standard eNodeB with three cells. Such approaches usually provide a very detailed picture of the cell traffic but are very hardware consuming. Thousands of hardware probes would be needed to monitor a nationwide LTE RAN network. The cost of the probe hardware would be equally high as the investment into the base stations itself, but also the deployment of the probe hardware would be a challenge as the cell sites are often exposed to severe environmental circumstances.

Centralized software–based probes (soft probes) covering a cluster of base stations are a solution to monitor and analyze nationwide RAN networks and to overcome the aforementioned issues of hardware-based probes. The challenge with this approach is to get access to the air interface information as all Uu protocols besides NAS and user plane are not natively accessible on LTE fixed line interfaces as discussed in the previous chapters.

Software-based probes receive a trace feed from the monitored network element and process it for aggregation and analysis. A monitoring port provides this trace feed and mirrors content and additional information of the network interface to be monitored. The receiving monitoring probe is called *Trace Collection Entity* (TCE). The concept of a monitoring port with receiving TCEs is defined by the 3GPP LTE standard in technical specification TS 32.421, TS 32.422, and TS 32.423.

Even though the monitoring port for the air interface is specified by the 3GPP, the content and especially the additional information vary between network equipment vendors. Besides the different provided content, the mechanism of how the data is delivered is divided into two general schemes: file mode and streaming mode, described in the next two sub-sections.

Main use cases for the trace feed are troubleshooting, optimization, and Minimization of Drive Test (MDT).

The trace port content provided by the eNodeB are control plane protocols of the three eNodeB surrounding interfaces Uu, S1_MME and X2. Figure 4.9 depicts the eNodeB trace port architecture and general backhaul concept. The same physical interface as the S1 and X2 interface is used to backhaul also the eNodeB trace port content. Most deployments use IP links; the trace port feed content is decoupled at a RAN routing aggregation point (IP router) and is routed to the designated TCE receiver or

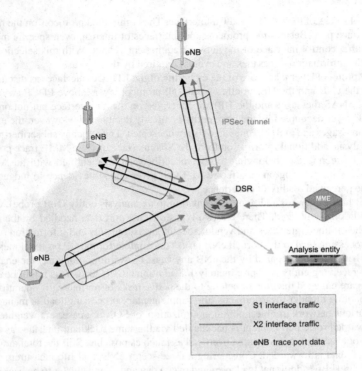

Figure 4.9 eNodeB trace port architecture and backhaul concept

pre-processor (see "file mode" in Section 4.1.6.1). This IP router option does IP decryption in case the link to the eNodeB uses IP security (encryption). Note that the IP backhaul architecture is subject to the deployment strategy of the network operator.

In order to keep the overhead traffic volume in a reasonable range, only parts of the control plane protocol stacks are available in the trace port content. As there is no possibility to access the Uu information on any fixed line interface, usually the main interest is the RRC protocol. However, the eNB trace port additionally provides the CP protocols (S1-AP and X2-AP) of the S1 and X2 interface. Figure 4.10 depicts

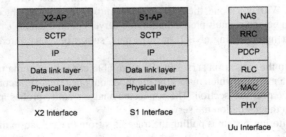

Figure 4.10 Control Plane stacks of the X2, S1, and Uu interface. Highlighted protocols are provided with the eNB trace port. Vendor specific information may provide additional information of other protocol layers

the CP protocol stacks of the S1, X2, and Uu interface. The highlighted protocols on the figure are delivered on eNB trace port. Besides the protocols and interfaces of interest, even specific messages can be configured with a control interface of the network equipment vendor. With this selection, the user can engage only the information (messages and events) required by the use case.

The PDCP protocol layer takes care of the ciphering of the LTE Uu interface (as discussed in Section 1.7) between the eNB and the UE. In other words, all protocol layers above PDCP (which is the RRC with embedded NAS messages and the UP) are encrypted on the air interface but not on the eNb trace port. The RRC layer decoupled by the trace entity is already deciphered. However, the embedded NAS layer in RRC messages and in S1-AP messages is not decrypted. Thus, for any subscriber-related analysis or handset analysis, additional information from the S6a interface or from MME trace ports is required. Nevertheless, UE connection belonging to one subscriber can be identified with the M-TMSI. After isolating an issue and filtering on an affected M-TMSI gives enough evidence to judge about the user impact and the perceived quality of experience.

As the eNB has to forward additional information to an analysis entity (soft probe), eNB resources have to be additionally utilized. There are two types of trace port data handled by the eNB: Protocol messages of the outlined interfaces and vendor specific measurements and information of lower layers. The on-top processing load to the normal eNB tasks is negligible for the CP protocol messages as those messages are handled and prepared by the eNB anyways. Forwarding (copying) the enabled protocols to a trace port receiving entity is approximately 1% additional CPU load. This might be different when measurements are engaged that do not belong to the native tasks of normal eNB operation. But even in the case of enabling broader measurements, the supplementary processing load is manageable and the benefits for various network troubleshooting, optimization, or SON use cases are weighted higher.

As the forwarded trace port content is backhauled via the same backhaul interface as the production network data, one wants to minimize the overhead as much as possible. Still the total amount is reasonable as only CP data is forwarded. Considering a three-sector eNB, real-life measurements show only approximately 360 kBps additional load averaged over a day and over a cluster to be backhauled at average loaded base stations. Peak data rates could reach 2.5 MBps for the busy hour of an entirely loaded eNB assuming all three sectors are fully loaded.

Multiple trace jobs with different content can be started in parallel. Those multiple parallel trace jobs enable different sets of content for various receivers or use cases.

There are two different basic trace port traffic concepts: the file or pull mode and the streaming or push mode. Both modes are described in the next two sub-sections.

4.1.6.1 File Mode/Pull Mode

One concept to deliver the trace port content to the analysis entity (network monitoring equipment) is the file or pull mode. With this mode, the eNBs are transmitting the engaged trace content to a mediation file server. There can be one or multiple mediation file servers in an operator's network. The file names follow a naming scheme including eNB IDs, time, date, subscriber trace reference number, and trace job number.

Files are written to the mediation server in a configurable cadence of some minutes. A common file duration interval is 15 minutes, which makes the file mode only near real-time compared to the streaming mode discussed in the next sub-section. The connected analysis entity probes in the same time periods for new trace files on the mediation file server.

The network monitoring system is pulling the trace files from the mediation file server via sftp to feed the analysis entity.

Figure 4.11 depicts a file mode eNB trace port setup with a mediation server, which stores and hosts the trace files from the eNBs and an analysis entity (soft probe), which pulls the trace files for further processing and storage.

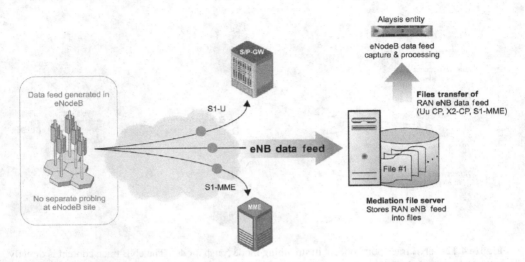

Figure 4.11 eNB trace port concept in file mode (pull mode). The eNB trace content is written into files at a mediation server and read-out (pulled) by the network monitoring analysis entities

4.1.6.2   Streaming Mode/Push Mode

Another concept is to transmit the trace port data in the streaming or push mode. The eNBs are streaming the trace port data directly to the analysis entity (soft probe). The used transport model is based on reliable TCP streaming with retransmissions. Network operations are taking care for a direct IP route and connectivity of the used TCP ports between eNBs and the analysis receiver.

The streaming concept allows real-time delivery of the data provided by the eNBs. Some analysis use cases desire a real-time feed. A disadvantage of streaming the data is that data will be lost in case of an interruption of the IP route or a temporal unavailability of the trace port receiver. Furthermore, generally only one receiver can receive the TCP stream. The file mode concept discussed in the previous sub-section could offer file access to multiple receivers (in this case, sftp clients).

Figure 4.12 depicts an eNB streaming trace port architecture. The TCP feed is decoupled at an RAN IP router and forwarded to the analysis probe.

4.2   Monitoring the Scheduler Efficiency

On the LTE radio interface, the most interesting aspect for radio quality and throughput of particular connections between the UE and network is inter-cell interference.

As in 3G UMTS, neighbor cells in LTE operate on the same frequency in UL and DL. Hence, in 3G UMTS, the DL signals of all cells received at a particular geographic position interfere with each other, while on the UL, each UE is an interferer to all other mobiles within a particular geographic area. This is a limitation of Code Division Multiple Access (CDMA) techniques in general that cannot be overcome.

In LTE, thanks to Frequency Division Multiple Access (FDMA), techniques are now being introduced that provide mechanisms to avoid inter-cell interference. Simply explained, in LTE, the base station (eNB) with very high periodicity collects information about the current interference situation in each cell.

Knowing which particular subcarriers of the available range are currently impacted by interference, the scheduler can assign only interference-free subcarriers to active connections as illustrated in Figure 4.13.

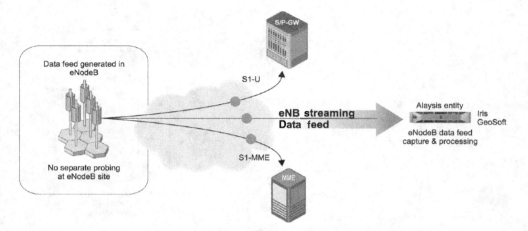

Figure 4.12   eNB trace port concept in streaming mode (push mode). The eNB trace content is directly streamed (pushed) to the network monitoring analysis entities

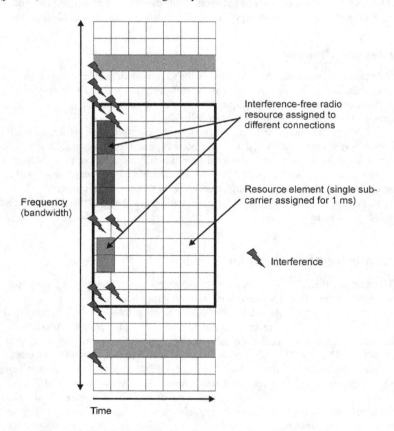

Figure 4.13   Scheduling of radio resources avoiding interference

A rescheduling of assigned resources is executed with a periodicity of 1 ms. In other words, within 1 second, the subcarriers used for a particular connection can change up to 1000 times.

For DL data transmission, the interference status of subcarriers is mostly derived from quality feedback sent by the UEs, for example, Channel Quality Indicator (CQI) and number of Hybrid Automatic Repeat Request (HARQ) retransmissions. To predict interference on the UL subcarriers, the neighbor eNBs exchange load information messages across the X2 interface with a maximum time granularity of 20 ms. For a system that reschedules radio resources every millisecond, this reporting granularity is certainly not sufficient. Hence, additional techniques are introduced to minimize interference impact in the UL scheduler, such as random frequency hopping.

In general, the scheduling of UL resources and hence the management of UL coverage and capacity are more difficult due to the fact that on the UL a set of neighbor subcarriers must be bundled together while on the DL any subcarrier is available for any connection. This "bundling" of a set of neighbor subcarriers for UL signal transmission of a particular connection is required to overcome the Peak-to-Average Power Ratio (PAR or PAPR) problem of Orthogonal Frequency Division Multiplex (OFDM) without introducing stronger and more power-consuming amplifiers in the handsets. An example for UL scheduling of three different subscribers is shown in Figure 4.14.

Note that this PAR problem of OFDM should be familiar to anybody who used to work in the Wireless Local Area Network (WLAN) environment. Here, even if your signal is weak, you can experience the DL throughput as quite acceptable while the upload of e-mails and other documents is wholly inadequate.

Now, if the mechanisms just described fail to avoid interference, this will have a significant impact on the radio connection quality. Degradation of throughput and, in the worst case, a loss of radio connection are the result.

In summary, we can say that the biggest challenge in LTE is to avoid interference from intelligent scheduling mechanisms. The algorithms implemented in the scheduler are not standardized and are highly proprietary. Hence, a comparison of scheduling algorithms implemented by different eNB vendors is key to optimizing the quality and capacity of the radio network.

To analyze the scheduling functions, it is necessary to visualize changes in both the frequency and time dimensions at a glance. This is the point where traditional measurement tools such as drive test equipment and network element counters are limited and fail to provide the necessary information for a complete analysis.

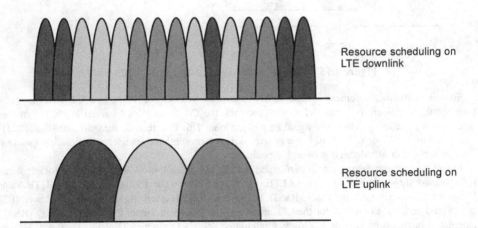

Resource scheduling on LTE downlink

Resource scheduling on LTE uplink

Figure 4.14   Resource scheduling for three different subscribers in same cell for downlink and uplink

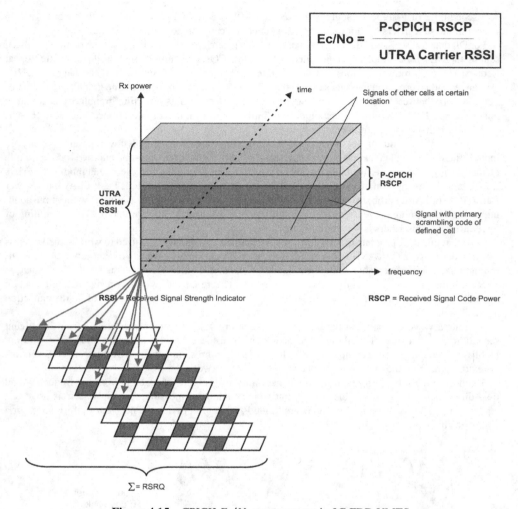

$$Ec/No = \frac{\text{P-CPICH RSCP}}{\text{UTRA Carrier RSSI}}$$

Figure 4.15   CPICH E_c/N_0 measurement in 3G FDD UMTS

To understand this statement, it is necessary to look back at 3G WCDMA. In WCDMA, the so-called pilot symbols are sent on a special control channel, the Common Pilot Channel (CPICH), in the DL direction *in parallel* with the user signaling and payload. This interference measured on the CPICH and expressed as E_c/N_0 or E_c/I_0 (chip energy over noise) represents the interference situation that is valid for any subscriber at a defined geographic position.

In fact, the same measurement is performed in LTE, but on each individual resource block that carries reference as shown in Figure 4.15. In LTE, the noise floor on the DL is named the E-UTRA carrier Received Signal Strength Indicator (RSSI), but instead the Received Signal Reference Power (RSRP) is reported and, based on this, for the LTE interface, the Received Signal Reference Quality (RSRQ) is calculated in the same way as E_c/N_0 was calculated for 3G Frequency Division Duplex (FDD) radio quality. The reference signals of a particular cell in LTE can be identified by the physical cell ID, which is in fact a scrambling code very similar to the primary scrambling code used on 3G FDD cells.

On the 3G UL, the noise floor is measured as the received total wideband power by node B. This measurement also applies to the complete bandwidth and is *in parallel* with all user connections. This means that, if the received total wideband power is high, all user connections will be affected.

Due to the fact that the 3G radio quality measurements in DL and UL are always in parallel with the dedicated channels used for the transmission of signaling and payload for particular UEs, the estimates of radio quality for these dedicated channels are quite accurate.

Looking now at the situation in LTE, it emerges that subscribers never use the complete bandwidth on the frequency range, but only particular subcarriers at a given time. Also the reference signals (LTE term for "pilot bits") are distributed in frequency and time using a predefined pattern as illustrated in Figure 1.44 (Section 1.8.6).

This pattern must be different for all antennas that overlap in a defined geographic area and it cannot cover the entire DL frequency range of a cell. This means that reference signals do *not* reflect the true DL radio quality of user connections, because these connections are scheduled on *other* resources. One can imagine that RSRP and RSRQ are measured in exactly the same way as their 3G companions RSCP (Received Signal Code Power) and E_c/N_0, but 4G DL radio quality is measured for each individual resource element shown in Figure 4.5. Then, the UE computes a statistical average that is later reported to the eNB.

In the end, the LTE DL radio quality parameters RSRP (3G equivalent: RSCP) and RSRQ (3G equivalent: E_c/N_0) provide only a fair estimation of the DL quality, but they are insufficient for root cause analysis of performance degradation and call drops if advanced scheduling techniques are used.

DL quality measurement reports sent by the UE to the eNB or logged by drive test mobiles are highly aggregated in the frequency and time dimensions to get one measurement result with a typical periodicity of 1 second. This allows a first view of the footprint of cells. Interference logs of Scanner (the mobile spectrum analyzer used in drive test campaigns) can help find geographic regions with permanent interference caused, for example, by radar stations, defective DECT handsets, or amplifiers of satellite dishes (known sources of interference from 3G networks), but they cannot help minimize inter-cell interference.

However, the most outstanding limitation of a drive test is that a drive test mobile can never measure the UL radio quality. As explained earlier, the UL quality and coverage are much more crucial for LTE quality than the DL.

Areas with coverage or interference problems are typically found on the cell edge. In LTE, it is possible to define a subset of subcarriers for DL transmission that are sent with a higher transmitted power than the others to serve subscribers located especially at the cell edge. This optimization strategy is called partial frequency reuse. Its principle is illustrated in Figure 4.16. Using this approach, it is quite simple to optimize the DL coverage and interference.

In turn, the UE cannot increase its transmitted power up to a certain level and interference prediction is much more difficult.

Note that, in LTE, the eNB is in charge of control of the UL TX power of the UE and to evaluate and optimize this power control loop, it is also necessary to monitor the UE signal received on the UL as well as transmit power commands sent by eNBs on the DL.

On the network side, radio quality parameters of 2G and 3G cells have been typically collected using histogram bins. The histogram is a statistical graph and each bin represents a range of discrete measurement results. Looking at the histogram shown in Figure 4.17, it is easy to identify abnormal behavior: outliers or "fat tails" to the left or right of the normal distribution function shown in the figure.

With this kind of histogram data and some statistical calculations, the UL or DL radio quality of a 2G or 3G cell can be relatively easy determined, because there is only one histogram per measurement type per cell required. If there are, for example, 1000 cells connected to a single RNC in 3G, then there are 1000 histograms to evaluate and using some common statistical calculations such as the average or median, it is not hard to identify the worst cell.

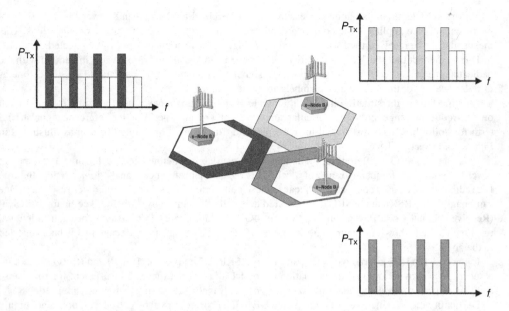

Figure 4.16 Partial frequency reuse to provide better signal quality for users at the cell edge on the downlink

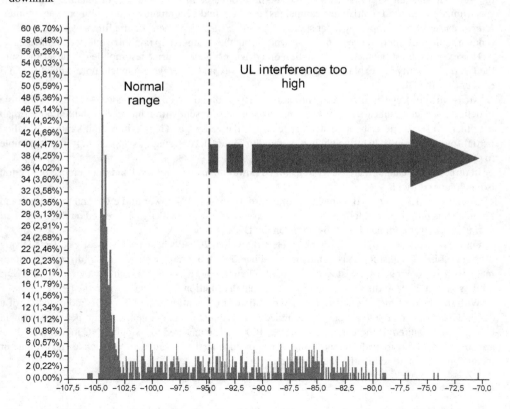

Figure 4.17 Statistical outliers in a histogram for received total wideband power

With LTE, this simple methodology is not applicable to most measurement types. The only exception is the received total wideband power, the UL noise floor in the cell's overall frequency band. The reason for the limited applicability is that the interference situation in the frequency domain of the LTE cell changes as fast as the neighbor cells reschedule their radio resource assignments: that is, 1000 times a second. The frequency band with the number of subcarriers in a single LTE cell is also rather large. In an LTE cell that operates at 20 MHz bandwidth, the number of subcarriers is 1200. To get meaningful results on the interference situation in this cell using the histogram approach, it would be necessary to write data for 1200 histograms per cell – for 300–6000 cells per MME cluster. No human eye can evaluate the thousands of histograms and there is also no mathematical algorithm to calculate meaningful statistics in such situations. Finally, the limited eNB's resources of processing power and RAM do not allow writing such huge amounts of statistical data. Remember that the eNB is a network element designed to switch connections and allocate radio resources. It is not a measurement instrument that gives insight into proprietary scheduling algorithms and radio quality parameters with best possible granularity in the frequency and time dimensions.

As a conclusion of these facts, the technical requirements for LTE radio network optimization measurement equipment are defined.

For data captured on the radio interface, it is necessary to have:

- full visualization of scheduling functions in the frequency and time domains for UL and DL signals, including visualization of radio resources assigned to single connections;
- visualization of absolute transmitting and receiving power in the frequency dimension;
- visualization of any possible deterioration of UL and DL radio signals, such as phase shift and amplitude errors in a simple view.

All these functions have to work in the real-time mode as well as in playback offline mode so that, offline, all particular actions of, for example, the scheduler can be tracked and analyzed.

In addition, it is necessary to capture and analyze data wired interfaces such as S1 and especially X2 and correlate this with the measurements from the air. This kind of important signaling analysis comprises:

- extract and quality feedback such as CQI and HARQ, Acknowledgment/Negative Acknowledgment (ACK/NACK) and its correlation with the particular UE that sent this feedback;
- extracted radio-related information about UL quality and resources from X2 load indication messages and its correlation with the sender/receiver eNB and comparison of this input to the reaction of the scheduler engines.

To address these specific requirements of LTE radio quality measurement and optimization, a radio interface tester must provide the following specific analysis functions.

4.2.1 UL and DL Scheduling Resources

To visualize the status of UL and DL scheduling resources and how they are dynamically assigned to the active UEs within a cell, it is necessary to plot a two-dimensional map of all available resource blocks in both the frequency and the time dimensions as was demonstrated in Figure 4.13.

Now each individual user that is assigned a resource block for a particular scheduling (time) interval is displayed using an individual color. Figure 4.18 shows how four different UEs are scheduled on the UL and the DL shared channel resources.

Looking at the distribution of color (in Figure 4.18 shown as gray scale) over frequency and time, it is possible to see which UE is dominant and which scheduling algorithms might have been implemented by the eNB vendor.

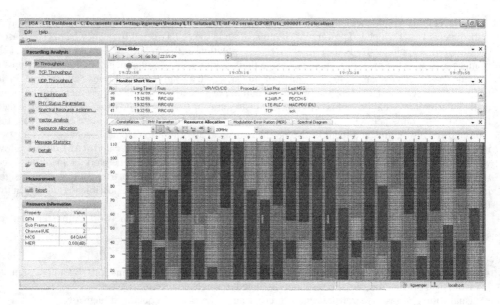

Figure 4.18 Visualized downlink scheduling for four individual subscribers

Since scheduling of the UE directly impacts the throughput of the connection, the throughput graphs for each UE are plotted on top of the scheduling scheme using the same color for each UE as below. The uppermost time plot for the UL and DL indicates the total throughput per cell.

In addition to the identity of the user, the color may also indicate the used modulation scheme by underlying a specific pattern or changing intensity of color. Within such a scheme, a dark green color may indicate that the "green UE" uses 64QAM while the light green color shows that the same "green UE" fell back to 16QAM and, hence, the green throughput graph declines while the total number of resource blocks assigned to the "green UE" remains unchanged.

A good air interface tester allows all these changes and measurements to be displayed in real time as well as in offline mode.

4.2.2   X2 Load Indication

An important input for the scheduler of a particular E-UTRAN cell comes from the load indication messages that are periodically sent on the X2 interface. Using the load indication mechanism, the neighbor cells inform each other about which UL resources are currently used. The intention is that a cell should ideally schedule its UL traffic on resource blocks that are not occupied by neighbor cells. The problem is that the scheduler works much faster than the information between the neighbor eNBs can be exchanged. So, the scheduler will change the allocation of radio resources every millisecond while the best possible time granularity for X2 load indication reports is limited to 20 ms. Also, the load indication message – as shown in Message Example 4.1 – does not contain scheduling information, but rather an abstract of the overall situation on the UL radio path.

Message Example 4.1: X2 Load Indication Message

```
+------------------------------------------------+------------------------+
|ID Name                                         |Comment or Value        |
+------------------------------------------------+------------------------+
|X2AP 3GPP TS 36.423 V8.4.0 (2008-12) (X2AP) initiatingMessage            |
(= initiatingMessage)                                                      |
|x2apPDU                                                                   |
|1 initiatingMessage                                                       |
|1.1 procedureCode                               |id-loadIndication       |
|1.2 criticality                                 |ignore                  |
|1.3 value                                                                 |
|1.3.1 protocolIEs                                                         |
|1.3.1.1 sequence                                                          |
|1.3.1.1.1 id                                    |id-CellInformation      |
|1.3.1.1.2 criticality                           |ignore                  |
|1.3.1.1.3 value                                                           |
|1.3.1.1.3.1 sequenceOf                                                    |
|1.3.1.1.3.1.1 id                                |id-CellInformation-Item |
|1.3.1.1.3.1.2 criticality                       |ignore                  |
|1.3.1.1.3.1.3 value                                                       |
|1.3.1.1.3.1.3.1 cell-ID                                                   |
|1.3.1.1.3.1.3.1.1 pLMN-Identity                 |'299000'H               |
|1.3.1.1.3.1.3.1.2 eUTRANcellIdentifier          |'00fce00'H              |
|1.3.1.1.3.1.3.2                                                           |
ul-InterferenceOverloadIndication                                         |
|1.3.1.1.3.1.3.2.1                               |low-interference        |
uL-InterferenceOverloadIndication-Item                                    |
|1.3.1.1.3.1.3.2.2                               |low-interference        |
uL-InterferenceOverloadIndication-Item                                    |
|1.3.1.1.3.1.3.2.3                               |high-interference       |
uL-InterferenceOverloadIndication-Item                                    |
|1.3.1.1.3.1.3.2.48                              |high-interference       |
uL-InterferenceOverloadIndication-Item                                    |
|1.3.1.1.3.1.3.2.49                              |low-interference        |
uL-InterferenceOverloadIndication-Item                                    |
|1.3.1.1.3.1.3.2.50                              |low-interference        |
uL-InterferenceOverloadIndication-Item                                    |
```

Besides the identity of the sending cell, the load indication message contains three different parameters:

- UL Interference Overload Information.
- UL High Interference Indication.
- Relative Narrowband Transmission Power (RNTP).

Message Example 4.1 shows only the structure of the UL Interference Overload Information. Here, for each UL resource block, the interference level experienced by the sending cell is reported using a scheme with three different values: "high interference," "medium interference," and "low interference." This interference level depends mostly on how often within an X2 load indication reporting interval LTE UEs are scheduled on the reported Physical Resource Block (PRB) and how much UL transmission power these UEs are using. For each UL resource block in the reporting cell, an interference level value is found. In the example, the numbering of information elements indicates that in the cell that sent this report, the total number of UL resource blocks is 50.

The UL High Interference Indication is encoded as a bit map. Each position in the bit map represents a PRB (first bit = PRB 0, and so on), for which value "1" indicates "high interference sensitivity" and value "0" indicates "low interference sensitivity." In the case of Message Example 4.1 – if included in the message – the bit map would consist of 50 bits. The maximum number of PRBs in an LTE cell is 110.

RNTP provides an indication of DL power restriction per PRB in a cell. Hence, the main difference between this information and the two previous pieces of information is that RNTP is input for the DL scheduler while the others are used as input for the UL scheduling algorithm.

Again, RNTP is reported as a bit map. Each position in the bitmap represents a PRB value (i.e., first bit = PRB 0, etc.). Instead of the full bit map, the value 0 might be transmitted to indicate "Tx not exceeding RNTP threshold" or value 1 indicates "no promise on the Tx power is given" by the reporting cell. The individual RNTP threshold values for the PRBs are encoded as an integer number covering the range from −11 to +3 dB.

Regarding aggregation and visualization of the X2 load indication measurements – and the same statement is true for the PRB usage reports discussed in the next section – it is obvious that simple counter or histogram data is an inapplicable approach. The best way to visualize what is periodically reported using bit maps is what was shown in Figure 4.18.

4.2.3   The eNodeB Layer 2 Measurements

The measurements described in this chapter are defined in 3GPP 36.314. The intention of this standard document is to define some common statistics for radio interface performance and in some cases to generate measurements to serve the X2 load indication reporting.

4.2.3.1   PRB Usage

The usage of PRBs can be reported by the eNB (to, e.g., the Operation and Maintenance Center (OMC)) for UL and DL in a particular cell in different ways:

1. *UL/DL total PRB usage:* The objective of the PRB usage measurements is to measure the use of time and frequency resources. One use case is cell load balancing, where PRB usage relates to information signaled across the X2 interface. Another use case is O&M performance observability.
2. *UL/DL PRB usage per traffic class:* This measurement is an aggregate for all UEs in a cell and is applicable to Dedicated Traffic Channels (DTCHs). The reference point is the SAP between MAC and L1. The measurement is done separately for DL DTCH, for each QCI, and UL DTCH, for each CQI.
3. *UL/DL PRB usage per Signaling Radio Bearer (SRB):* This measurement is applicable to Dedicated Control Channels (DCCHs). The reference point is the SAP between MAC and L1. The measurement is done separately for DL DCCH and UL DCCH.

4. *DL PRB usage for Common Control Channels (CCCHs):* This measurement is applicable to the Broadcast Control Channel (BCCH) and Paging Control Channel (PCCH). The reference point is the SAP between MAC and L1.
5. *UL PRB usage for CCCHs:* This is the percentage of PRBs used for CCCHs' Random Access Channel (RACH) and Physical Uplink Control Channel (PUCCH). Value range: 0–100%.

Basically, all PRB usage statistics can also be visualized using a radio interface tester as shown in Figure 4.7. The reporting format on the O&M interface between the eNB and OMC is not defined in international standards and, hence, is of proprietary nature. However, due to the limited processing resources available in the eNB for performance measurement and statistical tasks, it can be guessed that what is provided to the OMC is also rather highly aggregated in the time and maybe also frequency dimension. Compared to what can be measured and visualized using a radio interface tester, the OMC statistics may be sufficient to determine the average load in a cell on an hourly or 15 minute basis. These will not help to verify and optimize the scheduling algorithms as a radio interface tester can do.

4.2.3.2 Received Random Access Preamble

The number of received random access preambles is an important accessibility KPI for a cell. The more preambles the UE must send to get RACH resources assigned by the cell, the longer it takes to set up a call or to complete a handover to an E-UTRAN target cell. The worst cell is the cell with the highest number of received random access preambles. In the worst case, the UE never gets access to the RACH – an effect that is known from drive test campaigns as the "sleeping cell."

As a result, it emerges that the number of received random access preambles is a very important measurement, but experience with RAN vendors in 3G UTRAN has proved that this measurement (which also exists for UTRAN cells) was never enabled. Whether this will change in LTE RAN is a question that cannot be answered at the time this chapter was being written (Fall 2013).

4.2.3.3 Number of Active UEs

The number of active UEs is a simple gauge measurement that shows how many subscribers on average use the resources of the cell over a defined period of time.

This measurement is important for traffic and radio resource planning. It depends on the individual optimization task and which granularity is necessary on the time axis. Typical reporting intervals of OMC statistics are 15, 30, and 60 minutes. However, looking at Figure 4.19, which shows a much better time granularity, the peaks in subscriber activity (in the figure up to approximately 25, while the average value for 15 or 30 minutes could be something around 12–15) that may lead to a shortage of radio resources in the cell can be clearly identified.

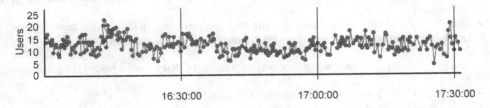

Figure 4.19   Number of active UEs ("users") over time

4.2.3.4   Layer 2 User Plane Measurements in the eNodeB

The eNode may provide the following measurement about user plane transport on layer 2: Packet Delay in the downlink per QCI; Packet Discard Rate in the downlink per QCI – this is the number of downlink PDCP packets discarded by the eNode; Packet Uu Loss Rate per QCI – number of PDCP packets lost during radio transmission as detected by the eNodeB; Scheduled IP throughput per QCI on the radio interface for uplink and downlink; Data Volume of PDCP SDUs transmitted on uplink and downlink

- Packet Delay in the downlink per QCI.
- Packet Discard Rate in the downlink per QCI – this is the number of downlink PDCP packets discarded by the eNode.
- Packet Uu Loss Rate per QCI – number of PDCP packets lost during radio transmission as detected by the eNodeB.
- Scheduled IP throughput per QCI on the radio interface for uplink and downlink.
- Data Volume of PDCP SDUs transmitted on uplink and downlink.

4.3   Radio Quality Measurements

There is a set of radio quality measurements specified by 3GPP. In particular, the definitions can be found in 3GPP 36.214 "E-UTRA Physical Layer Measurements." These measurements are split into E-UTRAN measurements that are provided by the eNB and UE measurements reported by the handset. Especially for the E-UTRAN measurements, the 3GPP standards must be seen as an option and there is no guarantee that eNB vendors will implement them. In any case there is room for proprietary implementations, because there are no standardized measurement reports defined for the S1 interface. It is expected that in most cases, the E-UTRAN measurement results will be sent to the OMC via O&M interfaces using a proprietary protocol. Also, the binning of E-UTRAN measurements is – in contrast to the 3G UTRAN standards – not defined by 3GPP.

The radio quality measurements can be split into UL and DL measurements. Looking at the UL measurements illustrated in Figure 4.20, the only measurement sent by the UE using an RRC measurement report is the UE Tx power, the power used by the handset to send the physical UL signal toward the eNB.

On the eNB side, the following parameters can be measured by the base station:

- *Downlink Reference Signal Power (DL RS TX power)* – This is the average power used to send the reference signals on the downlink. The DL RS TX power has an influence on network coverage and accuracy of geolocation results as described in Section 4.5.6.
- *Received Interference Power (RIP):* This is the UL noise floor for a set of UL resource blocks.
    Thermal noise power: This is the UL noise for the entire UL frequency bandwidth of the receiving cell without the signals received from LTE handsets.
    Timing advance: This is roughly the time it takes for the radio signal to travel from the UE to the eNB's receiver across the radio interface. Thus, it is equivalent to the distance between the UE and the cell's antenna.
- *eNB Rx – Tx time difference*: This is the estimated time difference in the eNodeB between receiving an UL frame and sending the next downlink frame of the same radio connection.

The following three measurements have been introduced by 3GPP for the purpose of geolocation:

- E-UTRAN GNSS Timing of Cell Frames for UE positioning.
- Angle of Arrival (AoA).
- UL Relative Time of Arrival ($T_{UL-RTOA}$).

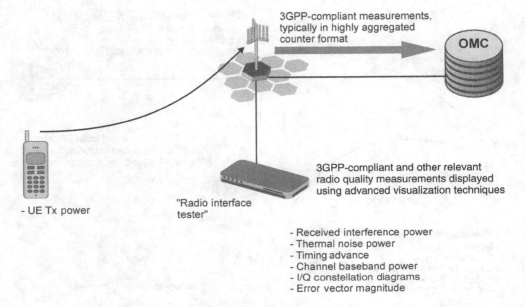

3GPP-compliant measurements, typically in highly aggregated counter format

OMC

- UE Tx power

"Radio interface tester"

3GPP-compliant and other relevant radio quality measurements displayed using advanced visualization techniques

- Received interference power
- Thermal noise power
- Timing advance
- Channel baseband power
- I/Q constellation diagrams
- Error vector magnitude

Figure 4.20 Uplink radio quality measurements and layer 2 measurements

An air interface tester can provide in addition:

- *Channel baseband power:* This is a measurement for the change in power amplitude of a particular physical channel in the time domain and can only be provided by an air interface tester.
- *I/Q constellation diagrams:* These are used to check the quality of the modulated symbols of the received radio signals and can only be provided by an air interface tester. The measurement principle of an I/Q constellation diagram is to compare the received symbol pattern and shape to the ideal constellation points (i.e., the expected shape and pattern of the symbols). Any differences are visualized in real time in the changing pattern of the picture.
- *Error vector magnitude:* This is a measurement related to the points in the I/Q constellation diagram. Basically, the Error Vector Magnitude (EVM) is the metric that indicates how far the points are from the ideal locations. This measurement is only available from air interface testers.

Both I/Q constellation diagrams and EVM measurements are available for UL and DL physical channels and a separate measurement for each type of physical channel in the cell. To measure the DL quality, the air interface tester must be used as a kind of drive test device and cannot remain connected to the CPRI of the eNB.

Also for the DL quality, a set of measurement tasks is performed by the UE (see Figure 4.21). Drive test equipment will perform the same measurement jobs but store the measurement results in its log files with higher granularity (typical time granularity for reference signal measurements: 1 second) and correlated with the true GPS location of the geographic measurement point. Regular subscriber handsets will send RRC measurement reports with reference signal measurements only in event-triggered mode, which means only if a predefined threshold is exceeded. Depending on the radio access capabilities of the handset and the availability of network coverage, the UE is able not just to receive reference signals

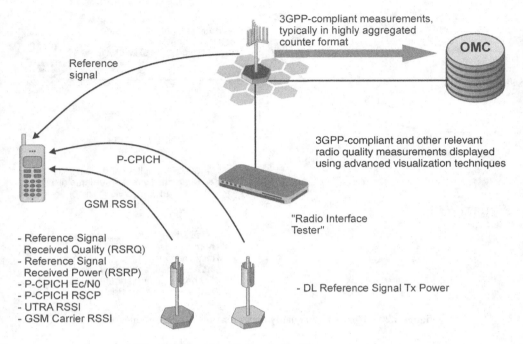

3GPP-compliant measurements,
typically in highly aggregated
counter format

OMC

Reference
signal

3GPP-compliant and other relevant
radio quality measurements displayed
using advanced visualization techniques

P-CPICH

GSM RSSI

"Radio Interface
Tester"

- Reference Signal
  Received Quality (RSRQ)
- Reference Signal
  Received Power (RSRP)
- P-CPICH Ec/N0
- P-CPICH RSCP
- UTRA RSSI
- GSM Carrier RSSI

- DL Reference Signal Tx Power

Figure 4.21   Downlink radio quality measurements

from LTE cells, but also to listen to 3G UMTS and GSM cells on demand. Hence, in the E-UTRAN, it is possible to monitor RRC measurements that contain the following parameters:

- *RSRP of LTE cells:* This is the absolute signal strength of the LTE reference signal related in dBm (absolute signal strength compared to a reference level of 1 mW). RSRP is the LTE equivalent of 3G UMTS RSCP.
- *RSRQ of LTE cells:* This is the DL signal-to-interference ratio in dB measured on the LTE reference signals. RSRQ is the LTE equivalent of 3G UMTS E_c/N_0.
- *P-CPICH E_c/N_0:* This is the Primary CPICH chip energy over noise, the signal-to-interference ratio measured on the DL reference channel of the 3G UMTS cell.
- *P-CPICH RSCP:* This is the absolute signal strength of the Primary CPICH (in dBm) measured on the UE's receiver antenna and compared to a reference level of 1 mW.
- *UTRA RSSI:* The RSSI is the DL noise level measured on the UE's radio receiver antenna for the entire bandwidth of the UTRAN or E-UTRAN cell.
- *GSM Carrier RSSI:* This is the RSSI (in dBm) for the signal level of the GSM cells measured on the UE's receiver antenna and compared to a reference level of 1 mW.

Furthermore, 3GPP also specifies the measurement capabilities for the UMTS TD-SCDMA standard that was launched in China. Here, the reference signal is the RSCP of the Primary Common Control Physical Channel (P-CCPCH). In those regions of China that have TD-SCDMA coverage, the P-CCPCH RSCP will be reported instead of the P-CPICH E_c/N_0 and RSCP that are defined for the more global UMTS FDD standard. On the network side, the eNB is equipped with the capability to measure and report the DL reference signal transmission power, which is the average absolute power level of all DL resource blocks that are used to send reference signals. This value is equivalent to the P-CPICH Tx power that was not measured in 3G UMTS cells, but configured as a fix value during cell setup or reset.

4.3.1 UE Measurements

4.3.1.1 RSRP

RSRP is used to measure the coverage of the LTE cell on the DL. The UE will send RRC measurement reports that include RSRP values in a binned format. The appropriate bin mapping is given in Table 4.1. The reporting range of RSRP is defined from −140 to −44 dBm with 1 dB resolution.

The main purpose of RSRP is to determine the best cell on the DL radio interface and select this cell as the serving cell for either initial random access or intra-LTE handover. The RRC measurement reports with RSRP measurement results will be sent by the UE if a predefined event trigger criterion is met (see Section 1.10.8.3).

There is certainly a correlation also between RSRP and the user plane QoS. As a rule of thumb for a cell in the outdoor environment, the RSRP measurement results can be categorized in three ranges. If RSRP > −75 dBm, excellent QoS can be expected as long as not too many subscribers struggle for the available bandwidth of the cell. In the range between −75 and −95 dBm, a slight degradation of the QoS can be expected, for example, throughput will decline by 30–50% if RSRP goes down from −75 to −95 dBm. Below −95 dBm, the QoS becomes unacceptable and throughput tends to decline down to zero at approximately −108 to −100 dBm. Under such radio conditions, call drops must be expected as the worst case.

Due to their limited footprint, in-house cells are normally able to deal with radio conditions that are worse compared to what is measured in the outdoor environment. Consequently, an acceptable QoS for RSRP values as low as −130 dBm can be assumed.

For city center cells with a foot print radius of 400–600 m the lower acceptable RSRP threshold is −120 dBm. However, starting from −115 dBm a significant rise of RRC accessibility issues can be observed.

4.3.1.2 RSRQ

Like RSRP, RSRQ is used to determine the best cell for LTE radio connection at a certain geographic location. However, while RSRP is the absolute strength of the reference radio signals, RSRQ is the signal-to-noise ratio. Like RSRP, RSRQ can be used as the criterion for initial cell selection or handover.

Table 4.1 RSRP measurement report mapping

Reported value	Measured quantity value	Unit
RSRP_00	RSRP < −140	dBm
RSRP_01	−140 ≤ RSRP < −139	dBm
RSRP_02	−139 ≤ RSRP < −138	dBm
...	...	...
RSRP_95	−46 ≤ RSRP < −45	dBm
RSRP_96	−45 ≤ RSRP < −44	dBm
RSRP_97	−44 ≤ RSRP	dBm

Source: Reproduced with permission from © 3GPP™.

Table 4.2   RSRQ measurement report mapping

Reported value	Measured quantity value	Unit
RSRQ_00	RSRQ < −19.5	dB
RSRQ_01	−19.5 ≤ RSRQ < −19	dB
RSRQ_02	−19 ≤ RSRQ < −18.5	dB
…	…	…
RSRQ_32	−4 ≤ RSRQ < −3.5	dB
RSRQ_33	−3.5 ≤ RSRQ < −3	dB
RSRQ_34	−3 ≤ RSRQ	dB

Source: Reproduced with permission from © 3GPP™.

Simply, the calculation of RSRQ can be expressed by the following formula:

$$RSRQ[dB] = 10 \lg \frac{RSRP}{RSSI} \tag{4.1}$$

The reporting range of RSRQ is defined from −19.5 to −3 dB with 0.5 dB resolution (Table 4.2).

When comparing the measurement results of RSRQ and RSRP that have been made at the same geographic location – in a protocol trace, they can be identified by the same timestamp – it is possible to determine if coverage or interference problems occur at this location. This is illustrated in Figure 4.22.

If a UE changes its location or if radio conditions change due to other reasons and RSRP (i.e., the absolute signal strength of the reference signals) remains stable or becomes even better than before while RSRQ is declining, this is an unambiguous symptom of rising interference. If, on the other hand, both RSRP and RSRQ decline at the same time/location, this clearly indicates an area with weak coverage. This kind of evaluation is very important for finding the root cause of call drops due to radio problems.

Similar to what was described for RSRP, for RSRQ also three quality ranges can be defined, but the numbers here are still very uncertain since no loaded network environment has yet been monitored, due to the fact that the number of calls and location of subscribers are limited during field trials. All in all, it seems that in general, RSRQ values higher than −9 dB guarantee the best subscriber experience. The range between −9 and −12 dB can be seen as neutral with a slight degradation of QoS, but overall customer experience is still at a fair level. Starting with RSRQ values of −13 dB and lower, things become worse with significant declines of throughput and a high risk of call drop.

4.3.1.3   Power Headroom

The Power Headroom (PH), expressed in decibels, is defined as the difference between the nominal UE maximum transmit power and the estimated power for the Physical Uplink Shared Channel (PUSCH) transmission. It is the power that can be added to UL data transmission if the UE moves toward the cell edge or requires a service with a higher guaranteed bit rate.

The PH reporting interacts with the UL scheduling function of the eNB. The reports can be sent either event-triggered or periodically. An event is triggered if, after a period of not allowing PH reporting, the appropriate timer in the UE expires, if the path loss change is higher than a predefined value. Otherwise, periodic PH reporting starts with configuration or reconfiguration of the PH measurement task.

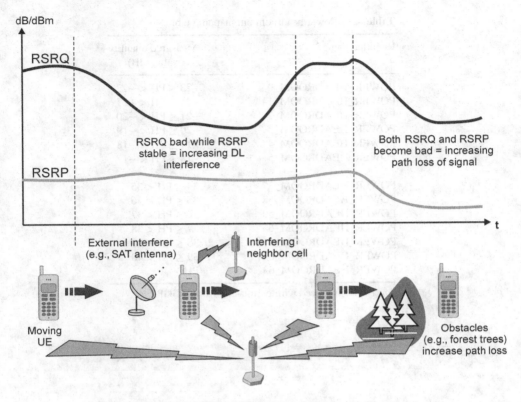

Figure 4.22   LTE coverage and interference problems

The PH reports are sent on the MAC layer, not in RRC measurement reports! However, the eNB is able to control the UE's maximum UL transmission power using the P-max parameter in RRC signaling.

The PH reporting range is from −23 to +40 dB. Table 4.3 defines the mapping for the report in binned format. The 64 possible values correspond to 6 bits of the PH control element in the MAC protocol definitions.

Since the PH reporting is not foreseen as an eNB statistical value to be reported via a northbound interface, it can only be analyzed and visualized by using a radio interface protocol tester.

4.3.1.4   UL Scheduling Requests

The UL scheduling request is used for requesting Uplink Shared Channel (UL-SCH) resources for new transmission. This is also not a measurement for statistical purposes, but mandatory for troubleshooting and optimizing the UL scheduling function of the eNB.

UL scheduling requests are sent on the MAC layer and can be tracked for measurement purposes by an air interface tester and visualized as shown in Figure 4.23. There are no scheduling request statistics provided by the eNB.

Table 4.3 Power headroom bin mapping table

Reported value	Measured quantity value (dB)
POWER_HEADROOM_0	$-23 \leq PH < -22$
POWER_HEADROOM_1	$-22 \leq PH < -21$
POWER_HEADROOM_2	$-21 \leq PH < -20$
POWER_HEADROOM_3	$-20 \leq PH < -19$
POWER_HEADROOM_4	$-19 \leq PH < -18$
POWER_HEADROOM_5	$-18 \leq PH < -17$
…	…
POWER_HEADROOM_57	$34 \leq PH < 35$
POWER_HEADROOM_58	$35 \leq PH < 36$
POWER_HEADROOM_59	$36 \leq PH < 37$
POWER_HEADROOM_60	$37 \leq PH < 38$
POWER_HEADROOM_61	$38 \leq PH < 39$
POWER_HEADROOM_62	$39 \leq PH < 40$
POWER_HEADROOM_63	$PH \geq 40$

Source: Reproduced with permission from © 3GPP™.

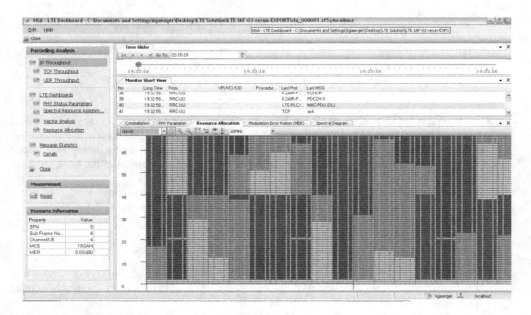

Figure 4.23   UL scheduling for four different UEs in same cell

4.3.1.5 Buffer Status Reporting

Another important input for the UL scheduling that is also sent on the MAC layer is the UL buffer status report of the UE. It is used to inform the serving eNB about the amount of data waiting for transmission in the UL buffers of the UE.

The Buffer Status Reports (BSRs) are either sent periodically or event-triggered. Typical event triggers are the following:

- UL data for a logical channel, which belongs to a logical channel group, becomes available for transmission in the Radio Link Control (RLC) or PDCP entity.
- UL resources are allocated and the UE detects that the number of padding bits is equal to or larger than the size of the BSR MAC control element. In such a case, the BSR is called a Padding Buffer Status Report.
- A serving cell change occurs or the retransmission timer for BSRs expires while the UE has data waiting for UL transmission. Here, the 3GPP specs use the name "Regular Buffer Status Report." A Regular Buffer Status Report to be sent to the eNB should trigger an UL scheduling request to be sent in parallel.

It is also necessary to distinguish between long and short BSRs (compare Figure 4.24 to Figure 4.25). The long report format is used if more than one logical channel group has data available for UL transmission in the TTI where the BSR is transmitted. In any other case, the short format is used.

For the reporting, a binned format (index) is used as defined in Table 4.4. Like other UE measurements reported on the MAC layer, there are no statistics for BSRs from the eNB defined by 3GPP. Hence, the availability of measurement results for troubleshooting and network optimization relies on proprietary implementations and the radio interface test equipment.

4.3.2 The eNodeB Physical Layer Measurements

4.3.2.1 Received Interference Power (RIP)

This is the UL noise floor for a set of UL resource blocks. Typically, the UL noise is generated by the UL signals of all UEs received in the particular frequency range of these resource blocks on a single

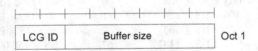

Figure 4.24 Short buffer status MAC control element. (*Source*: Reproduced with permission from © 3GPP™.)

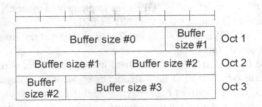

Figure 4.25 Long buffer status MAC control element. Reproduced with permission from © 3GPP™

Table 4.4 BSR bin mapping table

Index	Buffer Size (BS) value (bytes)	Index	BS value (bytes)
0	BS = 0	32	1132 < BS ≤ 1326
1	0 < BS ≤ 10	33	1326 < BS ≤ 1552
2	10 < BS ≤ 12	34	1552 < BS ≤ 1817
3	12 < BS ≤ 14	35	1817 < BS ≤ 2127
4	14 < BS ≤ 17	36	2127 < BS ≤ 2490
5	17 < BS ≤ 19	37	2490 < BS ≤ 2915
6	19 < BS ≤ 22	38	2915 < BS ≤ 3413
7	22 < BS ≤ 26	39	3413 < BS ≤ 3995
8	26 < BS ≤ 31	40	3995 < BS ≤ 4677
9	31 < BS ≤ 36	41	4677 < BS ≤ 5476
10	36 < BS ≤ 42	42	5476 < BS ≤ 6411
11	42 < BS ≤ 49	43	6411 < BS ≤ 7505
12	49 < BS ≤ 57	44	7505 < BS ≤ 8787
13	57 < BS ≤ 67	45	8787 < BS ≤ 10 287
14	67 < BS ≤ 78	46	10287 < BS ≤ 12 043
15	78 < BS ≤ 91	47	12043 < BS ≤ 14 099
16	91 < BS ≤ 107	48	14 099 < BS ≤ 16 507
17	107 < BS ≤ 125	49	16 507 < BS ≤ 19 325
18	125 < BS ≤ 146	50	19 325 < BS ≤ 22 624
19	146 < BS ≤ 171	51	22 624 < BS ≤ 26 487
20	171 < BS ≤ 200	52	26 487 < BS ≤ 31 009
21	200 < BS ≤ 234	53	31 009 < BS ≤ 36 304
22	234 < BS ≤ 274	54	36 304 < BS ≤ 42 502
23	274 < BS ≤ 321	55	42 502 < BS ≤ 49 759
24	321 < BS ≤ 376	56	49 759 < BS ≤ 58 255
25	376 < BS ≤ 440	57	58 255 < BS ≤ 68 201
26	440 < BS ≤ 515	58	68 201 < BS ≤ 79 846
27	515 < BS ≤ 603	59	79 846 < BS ≤ 93 479
28	603 < BS ≤ 706	60	93 479 < BS ≤ 109 439
29	706 < BS ≤ 826	61	10 9439 < BS ≤ 128 125
30	826 < BS ≤ 967	62	12 8125 < BS ≤ 150 000
31	967 < BS ≤ 1132	63	BS > 150 000

Source: Reproduced with permission from © 3GPP™.

Rx antenna. The measurement is defined by 3GPP to be implemented in the eNB, but can also be provided by an air interface tester that is connected to the eNB's CPRI.

The reporting range for RIP is from −126 to −75 dBm. The values are reported in binned format according to the definitions in Table 4.5.

Besides the UL load in the cell that is determined by the number of active subscribers, the exceptional values of RIP can also be caused by high-frequency signal sources outside the LTE radio network. In 3G UMTS networks, typical sources of interfering high-frequency signals have been identified as old-fashioned DECT phones, radar systems of airports and ships, and amplifiers of satellite dishes used to receive satellite TV at home. The same kind of interference caused by external signal sources can be

Table 4.5   Received interference power – reporting range and bin mapping table

Reported value	Measured quantity value	Unit
RTWP_LEV_000	RIP < −126.0	dBm
RTWP_LEV_001	−126.0 ≤ RIP < −125.9	dBm
RTWP_LEV_002	−125.9 ≤ RIP < −125.8	dBm
...	...	...
RTWP_LEV_509	−75.2 ≤ RIP < −75.1	dBm
RTWP_LEV_510	−75.1 ≤ RIP < −75.0	dBm
RTWP_LEV_511	−75.0 ≤ RIP	dBm

Source: Reproduced with permission from © 3GPP™.

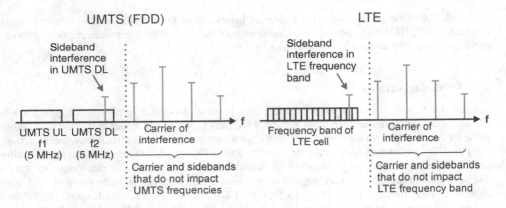

Figure 4.26   Impact of external interference in UMTS and LTE cells

expected to be found also in LTE cells, but as shown in Figure 4.26, the impact on the LTE cell is much less dramatic compared to what happens in the UMTS cell.

In UMTS, if a single sideband of an external high-frequency signal source strikes the UL or DL frequency band of the cell with a certain power, the entire bandwidth of the cell is impacted and in the worst case, the cell becomes unusable for transmission of UMTS radio signals, because all connections are distributed over the entire frequency band and diversity between different connections is only available in the power domain (amplitude of UMTS signals). In LTE, the scheduling grid that gives diversity in both the frequency and time domains reduces the impact of the interfering sideband to a minimum. If the interfering carrier is permanently present, no more than just a few subframes of the entire frequency band become unusable and in the case of, for example, radar beams that strike the cell with a certain periodicity, the impact on the subcarriers is further limited to a few sub-slots of a particular frequency in the time domain.

All in all, this comparison shows that LTE is much less interference sensitive than UMTS FDD and, hence, it can be expected that also "interference hunting" will have much less importance during the deployment phase of the networks compared to what was done during UMTS FDD rollout.

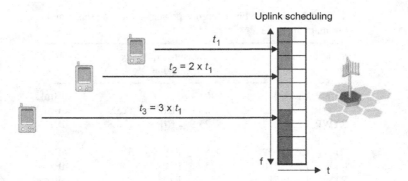

Figure 4.27   Timing advance principle

4.3.2.2   Thermal Noise Power

This is the UL noise for the entire UL frequency bandwidth of the receiving cell without the signals received from LTE handsets. So it is the UL noise without the LTE traffic. The measurement is optionally provided by the eNB and can also be provided by a radio interface tester.

4.3.2.3   Timing Advance

Actually, the timing advance is not measured for the purpose of statistics. Rather, it is required to synchronize the transmission and reception of UL radio signals between the UE and eNB in the time domain of the air interface. The timing advance is the estimated time that a particular UL subframe needs to travel from the UE's Tx antenna to the cell's Rx antenna. This is illustrated in Figure 4.27 where three different UEs have been scheduled for the same UL sub-slot on the time domain, but due to the fact that the distance between the UE and receiving cell is different for each handset, the radio signal of UE3 needs three times more time to travel all the way from the UE to the cell. Hence, the UL signal of UE3 must be sent earlier compared to the signal of UE1 if both signals are to arrive at the same sub-slot of the time domain in the cell. By sending the timing advance command, the eNB adjusts the proper arrival time of all three UL radio signals individually.

The initial timing advance command is sent together with the random access response encoded in an 11-bit timing advance value T_A. The 11 bits define a range of possible random access timing advance values represented by integer index values of $T_A = 0, 1, 2, \ldots, 1282$. The step size of the timing advance value is expressed in multiples of $16T_s$, where T_s is the basic time unit of the LTE radio interface and defined in 3GPP 36.211 as follows:

$$T_s = 1/(15\,000 \times 2048)\mathrm{s} = 1/30\,720\,000\,\mathrm{s} = 32.552083\,\mathrm{ns}$$

Thus, one step timing advance (T_A) in the time domain of the radio signal is

$$T_A = 16T_s = 16 \times 32.552083\,\mathrm{ns} = 0.52\,\mu\mathrm{s}$$

Considering that the radio waves travel at speed of light c or $300\,000$ km/s, the geographic distance for a single timing advance step can be calculated using the following formula where r represents the radius of the cell at a distance equal to T_A:

$$r = c \times 16T_s = 300\,\mathrm{m} \times 0.52 = 156\,\mathrm{m}$$

As T_A is a round-trip time measurement, the step distance needs to be divided by factor 2. In other words, the distance between UE and cell's antenna is only half of the TA measurement result. Thus, the true distance steps are increasing in multiples of 78 m.

If the maximal timing advance index value of 1282 is seen during the random access response, this means that the UE is $1282 \times 78\text{ m} = 99\,996\text{ m}$ (roughly 100 km) away from the cell's antenna when sending the random access preamble.

After establishment of the RRC connection, the 6-bit timing advance values are sent on the MAC layer using the Physical Downlink Shared Channel (PDSCH) whenever the distance between the UE and cell changes significantly. The 6-bit field allows a range of MAC timing advance command values between 0 and 63. These timing advance command values sent during the active call are relative to the current UL timing, which means that they do not correspond to the total distance between the UE and cell, but only to the change in distance since the last timing advance command was sent.

Thus, whenever a UE receives a timing advance command from the eNB, it needs to calculate the new timing advance using the formula

$$N_{T_A,\text{new}} = N_{T_A,\text{old}} + (T_A - 31) \times 16$$

The best possible granularity for timing advance commands is 2 Hz – when a new timing advance command is sent every 500 ms. Using the 64 index values, a distance of 64 x 78 m = 4992 m (roughly 5 km) is covered and, hence, the timing advance can be properly adjusted when the UE changes its position relative to the eNB by ±5 km within 500 ms. Theoretically, this would be sufficient in cases where the UE moves at 3600 km/h. However, it must be taken into account that changes in the air interface radio way are not just due to subscriber mobility, but also the longer ways in the air due to reflection of radio signals, especially in city center environments like New York City.

It should also be noted that there is a delay between the reception and execution of a timing advance command inside the UE. Normally, a timing advance command received at the UE is executed for the UL subframe that begins six subframes later. Errors in timing advance measurement, transmission, and execution will cause in the worst case the loss of UL radio frames, which seriously deteriorates the user's quality of experience, especially for real-time services like Voice over IP (VoIP). To detect defects in the timing advance procedure, these measurements should not be just measured per call, but also collected in an aggregated format per handset type and cell to benchmark the equipment of different UE manufacturers and cells that cover different geographic areas against each other. Besides correlation with UL radio quality measurements, the timing advance information can also be helpful in estimating the UE's geographic position in cases when GPS methods are not available.

The timing advance measurements are not defined by 3GPP as part of the eNB statistical measurements. Rather, it is a task for a protocol tester to decode and store the timing advance command values in a trace file. It is possible to capture the timing advance commands directly on the radio signal stream that is monitored at the CPRI. Alternatively, eNBs may have a monitoring port to allow the capture of radio protocol traces.

4.3.3   Radio Interface Tester Measurements

4.3.3.1   Channel Baseband Power

The channel baseband power measurement is used to track the changes in the power amplitude of physical channels over time. This measurement is available for both the receiver and transmitter sides of a particular physical channel. Thus, in an ideal measurement scenario, a radio interface tester should be located at each side of the connection, at the UE and eNB.

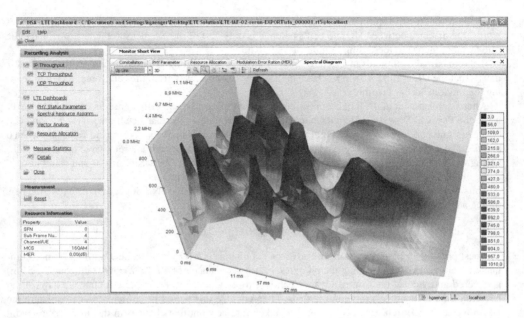

Figure 4.28   Channel baseband power measurement graph

Under these conditions, the baseband power measurements for all physical channels are available in a graphical format as shown in Figure 4.28. In particular, these measurements are used to evaluate the amplitude of sent and received signals for the following:

- Physical Downlink Shared Channel (PDSCH).
- Physical Downlink Control Channel (PDCCH).
- Primary Synchronization Channel (P-SCH).
- Secondary Synchronization Channel (S-SCH).
- Physical Broadcast Channel (PBCH).
- Physical Uplink Shared Channel (PUSCH).
- Physical Uplink Control Channel (PUCCH).

4.3.4   I/Q Constellation Diagrams

A constellation diagram is a scatter diagram used to visualize the distribution of symbols of a modulated signal in the so-called complex plane. For each common modulation scheme such as Phase Shift Keying (PSK) and Quadrature Amplitude Modulation (QAM), the ideal distribution pattern of symbols in the complex plane is known from signal theory. Now, the measured pattern can be visually compared to the expected ideal pattern.

In the ideal pattern, each symbol dot is laser focused on a particular fixed position of the constellation diagram. The real-time measurement as shown in Figure 4.29 shows the symbol measurement samples "dancing" around their ideal positions and the further they are from the ideal position as plotted, the more the signal was corrupted.

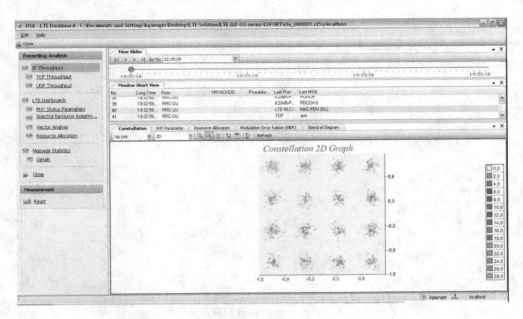

Figure 4.29   Two-dimensional I/Q constellation diagram

For the purpose of analyzing received signal quality, some types of corruption are very evident in the constellation diagram. Typical radio transmission problems can be easily recognized as follows:

- Gaussian noise becomes evident as fuzzy constellation points.
- Non-coherent single frequency interference manifests as circular constellation points.
- Phase noise leads to rotationally spreading constellation points.
- Amplitude compression causes the corner points to move toward the center.

The I/Q constellation diagram can be measured on the transmitter side – then it shows the quality of signal modulation before transmission over the air interface. However, this use case is seen more in the lab than in live networks. The typical use case for live networks is to measure the modulation quality of a received signal. A high-quality air interface tester should provide the particular I/Q constellation diagrams for the following physical channels:

- Physical Downlink Shard Channel (PDSCH).
- Physical Downlink Control Channel (PDCCH).
- Primary Synchronization Channel (P-SCH).
- Secondary Synchronization Channel (S-SCH).
- Physical Broadcast Channel (PBCH).
- Physical Uplink Shared Channel (PUSCH).
- Physical Uplink Control Channel (PUCCH).
- Sounded Reference Symbols (SRSs) for UL.

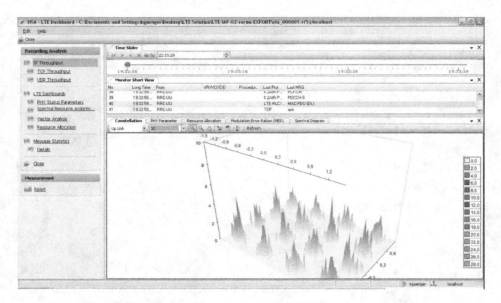

Figure 4.30   Three-dimensional I/Q constellation diagram

The 2D constellation diagram can be enriched with the individual signal strength amplitude for each received and demodulated symbol. The result is a 3D constellation diagram as shown in Figure 4.30.

4.3.5   EVM/Modulation Error Ratio

The EVM is sometimes also called the Receive Constellation Error (RCE). It is a measurement used to express the difference between measured constellation points and the ideal constellation point in the complex plane; in other words, how far the measured points are from the ideal location.

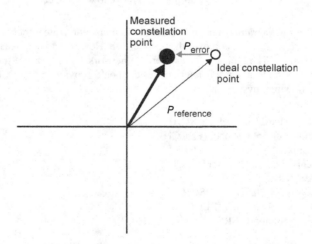

Figure 4.31   Measurements to compute EVM/MER

Like I/Q constellation diagrams, the EVM can be measured with radio interface testers and cannot be provided by any drive test or eNB statistics.

The principle of determining the EVM is shown in Figure 4.31. Due to radio transmission error, the measured constellation point drifted away from the ideal location and, starting from the center of the I/Q constellation diagram, a power vector for each point – ideal and measured – can be calculated. These vectors are P_{error} and $P_{reference}$.

To compute the EVM, P_{error} and $P_{reference}$ are set in relation to each other so that the EVM is the ratio of the power of the error vector to the root mean square of the reference power. The result can be expressed as a percentage when using Equation 4.1:

$$EVM(\%) = \sqrt{\frac{P_{error}}{P_{reference}}} \times 100\% \tag{4.2}$$

The measurement can also be defined in decibels when Equation 4.2 is used with identical values for error vector power and reference vector power:

$$EVM(dB) = 10 \log_{10}\left(\frac{P_{error}}{P_{reference}}\right) \tag{4.3}$$

Now, within LTE cells, the power and modulation scheme assigned to particular subcarriers over time may change frequently and, to give an overview of the radio transmission quality of the entire frequency spectrum, an aggregation of measurement results is required. The aggregated view with one MER (Modulation Error Ratio) bar for each subcarrier of the frequency spectrum is shown in Figure 4.32. By filtering on radio frames belonging to a particular UE, it is even possible to visualize in this view the changes in EMV/MER during a single call.

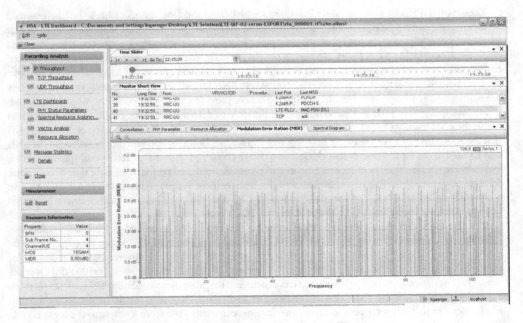

Figure 4.32   Modulation error ratio for LTE spectrum

4.4   Control Plane Performance Counters and Delay Measurements

A set of common performance counters for accessibility, retainability, and mobility KPIs is generated by network elements like eNB and MME. Hence, these performance counters are also named network element counters or OMC counters.

The disadvantage of such network element counters is that their trigger point definitions often follow proprietary standards so that, for example, a set of Nokia Siemens Network MME counters will never be 100% identical to a set from an Ericsson MME. The differences are growing in the field of KPI formula definition that is in large portions not covered by any international standard although a comprehensive set of KPIs for the E-UTRAN is defined in 3GPP 32.450.

The purpose of these 3GPP-defined KPIs is to provide insights to the user experience, while the Layer 2 user plane measurements according to 3GPP 36.314 are designed to detect problems in the radio transmission and in the eNodeB itself when handling the payload. The 3GPP 32.450 KPI set comprises:

- a combined formula for E-RAB Accessibility that takes successful establishment of the RRC signaling connection into account;
- an E-RAB Retainability KPI that gives the number of E-RAB drops over time of E-RAB activity. The drop counter of this KPI only increases if there was payload data in the buffer waiting to be transmitted.
- The E-UTRAN IP Throughput measures how fast the payload in the buffer is forwarded in uplink or downlink direction. Time periods of an active radio connection or UE context without having payload in the buffer are not considered in this throughput measurement, so the real speed of the radio data transmission is measured.
- The E-UTRAN Cell Availability provides information about network element outages that have an impact on subscriber experience.
- The E-UTRAN Mobility Success Rate shows the percentage of all successfully executed intra-LTE handovers per QCI.

The set of 3GPP 32.450 standard KPIs is quite sophisticated and almost impossible to be implemented by external measurement equipment, because check of user plane buffer status is mandatory and external measurement equipment cannot check the user plane buffer inside the network elements. However, even implementation inside the network element is not easy and significant processing resources are required to compute such sophisticated KPIs in the network elements. For this reason, NEMs will hesitate to move forward with the implementation as proposed by 3GPP. Also, any network element performance counters are aggregated at a single measurement point and do not provide end-to-end visibility, which means that if a failure is found in eNB statistics, it might be possible to determine that the origin of this failure was not in the eNB itself, but somewhere in the core network. However, it is impossible to see from the eNB that the root cause of a core network problem that triggered a chain reaction including failure in the eNB was located in the Public Data Network Gateway (PDN-GW). To follow a call across multiple interfaces, synchronize the timestamps of measurements and protocol events, and provide a single user access point for data analysis are the job of a classical network management system. Although it is not easy to give a clear definition of a "call" in the all-IP world of LTE and EPC, the general measurement tasks and aggregation levels are almost the same as in earlier 2G/3G networks.

It is also clear that the air interface measurements that have been discussed in the previous sections in deployment scenarios with 300–6000 eNBs connected to a single MME cannot be expected to be available from all base stations of a network at any time. Nor is it technically required to monitor, for example, the scheduler efficiency for all sites 24/7 (means: 24 hours, 7 days a week), neither does it make sense from a cost-per-site perspective when it is considered that already a single air interface tester is more expensive than a complete eNB with its hardware and software. Further, it would be necessary to build an architecture that allows access to all air interface measurement data from a single user access point.

Knowing these limitations, the optimal way to troubleshoot and optimize the LTE radio access network is to use a set of rather simple control plane and user plane performance counters and measurements that are aggregated on a per cell or per site basis to allow rapid identification of problematic areas. The details of such problems – as far as they are related to air interface parameters and radio transmission conditions – can then be analyzed in an on-demand scenario using the highly sophisticated measurements and statistics of portable air interface testers. Another solution of gaining access to at least a subset of radio interface statistics is the implementation of trace ports in the eNB that mirror the RRC traffic together with some key parameters of the MAC layer (such as the Radio Network Temporary Identifier, RNTI), and thus, this may become the most common solution for performance measurement and radio network optimization.

The following subsections contain a detailed description of the most common signaling KPIs, their meanings, and options for troubleshooting and optimization.

4.4.1   Network Accessibility

Network accessibility in LTE requires different performance counters than those known in 2G and 3G radio access networks. Only RRC connection setup counters are almost identical to what is known from 3G UMTS.

When the term "call establishment" is used, it needs to be discussed with respect to the all-IP environment of E-UTRAN/EPC: what is the definition of a "call" in this network environment? Actually, the term "call" is no longer used by 3GPP standard documents that deal with the E-UTRAN and EPC network architecture and concept, but it is still commonly used in the field of drive test, network operation and maintenance, and performance measurement. Typically, a "call" means all signaling messages and payload belonging to a single subscriber. The function of a call trace application is to filter all messages and payload packets belonging to such a particular user connection and display these messages and packets in the correct time sequence. Knowing this, a call can be defined as a single RRC connection between the UE and the network. This also fits the definition of a "call drop" in the radio access network environment, which means that the connection between the UE and network was lost. Surely, within such a "call," multiple different service flows and bearers are active. Each bearer is characterized by an individual set of QoS parameters and nine different QCIs have been defined by 3GPP that fit the most common service requirements. At the end, it is meaningful to aggregate all accessibility performance counters on two different levels: the subscriber level and the service level. While the subscriber level allows computing "call setup" KPIs that reflect the user experience, the service level reflects more the network-centric view of a network manager who asks the question: "How are the services in my network performing?" In addition, it makes sense to aggregate accessibility counters on a "per handset" dimension. This will be useful for checking the robustness of new handset models and for benchmarking handsets of different brands against each others. That the differences on the handset level can be significant has been observed in 3G UTRAN when the Apple iPhone was introduced: early models of the iPhone had a tendency to trigger inter-RAT handover from 3G to 2G by factor three times faster compared to other 3G phones with the same UE capabilities. The root cause of this behavior was a minor defect in the RF receiver's chipset of the early iPhone. Problems like this can occur at any time using any radio access technology and, hence, there should be measurement tasks and reports defined that address such use cases.

4.4.1.1   Random Accessibility Failures

Due to radio interface testers and eNB trace ports, it will become quite easy to monitor the random access procedure. This was not possible in 2G and 3G RAN monitoring scenarios.

Now a random access success ratio KPI can be defined as follows:

$$\text{Random access success ratio} = \frac{\sum \text{RRC\_Connection\_Request}}{\sum \text{MAC\_Random\_Access\_Response}} \times 100\% \qquad (4.4)$$

Low percentage values of this KPI will help to identify cells with serious problems on the radio interface, because the message defined for raw counters in this formula are the first messages sent on the UL-SCH and Downlink Shared Channel (DL-SCH) of a cell. If there are radio transmission problems in a cell, the random access procedures will be the ones most impacted. Failed random access also has the highest user impact, because missing network access is immediately recognized.

4.4.1.2   RRC Accessibility Failures

SRBs are established when an RRC connection between the UE and eNB is set up. The possible options of successful and unsuccessful message flow scenarios for RRC connection setup are shown in Figures 4.33 and 4.34. An RRC connection request message is sent to the eNB after the UE has performed initial cell selection and detected the best suitable cell for radio access. Besides the UE identity, this RRC connection request message also contains an establishment cause that allows distinguishing between mobile originating signaling, mobile originating data calls, mobile terminating connections (these are all data calls), emergency calls, and high-priority access calls. A further indication of the service that is intended to be used is not possible. Hence, the idea to aggregate RRC accessibility counters on the service level is obsolete.

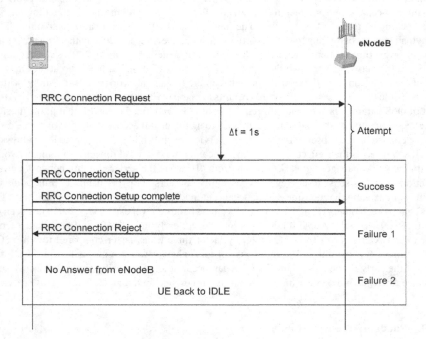

Figure 4.33   RRC connection setup procedure

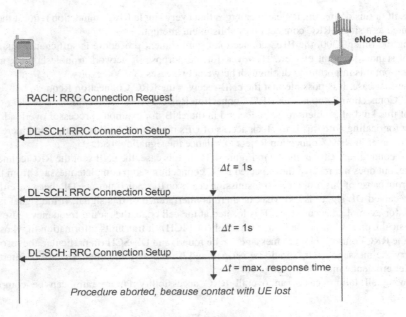

Figure 4.34   RRC setup timeout on UE side

Once the RRC connection request message is received, the eNB should respond by sending an RRC connection setup message. In case of successful RRC connection establishment, the UE will return an RRC connection setup complete message to the eNB.

Using a simple equation, the RRC connection setup success ratio can be computed per cell as follows:

$$\text{RRC connection setup success ratio} = \frac{\sum \text{RRC\_Connection\_Setup\_Complete}}{\sum \text{RRC\_Connection\_Request}} \times 100\% \qquad (4.5)$$

However, instead of the RRC connection setup message, the eNB may block the UE's connection request by sending RRC Connection Reject and an RRC blocking ratio can be computed per cell by using Equation 4.6:

$$\text{RRC blocking ratio} = \frac{\sum \text{RRC\_Connection\_Setup\_Reject}}{\sum \text{RRC\_Connection\_Request}} \times 100\% \qquad (4.6)$$

Different from the same message in 3G UTRAN, there will be no cause value signaled to the UE with RRC Connection Reject, but a wait time value of 1–16 seconds that can help to protect the eNB from overload due to rising numbers of RRC Connection Requests. When RRC Connection Request is sent, the timer T300 is started in the UE: if the RRC connection setup message is not received before timer T300 expires on the UE side, there will be no retransmission of the RRC connection request message as was seen in 3G UTRAN. Instead, the UE will reset the MAC, release the MAC configuration, and inform upper layers about the failed RRC connection establishment attempt. It is then the job of these upper layers (e.g., NAS signaling) to decide if and when a new RRC Connection Request should be sent.

As a result of this definition, it clearly emerges that every single RRC connection request message must be counted as a single RRC connection establishment attempt.

Using the timer T300, the RRC connection establishment procedure is sufficiently guarded from a protocol standard's point of view. However, for the purpose of network troubleshooting and network optimization, it is important to distinguish between two cases of T300 expiry.

In the first case, it is possible that the eNB receives the RRC Connection Request, but sends neither an RRC Connection Setup nor an RRC Connection Reject before T300 expires. It is clear that the root cause of this kind of problem must be located in the eNB, for example, processor overload. The proper strategy for dealing with this is to block access of UEs if a critical limit is reached and increase the wait time parameter in RRC Connection Reject to balance the signaling load.

The second failure case is shown in Figure 4.34. In this case, the eNB sent the RRC connection setup message, but does not receive the expected RRC connection setup complete message from the UE. The typical root cause of this problem is transmission errors on the radio interface DL, because either the level of the received DL signal is too weak or there is interference of this signal. If the signal is not strong enough, for example, because the UE is located at the cell edge, the partial frequency reuse settings for the Physical Control Format Indicator Channel (PCFICH) (it transmits information in whose resources blocks the RRC connection setup message can be found) and DL-SCH (in particular, the resource blocks that carry the message itself) should be verified and if necessary modified. In case of interference, the scheduler efficiency must be investigated.

Knowing all failure cases, an overall RRC accessibility failure ratio can be computed as in Equation 4.7:

$$\frac{\sum \text{RRC\_Connection\_Setup\_Reject} + \sum \text{RRC\_Setup\_timeout\_eNodeB} + \sum \text{RRC\_Setup\_timeout\_UE}}{\sum \text{RRC\_Connection\_Request}} \times 100\% \qquad (4.7)$$

4.4.1.3 Call Setup Failures

If the definition can be agreed that a "call" is a single radio connection between the UE and the network that is used to transmit payload using multiple bearers and service flows, then all failures that prevent a UE from attaching to the network and enabling Public Data Network (PDN) connectivity can be summarized by the umbrella term "call setup failures."

The strong link between attach to network and PDN connectivity is also reflected by the standards, especially 3GPP 24.301 "NAS Protocol for Evolved Packet System (EPS)." Here, it is defined that in case of an unsuccessful attach, an explicit PDP connection reject message should be sent to the UE to ensure proper transitions of EPS Mobility Management (EMM) and ESM (EPS Session Management) states in the UE's NAS signaling entity.

Figure 4.35 shows an attach rejected by the network due to problems in the EPC. The most common problems are that one of the GPRS Tunneling Protocol (GTP) tunnels for the default EPS bearer cannot be established or the location update procedure between the MME and Home Subscriber Server (HSS) fails. Also, failures in Stream Control Transmission Protocol (SCTP) transport as described in Section 1.10.3 belong to this category. However, it can be expected that errors on the SCTP level will have a more serious impact on the network performance than just the failed attach of a single subscriber.

A more detailed view of a typical update location failure is shown in Figure 4.36. Such a failed update location does not need to be necessarily a network issue. In fact, there are a couple of cases where it is meaningful that access to the network is restricted to particular subscribers or groups of subscribers. For instance, if the home network operator of the subscriber does not have a roaming agreement with the operator of the visited network, then the subscriber cannot be charged for the roaming services and,

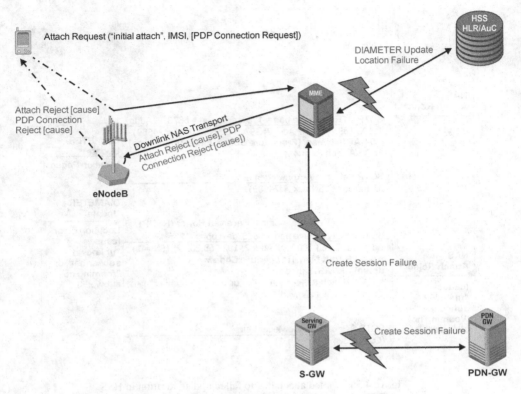

Figure 4.35   Attach failure due to failures in EPC

hence, it is good not to reject the attach request. In such cases, the cause value signaled on DIAMETER or MAP from HSS will be the same as in Attach Reject.

However, it is also quite likely that, especially in roaming scenarios, other failures in the communication between visited MME and the home network's HSS will occur. It was known from 2G and 3G networks that sometimes the latency on the link between MME and HSS is too high and, hence, the Update Location Response (DIAMETER or MAP) does not arrive in time. If this happens, typically, the cause value "network failure" is signaled in the attach reject message as shown in Figure 4.37, but not every "network failure" must have its origin in the core network. The "network failure" cause is also used if security functions in the E-UTRAN cannot be activated as requested. This generates the requirement for an intelligent network management/troubleshooting tool to distinguish and report unambiguously the true location and root cause of "network failures."

Since the initial attach procedure is linked with many other signaling procedures in the E-UTRAN and EPC, failures that occur in a particular network element or on a particular signaling link often trigger a chain reaction. This is also true in case the initial context setup procedure on S1 between the MME and eNB fails, as shown in Figure 4.38.

In the example, the eNB is not able to set up the initial context, which triggers in turn Attach Reject (with cause "network failure"? – which depends on the implementation in MME software) and PDP Connection Reject. In addition, due to the failed attach, the GTP tunnels on S1-U and S5 that have already been established need to be deleted. For this purpose, the GTP-C delete session procedure will

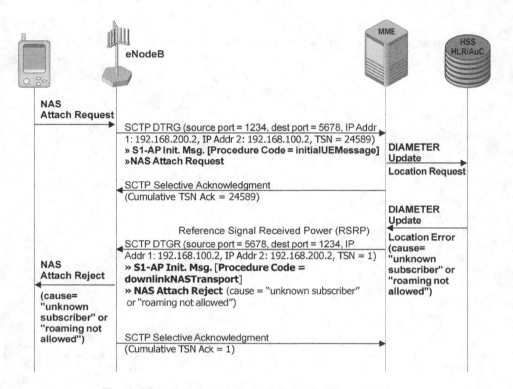

Figure 4.36   Failed attach due to failed update location in HSS

be used on the S11 signaling link between the MME and Serving Gateway (S-GW) and on S5 between S-GW and PDN-GW. Since the UE's new location was already successfully updated in the HSS, it is now necessary to delete this entry and mark the UE in the HSS database as "not reachable," because a UE that is not attached to the network should not be paged. The signaling procedure that is used to notify the HSS about this new state of the connection is the purge UE procedure.

Even if the UE can successfully attach to the network, there is another potential error with an impact on the PDN connectivity. In this case, after the successful attach, the activation of the default bearer fails. In most cases, the origin of this failure is expected to be found in the UE itself. Actually, all handsets have to undergo various load and stress test scenarios in the lab before they become available on the market. Thus, the likelihood of seeing such failures is rather small. Nevertheless, it may happen and is for good reason defined in the new LTE NAS protocol (3GPP 24.301). Figure 4.39 illustrates the signaling pattern of such failures. As was explained in Chapter 2, Activate Default EPS Bearer Request is sent by the MME to the UE together with Attach Accept. While Attach Accept is a mobility management (EMM) message, Activate Default EPS Bearer Request belongs to the category of session management (ESM) messages. If for some reason the UE is not able to confirm the successful activation of the default EPS bearer, it will respond with an activate default EPS bearer failure message. As a result, the UE will remain attached to the network (NAS state: EMM Registered) while it does not have an active bearer (NAS state in UE: ESM Bearer Context Inactive). The reaction of the MME when receiving activate default EPS bearer failure is to repeat sending Activate Default EPS Bearer Request for a maximum of four times.

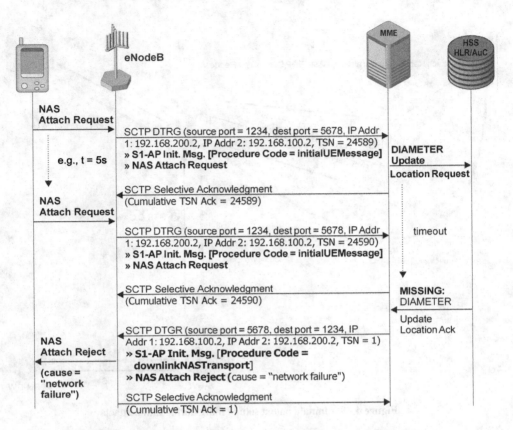

Figure 4.37 Network failure in attach procedure due to EPC latency

If this does not help to activate the bearer context on the UE side, the MME may detach the UE so that a new initial registration to the network is required.

4.4.1.4 Dedicated Bearer Setup Failure

Whenever a dedicated bearer cannot be established, as shown in Figure 4.40, access to an individual service and its contents is blocked for the subscriber. With regard to measuring user experience, this kind of failure should be aggregated on a subscriber ID that can later be used to generate customer-centric reports. However, the aggregation of performance counters for dedicated bearer setup on the service level (defined by the embedded QoS parameters in the bearer setup request message) and on the location level (defined by tracking area and E-UTRAN cell ID) is also highly recommended.

The interesting thing about the bearer setup failure is that its message name is "Successful Outcome." We can see here the same concept that was introduced with the RANAP outcome message in the UTRAN and also used to confirm successful setup and/or failed setup for a list of multiple Radio Access Bearers (RABs). This is the reason why the message name in S1AP is now "Successful Outcome" even if it reported that the required E-RAB was not going to be established.

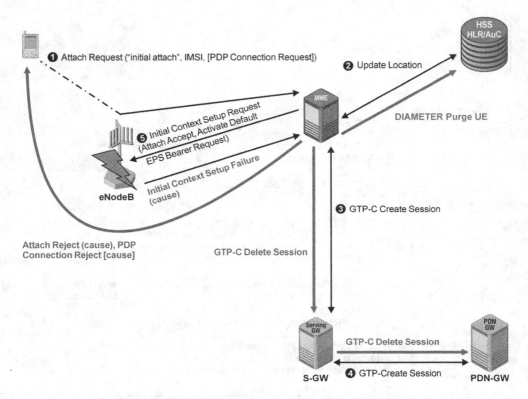

Figure 4.38   Initial context setup failure during initial attach

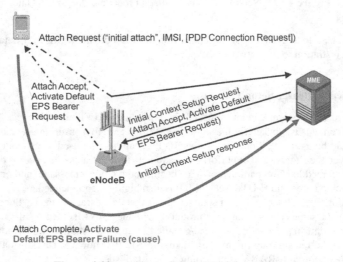

Figure 4.39   Activate default EPS bearer failure

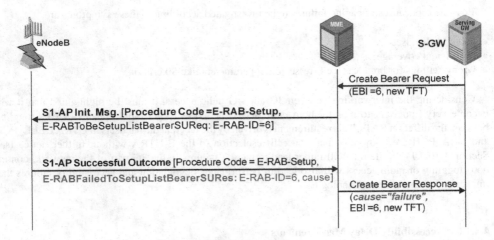

Figure 4.40   Dedicated bearer setup failure

The most typical reasons for failed E-RAB setup are drops of the radio connection before RAB establishment is completed. The eNB may also reject the setup of a particular bearer, because it is not able to provide the necessary resources for the requested service, for example, it cannot support the desired bit rates. Often the leak of resources is not necessarily related to the availability of subcarriers and time slots on the radio interface, but simply to the limited processing power and the system memory in the eNB. Processors and memory to serve a certain number of connections are found on the so-called channel cards, boards that allow scalability of hardware (and in turn scalability of the hardware price) according to the amount of expected traffic. If the eNB's hardware is inadequate, this will result in error messages with cause values like "not enough user plane processing resources," especially during peak hours of traffic.

4.4.1.5   Paging Failures

The analysis of paging failures requires a distinct workflow rather than just a simple KPI. Due to the tracking area concept of the E-UTRAN, the last known location of a handset that does not respond to paging messages sent by the MME can be more tightly encircled. In a network with one cell representing a single tracking area, the UE will be paged in a few cells only, whereas in 2G and 3G RAN, paging messages are typically sent to all cells of a particular Location Area Code (LAC) – a few hundred cells compared to a few dozen in the E-UTRAN. However, a paging failure ratio KPI on cell aggregation level will always have a questionable value due to the fact that a UE can only camp in a single cell and, hence, from most cells to where the paging was sent there will be no paging response.

It, therefore, makes sense to store the subscriber ID of UEs that have not answered the paging and investigate what may have happened to them before the paging failure and where they may have gone to. The most likely reasons for paging failures are as follows:

- The UE is defective and went out of order without being able to detach from the network.
- The UE while in IDLE mode performed a new cell (re)selection and is camping now on a 3G or 2G cell without informing the network properly about the new location that is geographically the same as before, but covered by a different RAT.

Possible root causes of paging failures to be investigated according to these symptoms are:

- defective handsets;
- insufficient coverage;
- wrong settings for broadcast cell (re)selection parameters like S0 criteria.

Considering the interworking between 3G and 4G radio access, it must be highlighted that it will become very important to monitor both radio access networks to get rapid updates of UEs toggling between the different RATs. This requirement is driven by the fact that the handsets in UTRA CELL_PCH and URA_PCH are allowed to perform cell reselection to the E-UTRA without further notice (see Section 1.10.7.1). While this will rather lead to paging failures in the 3G part of the network, continuous toggling of radio access will also lead to performance degradation and accessibility failures that impact both radio access technologies.

4.4.1.6 Accessibility Delay Measurements

An accessibility problem cannot always be detected on behalf of a dedicated failure message sent by an involved protocol entity. Some problems may not block the progress of the accessibility procedures, but cause an unacceptable delay. Bearing in mind that minimizing the access delay for the subscriber is one of the key targets in the 3GPP specifications, it appears that a couple of accessibility delay measurements are crucial for benchmarking the E-UTRAN performance. These delay measurements are:

- *Random Access Time:* Δt MAC Random Access Preamble $\rightarrow$ RRC Connection Request.
- *RRC Connection Setup Time:* Δt RRC Connection Request $\rightarrow$ RRC Connection Setup Complete.
- *NAS Attach Delay:* Δt NAS Attach Request $\rightarrow$ NAS Attach Accept.
- *Activate Default EPS Bearer Delay:* Δt NAS Activate Default EPS Bearer Request $\rightarrow$ NAS Activate Default EPS Bearer Accept.
- *Initial Context Setup Delay:* Δt S1AP Initial Context Setup Request $\rightarrow$ S1AP Initial Context Setup Response.
- *E-RAB Setup Delay:* Δt S1AP E-RAB Setup Request $\rightarrow$ S1AP E-RAB Setup Response.
- *Service Request Delay:* Δt S1AP Initial UE Message (NAS Service Request) $\rightarrow$ S1AP Initial Context Setup Response.
- *Paging Response Time:* Δt S1APPaging $\rightarrow$ S1AP Initial UE Message (NAS Service Request).

All delay measurements that involve the handset, especially RRC Connection Setup Time and Paging Response Time, should be correlated not just with the handset type, but also with the initial timing advance measurement from the random access response. This allows efficient benchmarking of different UE types that work under similar radio conditions (same distance between the cell antenna and UE).

4.4.2 Network Retainability

4.4.2.1 Call Drops

As previously stated, a "call" in the environment of the all-IP always-on E-UTRAN can only be defined as a single radio connection between a UE and the network. When this connection is interrupted due to a suddenly occurring exception (e.g., signal lost on the radio interface), the definition of "call drop" is fulfilled.

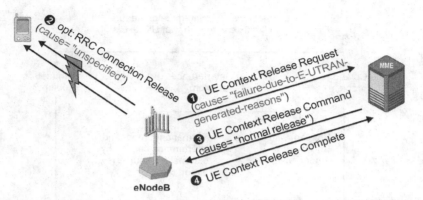

Figure 4.41   Call drop due to transmission failures on the radio interface

The drop events can be found in the S1AP UE context release request message sent by the eNB to the MME (see Figure 4.41). When this message is sent, the radio connection with the UE is already terminated on the RRC layer, so UE and eNB go back to the E-UTRA RRC IDLE state. However, the PDN connection between the UE and the server hosting application contents on the IP network will typically remain active.

The KPI formula for a call drop ratio on the cell aggregation level can be defined as follows:

$$\frac{\sum \text{UE Context Release Request (abnormal cause)}}{\left(\sum \text{S1Ap Initial Context Setup} + \sum \text{S1Ap Handover Notify} - \sum \text{UE Context Release [cause ``successful handover'']}\right)} \times 100\% \qquad (4.8)$$

In this equation, the total number of active calls is defined as the number of initial contexts successfully established plus the number of incoming handovers (identified by the S1AP handover notify message) minus the number of outgoing handovers (UE Context Release due to "successful handover").

The "call drop" in the all-IP world of E-UTRAN and EPC will not always be perceived as a dropped connection by the subscriber. For non-real-time services like web browsing or e-mail, user perception is described as rather a temporary interruption of data transport, a delay in accessing the next website, or a significant downturn of the data transmission rate of an ongoing download. If the network can re-establish the lost radio connection fast enough (and this is an important KPI for RRC retainability), the drop of the radio connection will not be recognized by the subscriber.

It is a different situation if real-time services are used by the subscriber. In this case, the user will immediately recognize the loss of connection, because the ongoing conversation with the peer party, for example, in a VoIP call is suddenly interrupted and it would require extremely fast RRC re-establishment procedures (successful re-establishment within 1 or 2 seconds) to save the situation.

Knowing this difference, it appears that it is mandatory not just to compute a call drop ratio per cell, but also to have a call drop ratio per service (per QCI is sufficient) within the cell to measure the user-perceived QoE.

The root causes for call drops are varied and cannot be unambiguously identified just by looking at the cause value in the S1AP UE context release request message. In fact, the identification of the root cause requires an in-depth analysis of all transport and signaling protocols involved in the call, with a focus on call state transitions and changing radio quality parameters. Figure 4.42 gives an overview of

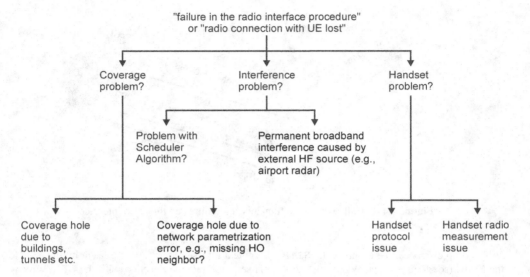

Figure 4.42   Possible root causes for an E-UTRAN call drop

which different root causes can be revealed from common call drop S1AP causes like "failure in the radio interface procedure" and "radio connection with UE lost."

However, different from 3G, the "missing HO neighbor" would indicate a problem with SON Automatic Neighbor Reporting functions.

4.4.2.2   RRC Re-Establishment

As discussed in Section 3.2, it is mandatory for the UE to try a RRC re-establishment before giving up the call as dropped and going back to IDLE. RRC re-establishment can help to recover connection problems in areas of quickly changing radio conditions. In turn, counting the number of RRC re-establishment messages and computing a meaningful KPI for this procedure per cell will help to identify such problem areas in the network.

Looking at the messages shown in Figure 3.7, an RRC re-establishment success ratio can be defined as follows:

$$\text{RRC re-establishment success ratio} = \frac{\sum \text{RRC\_Connection\_Re-Establishment\_Complete}}{\sum \text{RRC\_Connection\_Re-Establishment\_Request}} \times 100\%$$

(4.9)

4.4.3   Mobility (Handover)

Basically, three major groups of problems due to mobility can be identified when monitoring the E-UTRAN:

- Handover preparation failures.
- Handover execution failures.
- Data forwarding failures.

Handover preparation is the phase in which the target cell assigns the necessary radio resources for taking over the connection and sending back a handover command message containing the new radio parameters to the source cell.

Root causes for handover preparation failures are insufficient resources in the desired target cell, signaling transport between the source and target cell, protocol errors in one of the involved peer entities, and parameterization errors in the network configuration, for example, errors in IP/ATM routing tables.

The handover execution phase starts when the previously received handover command message is sent to the UE and successfully finished after the UE has arrived at the target cell.

Handover execution can fail because the UE refuses to execute the handover by sending an RRC reconfiguration failure message or if radio contact with the UE is lost during handover execution on the radio interface.

The various handover procedures in the E-UTRAN can also be classified by their type:

- X2 handover/S1 path switch = Inter-eNB handover with X2.
- S1 handover = Inter-eNB handover without X2. Special cases of this scenario are the inter-MME handover with/without S-GW and/or PDN-GW relocation.
- Inter-RAT handover to 3G UTRAN.
- Inter-RAT handover to GERAN.
- Other inter-RAT handover types, for example, to TD-SCDMA or CDMA2000 radio access technology (not discussed in more detail in this chapter).
- Inter-RAT handover from 3G UTRAN to E-UTRAN.
- Inter-RAT handover from GERAN to E-UTRAN.

The handover type can be distinguished according to the interface (X2 or S1) and for S1 according to the different containers embedded in the S1AP handover required message. In the case of S1 intereNB handover, the *Source eNB to Target eNB Transparent Container* is included. Inter-RAT handover to 3G UTRAN requires the *Source RNC to Target RNC Transparent Container* and inter-RAT handover to GERAN the *Source BSS to Target BSS Transparent Container*.

Furthermore, it is necessary to distinguish between incoming and outgoing handover legs in the case of intra-system S1 handover. This differentiation is possible due to the S1AP mobility elementary procedures. While the S1AP handover preparation procedure is found on the outgoing leg, the incoming leg deals with the S1AP handover resource allocation procedure.

3GPP 36.300 defines three different HO failure types according to their root causes to be detected automatically by eNodeB SON functions:

- The "Too Late Handover" is seen if a radio link failure occurs after the UE has stayed for a long period of time in the cell and the UE sends RRC Re-Establishment Request message to a different cell than the original serving cell. If the radio connection cannot be re-established, this "Too Late Handover" can seem as a special root cause of call drops.
- The "Too Early Handover" conditions are fulfilled when a radio link failure occurs shortly after a successful handover from a source cell to a target cell or if a handover failure occurs during the handover procedure and the UE sends an RRC Re-Establishment Request message to the previously serving cell. This case is similar to the "reconfiguration failure" example shown in Figure 3.27.
- The "Handover to Wrong Cell" case is characterized by the fact that a radio link failure occurs shortly after a successful handover from a source cell to a target cell or a handover failure occurs during the handover procedure and the UE sends an RRC Re-Establishment Request message to a cell other than the source cell and the target cell involved in the handover procedure.

"Successful handover" in the context of the 3GPP standards means at least the random access procedure in the target cell was successfully completed.

The "Too late HO" and "HO to wrong cell" occurrences shall be explicitly signaled in the non-call-related X2AP Handover Report message, but by the time of writing this chapter (October 2013), these messages have not been seen yet in any environment.

4.4.3.1 Handover Preparation

Consequently, the following KPIs can be defined for handover preparation and the failure report should display the cause values found in handover failure messages[1]:

$$\text{X2 handover preparation failure ratio} = \frac{\sum \text{X2\_Handover\_Preparation\_Failure}}{\sum \text{X2\_Handover\_Request}} \times 100\%$$

$$(4.10)$$

$$\text{S1 outgoing handover preparation failure ratio} = \frac{\sum \text{S1AP\_Handover\_Preparion\_Failure}}{\sum \text{S1AP\_Handover\_Required}} \times 100\%$$

$$(4.11)$$

$$\text{S1 incoming handover preparation failure ratio} = \frac{\sum \text{S1AP\_Handover\_Failure}}{\sum \text{S1AP\_Handover\_Request}} \times 100\% \qquad (4.12)$$

The message names used in 3GPP standard documents and in the aforementioned KPI formulas need to be mapped to the ASN.1 encoded message format used to construct a particular X2AP/S1AP message based on the X2AP/S1AP message type and associated procedure code:

```
handoverPreparation X2AP-ELEMENTARY-PROCEDURE ::= {
  INITIATING MESSAGE        HandoverRequest
  SUCCESSFUL OUTCOME        HandoverRequestAcknowledge
  UNSUCCESSFUL OUTCOME      HandoverPreparationFailure
  PROCEDURE CODE            id-handoverPreparation
  CRITICALITY               reject
}
handoverPreparation S1AP-ELEMENTARY-PROCEDURE ::= {
  INITIATING MESSAGE   HandoverRequired
  SUCCESSFUL OUTCOME   HandoverCommand
  UNSUCCESSFUL OUTCOME      HandoverPreparationFailure
  PROCEDURE CODE            id-handoverPreparation
  CRITICALITY       reject
}

handoverResourceAllocation S1AP-ELEMENTARY-PROCEDURE ::= {
  INITIATING MESSAGE   HandoverRequest
  SUCCESSFUL OUTCOME   HandoverRequestAcknowledge
  UNSUCCESSFUL OUTCOME      HandoverFailure
  PROCEDURE CODE            id-HandoverResourceAllocation
  CRITICALITY       reject
}
```

---

[1] Check the embedded container to distinguish between intra-system and inter-RAT handover preparation.

4.4.3.2 Handover Execution

KPIs for the handover execution phase can be defined as follows:

$$\text{X2 handover execution success ratio} = \frac{\sum \text{X2\_SN\_Status\_Transfer}}{\sum \text{X2\_Handover\_Request\_Acknowledge}} \times 100\% \quad (4.13)$$

$$\text{S1 path switch failure ratio} = \frac{\sum \text{S1AP\_Path\_Switch\_Failure}}{\sum \text{S1AP\_Path\_Switch\_Request}} \times 100\% \quad (4.14)$$

S1 outgoing handover execution success ratio

$$= \frac{\sum \text{S1AP\_UE\_Context\_Release\_("successful handover")}}{\sum \text{S1AP\_Handover\_Command}} \times 100\% \quad (4.15)$$

The handover type (intra-system or inter-RAT) must be detected according to the previously monitored Handover Required message:

$$\text{S1 incoming handover execution success ratio} = \frac{\sum \text{S1AP\_Handover\_Notify}}{\sum \text{S1AP\_Handover\_Request\_Acknowledge}} \times 100\%$$
$$(4.16)$$

The handover type (intra-system or inter-RAT) must be detected on the source ID found in *Source eNB to Target eNB Transparent Container* in the previously monitored handover request message.

One interesting information element is the *UE History Information* that is included as a mandatory parameter in the *Source eNB to Target eNB Transparent Container* and contains a list of up to 6 cells maximum visited by the UE before the handover attempt to the target cell. The history can include E-UTRAN cells as well as 3G UTRAN cells but is not defined for cells of any other radio access technology.

4.4.3.3 Handover Delay Measurements

These measurement are:

- *X2 Handover Delay* = Δt X2AP Handover Request → X2AP Sequence Number (SN) Status Transfer.
- *S1 Path Switch Delay* = Δt S1AP Path Switch Request → S1AP Path Switch Request Acknowledge.
- *S1 Outgoing Handover Preparation Delay* = Δt S1AP Handover Required → S1AP Handover Command.
- *S1 Outgoing Handover Total Delay* = Δt S1AP Handover Command → S1AP UE Context Release ("successful handover"). Although this delay measurement is not meaningful for the user experience, it is important to measure how long radio resources in the source eNB are blocked after a handover command was sent to the UE.
- *eNB Handover Command Latency* = Δt S1AP Handover Command → RRC Handover Command.
- *S1 Incoming Handover Total Delay* = Δt S1AP Handover Request Acknowledge → S1AP Handover Notify.
- *eNB Handover Success Latency* = Δt RRC Handover Confirm → S1AP Handover Notify.

4.4.3.4 Data Forwarding

Data forwarding applies to the X2 handover procedures and can start after the X2AP handover request acknowledge message is received by the source eNB.

In order to assist the reordering function in the target eNB, the S-GW should send one or more "end marker" packets to the source eNB using the old S1-U path immediately after switching the path for each E-RAB of the UE. The "end marker" packets do not contain any user data and are marked with a special flag in the GTP header. After sending the "end marker" packets, the S-GW should not send any further user data packets via the old path.

Upon receiving the "end marker" packets from the S-GW via the old S1-U path, the source eNB should, if forwarding is activated for that bearer, forward the packet toward the target eNB.

The PDCP SN of forwarded Service Data Units (SDUs) is carried in the "PDCP PDU number" field of the GTP-U extension header. The target eNB should use the PDCP SN if it is available in the forwarded GTP-U packet.

Important KPIs for data forwarding are:

- number of forwarded SDUs;
- number of lost SDUs;
- time difference lt X2AP Handover Request Acknowledge → first UP packet forwarded on X2;
- time difference lt first "end marker" packet on old S1-U path → first UP packet forwarded on X2.

4.5   User Plane KPIs

User plane KPIs are typically measured on a per call basis. However, this does not mean that they always have to be aggregated and stored per call. There are various ways to store and display user plane performance measurements. Which approach is chosen depends on the purpose of the measurement. What this means will be demonstrated hereafter using the example of throughput measurements.

To get a rough idea of the user plane load, it is sufficient to collect the data volume for a longer time period like 15, 30, or 60 minutes. Based on this time interval, an average throughput can be computed, but typically this is not done since the data volume fulfills the requirement to have a user plane metric for the busiest cells. However, to describe the user experience in these cells, this kind of measurement is not meaningful.

A better way to measure the user experience of throughput is to collect throughput measurement samples during each call that is active in a particular cell and store and count the results in a bin histogram table. The bin distribution of throughput measurement samples allows a good evaluation of the subscriber's throughput quality experience. However, the applications used most commonly in today's mobile networks are not very throughput sensitive. The throughput graphs in connections that are mainly used for e-mail and web-browsing services show a profile with high volatility due to the traffic profile itself. As a result, the distribution of samples in the bin histogram reflects the nature of the traffic rather than the experience of the user. Even if the throughput is lower during downloading elements of, for example, a website, this is typically not recognized by the subscriber as a problem with mobile IP access, because the same kinds of issues have been experienced many times before in fixed and WLANs for other reasons. Knowing this, it seems that the throughput bin histogram approach is not a universal one. It is meaningful to evaluate the user experience of throughput-sensitive real-time and non-real-time services like streaming video or File Transfer Protocol (FTP), but for a generic measurement of user experience and the quality of a cell, it does not work. Another disadvantage is that measuring this kind of throughput requires a lot of hardware. Hence, to deploy it either embedded in network elements or in using external measurement equipment is very costly and the benefits one gains must be carefully considered versus the expenses.

Another way to gain an impression of the user's throughput experience is to measure the throughput of a particular connection and store the maximum, minimum, and mean results of this measurement in CDRs. This method works quite well in the lab, where the number of connections and cells to be

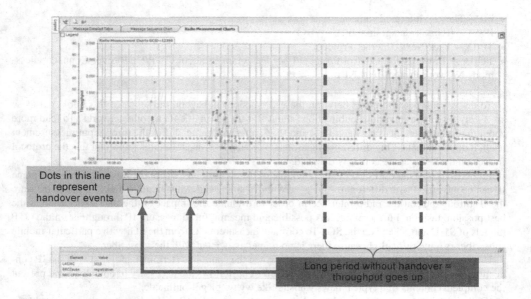

Figure 4.43   Throughput measurement graph of a single connection correlated with occurrence of handover events

monitored is rather small. The use case is to quickly identify under-performing calls from a set of connections with a known, determined traffic profile.

The most detailed way of throughput measurement presentation is to compute a set of measurements that have been sampled at a high sampling rate for the duration of an ongoing connection. The throughput measurement results are stored together with their timestamps in a database to be later presented as a table or as a graph along the time axis of a measurement diagram. As shown in Figure 4.43, the greatest benefit of this presentation format can be reached if the throughput measurement results are shown graphically in correlation with other measurements or with the occurrence of signaling events (in the figure: handover events) that impact the speed of data transmission. In fact, the graph shown in this figure is based on a 3G UTRAN Iub High-Speed Downlink Packet Access (HSDPA) recording, but handover procedures will have the same impact on throughput in 4G E-UTRAN.

In the end, it depends on the individual use case, what kind of measurement granularity and which reporting format should or can be chosen. What was discussed here for throughput measurements also applies to most other user plane performance measurements.

4.5.1   IP Throughput

The IP throughput is defined as the data volume of IP frames transmitted within a defined time period in the UL or DL direction. There are different versions and derivatives of the IP, so a suite of protocols should be supported by this measurement, especially as we find in mobile IP networks:

- IPv4 – Internet Protocol version 4 (32-bit addresses).
- IPv6 – Internet Protocol version 6 (128-bit addresses).

- PIP – The "P" Internet Protocol.
- IP/ST – IP in ST datagram mode.
- TP/IX – The "Next" Internet Protocol.
- TUBA – Transmission Control Protocol/User Datagram Protocol (TCP/UDP) over Connectionless-mode Network Protocol (CLNP).

To discuss all the differences between these IP versions is beyond the scope of this book, but these are surely discussed in various publications about IP services in 3G/4G mobile networks. To read more about the basic differences between IPv4 and IPv6, refer to Chapter 1. For the throughput measurement itself, it does not make a big difference regardless of which IP is used. But it is important that the protocol frames on the IP layer are correctly decoded.

In the E-UTRAN and EPC, there are a couple of different points where the throughput of user plane IP can be measured with different results even within a single call. Most of all, it makes sense to measure the IP throughput on the radio interface in order to evaluate the impact on the user experience in the best possible fashion. Furthermore, it is possible and meaningful to measure IP throughput on the GTP tunnels of S1-U and S5 and on the SGi. To correlate measurements from the SGi with a particular mobile subscriber is quite difficult, because there is no tunnel associated with the subscriber.

Dozens of TCP/UDP flows are exchanged between the endpoints (terminal and server) of the IP connection. To give an example, approximately 30 different IP flows are necessary to download and present the complete contents of a typical news website like www.spiegel-online.de.

While it is easy to detect the sender and receiver of a particular IP packet based on the source address and destination address in the IP header, it is not so easy to determine which of these addresses belongs to the subscriber. This knowledge is necessary to enable separate measurements for the UL and DL direction. The best way to sort UL and DL frames is to track the assignment of temporary IP addresses during the attach procedure or to track assignment of GTP tunnel endpoint identifiers in the tunnel management procedure of GTP-C and S1AP and associate this context information with the throughput measurements.

The volume of IP data can be easily measured using the length field of the IP header that indicates the size of the total IP frame: header plus data. An example is given in Figure 4.44: 1480-byte data + 20-byte header = total length of 1500 bytes.

Figure 4.44 IP frame header

If the total size of the cumulatively counted header length (= data volume!) for a particular call and direction is stored with a 1-second granularity, this value automatically reflects the throughput in bytes per second to be converted into kilobits per second (kbps) or kilobytes per second using a simple post-processing calculation.

4.5.2   Application Throughput

The application throughput for non-real-time services is the data volume measured on the layer 7 protocol of TCP/IP user plane protocol stacks divided by time. Basically, the measurement approach is the same as explained earlier in the case of IP throughput and the different layer 7 applications can be distinguished in terms of UDP or TCP port numbers.

There are two challenges for the design of an application that measures application throughput: find out the correct amount of application data volume itself, and find the correct start and stop trigger points for the measurement.

To understand the challenge of determining the correct application layer data volume, it is useful to look at an example, namely the FTP frame shown in Figure 4.45.

The FTP frame contains pure data and there is no length field in TCP, just a data offset information field. This data offset indicates the first bit of the TCP frame where the payload field of the TCP frame begins. Since the whole TCP frame is organized in multiples of 32 bits (4 bytes), it is clear that 5×4 bytes = 20 bytes. This amount needs to be subtracted from the total TCP frame length to get the

ID Name	Comment or Value	BITMASK
Delay	Normal	---0----
Throughput	Normal	----0---
Reliability	Normal	-----0--
Reserved	'00'B	------00
Total Length	1500	***B2***
Identification	cca0	***B2***
Reserved	'0'B	0-------
DF	Don't Fragment	-1------
MF	Last Fragment	--0-----
Fragment Offset	0	**b13***
Time to Live	119	01110111
Next Header/Protocol	TCP Transmission Control [RFC793]	00000110
Header Checksum	ca24	***B2***
Source Address (IPv4)	4.23.88.219	***B4***
Destination Address (IPv4)	10.90.0.11	***B4***
Data	00 14 06 16 d7 9d 8b 88 1a 18 b1 9d 50 10 ff...	*B1480**
TCP - Transmission Control Protocol, RFC793 (+RFC1072/1323/2... (TCP) ack (= acknowledge)		
acknowledge		
Source Port	ftp-data - File Transfer [Default Data]	***B2***
Destination Port	xingmpeg	***B2***
Sequence Number	3617426312	***B4***
Acknowledgment Number	437825949	***B4***
Data Offset	5	0101----
Reserved	'000000'B	***b6***
URG	Urgent Pointer field not significant	--0-----
ACK	Acknowledgement field significant	---1----
PSH	No Push Function	----0---
RST	Don't reset the connection	-----0--
SYN	Don't synchronize sequence numbers	------0-
FIN	More data from sender	-------0
Window	65535	***B2***
Checksum	e6b5	***B2***
Urgent Pointer	0	***B2***
data	91 e8 3f ff 9d e3 bf a0 d0 06 60 00 b6 10 00...	*B1460**
FTP-DATA - File Transfer Protocol (Data), RFC959 (IP_FTP_DATA) data (= Data)		
Data		
Data	91 e8 3f ff 9d e3 bf a0 d0 06 60 00 b6 10 00...	*B1460**

Figure 4.45   TCP and FTP data frame

```
                                                                            Frame View
              ID Name                              Comment or Value              BITMASK
UDP, RFC 768 08.80 (UDP_HIGH)   DTGR (= Datagram)
Datagram
Source Port                            - unknown / undefined -                 ***B2***
Destination Port                       Domain Name Server (DNS)                ***B2***
Length                                 69                                      ***B2***
Checksum                               32203                                   ***B2***
UDP contents                           1a 10 01 00 00 01 00 00 00 00 00 00 09 75 73...  **B61***
DNS - Domain Name Service (RFC1035) (DNS)  qry (= query)
query
header
id                                     1a10                                    ***B2***
qr                                     query                                   0-------
opcode                                 standard query                         -0000---
```

Figure 4.46   UDP datagram with length indicator

payload field size. Now the problem is also that the previous IP frame that carried another TCP block does not contain any information about the TCP payload size, just the total length of the IP frame including the header. Often the IP header is 20 bytes long, but we cannot count on this fixed value because, due to a number of options, IP headers may become extremely long – so long that special algorithms for IP header compression have been developed.

To calculate the FTP data volume using the information available in what is shown in Figure 4.41 as the BITMASK column would be easy, but this BITMASK information is delivered by the decoder of the measurement unit, not by the protocol itself, and to refer to this decoder-specific information, it needs a more sophisticated measurement application than one that just adds up values of length fields.

In the case of UDP, it might be a little bit easier, because the UDP header has a fixed size of 8 bytes that can be subtracted from the length field that indicates the size of the whole UDP frame: that is, header plus UDP contents. For example, see Figure 4.46.

For real-time services, the data volume of Real-Time Transport Protocol (RTP) packets is a good metric to determine the throughput, but since there is no length filed in the RTP header, the size of RTP blocks must be determined using a little algorithm as is typically implemented in decoder engines. One possible approach is to take the length field from the UDP header if the UDP ports indicate that RTP packets are transported and to plot the length of the payload field (UDP length minus UDP header size) in a trailer that comes along with the plain decoder output. By reading this trailer information and summing the size of RTP blocks over time, it is easy to calculate the RTP throughput.

As already explained, an IP connection consists of multiple TCP and UDP flows that may run subsequently or in parallel depending on the application. Also, on the IP layer, the server addresses are changing depending on the source of contents, for example, a website. Thus, on the IP layer also multiple flows can be identified as shown in the table in upper part of Figure 4.47, which also gives a nice example of data volume and average throughput measurement results in a tabular format that allows easy detection of the most active IP data flow of this connection.

To see TCP and UDP flows that allow one to distinguish between different application layers according to the port numbers, it is necessary to look one layer deeper into the IP flow contents. Here, one can see that UDP immediately transmits application data while TCP needs a startup procedure for its AM transmissions. This means that the application throughput on UDP can start straightaway with the application data volume of the first UDP frame that includes the application's specific UDP port numbers.

In the case of TCP, the startup messages of the three-way handshake "syn"–"syn-ack"–"ack" are not used to transmit any application layer data, although they include the port number of the application. Hence, the throughput measurement start trigger should be the first application layer message, for example, HTTP GET in the case of downloading a website stream, and the stop trigger for this measurement is the HTTP 200OK message.

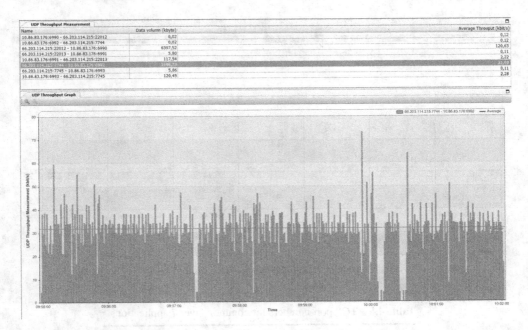

Figure 4.47 UDP throughput of a connection between two terminal endpoints as measured on S1-U interface

Last but not least, a strong limitation on measuring the application layer KPIs should be mentioned. In the case of IPsec, which is commonly used for mobile VPN connections, a decoding of TCP/UDP frames is not possible due to the ciphering algorithms used by IPsec. One must bear in mind that in 3G networks IPsec, at the time of writing this chapter (Spring 2010), already makes up between 30% and 60% of the overall IP data volume transmitted in mobile networks. This percentage is expected to rise.

4.5.3 TCP Startup KPIs

For non-real-time services such as file transfer, web browsing, and e-mail, a set of service startup failure ratio, service startup time, IP service startup failure ratio, and IP service startup time KPIs is defined in ETSI TS 102.250 together with other QoE KPIs such as UL/DL throughput for individual services.

Originally, ETSI TS 102.250 was defined to measure QoS and QoE KPIs with drive test equipment. This explains, for example, the different definitions for service setup time and IP service setup time. The difference is that service setup time refers to start trigger point of the first real application frame sent, while IP service setup sees the trigger points for the start and stop of delay measurements in the TCP layer. In fact, the service setup time definition also has a fallback to TCP layer-based trigger points defined in case the trigger point cannot be provided from the application layer by the measurement equipment. In the end, service setup and IP service setup trigger points can in fact be identical.

Figure 4.48 shows an example how the (IP) service setup time is measured in the case of FTP service. It is the time difference for the TCP three-way handshake (see Section 1.10.12). In this example, the delay between the TCP syn (frame number 1983) and the appropriate first TCP ack (frame number 1992)

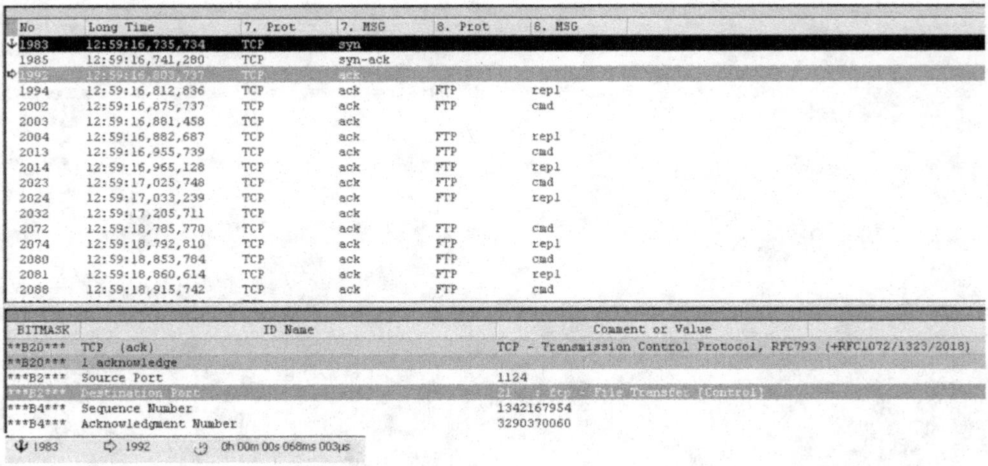

No	Long Time	7. Prot	7. MSG	8. Prot	8. MSG
1983	12:59:16,735,734	TCP	syn		
1985	12:59:16,741,280	TCP	syn-ack		
1992	12:59:16,803,737	TCP	ack		
1994	12:59:16,812,836	TCP	ack	FTP	repl
2002	12:59:16,875,737	TCP	ack	FTP	cmd
2003	12:59:16,881,458	TCP	ack		
2004	12:59:16,882,687	TCP	ack	FTP	repl
2013	12:59:16,955,739	TCP	ack	FTP	cmd
2014	12:59:16,965,128	TCP	ack	FTP	repl
2023	12:59:17,025,748	TCP	ack	FTP	cmd
2024	12:59:17,033,239	TCP	ack	FTP	repl
2032	12:59:17,205,711	TCP	ack		
2072	12:59:18,785,770	TCP	ack	FTP	cmd
2074	12:59:18,792,810	TCP	ack	FTP	repl
2080	12:59:18,853,784	TCP	ack	FTP	cmd
2081	12:59:18,860,614	TCP	ack	FTP	repl
2088	12:59:18,915,742	TCP	ack	FTP	cmd

BITMASK	ID Name	Comment or Value
B20*	TCP (ack)	TCP - Transmission Control Protocol, RFC793 (+RFC1072/1323/2018)
B20*	1 acknowledge	
B2	Source Port	1124
B2	Destination Port	21 : tcp - File Transfer [Control]
B4	Sequence Number	1342167954
B4	Acknowledgment Number	3290370060

↓ 1983   ⇨ 1992   ⏱ 0h 00m 00s 068ms 003µs

Figure 4.48   FTP (IP) service setup time

Table 4.6   TCP port numbers for common layer 7 applications

TCP port number	Layer 7 application number
20, 21	File Transfer Protocol (data and control)
80	Hyper Text Transfer Protocol (web-browsing)
110	Post Office POP 3 (e-mail)
143	Internet Message Access Protocol (e-mail)
2848	Wireless Application Protocol (WAP) push MMS

message is 68 ms. The key for correlating this plain TCP delay measurement with the application service is to look at the source and destination ports of the TCP header. The port numbers for the most common non-real-time services currently seen in mobile networks are listed in Table 4.6.

To compute the service startup failure ratio is also quite easy using the same protocol trigger points. Here, each TCP syn message is counted as a service attempt and the first ack message of the TCP flow is counted as a success event.

4.5.4   TCP Round-Trip Time

The TCP Round-Trip Time (RTT) is the delay between sending a TCP packet identified on behalf of an individual TCP SN in a particular TCP flow and acknowledgment of this packet in the same TCP flow as shown in Figure 4.49.

When implementing this measurement, it must be considered that TCP has some options to work with selective acknowledgments as defined in RFC 2018. In this case, the acknowledgment number refers not just to a single sent SN, but also to a set of previously sent packets.

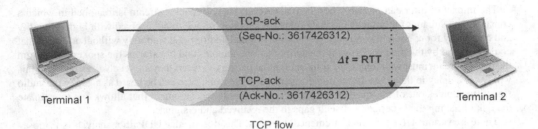

Figure 4.49   TCP round-trip time measurement principle

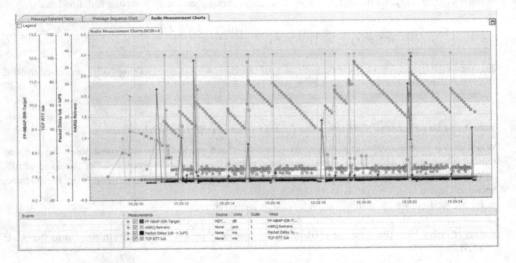

Figure 4.50   HARQ retransmissions cause TCP round-trip time peaks

The TCP RTT is heavily influenced by radio quality and radio interface procedures. Figure 4.50 shows how HARQ retransmissions on the Enhanced Dedicated Channel (E-DCH) of the 3G UTRAN trigger peaks of TCP RTT and how the power control increased the UL SIR (Signal-to-Interference Ratio) target to prevent further deterioration of user plane QoS. Once increased after the occurrence of HARQ retransmissions, the UL SIR target is decreased in quick small steps to ensure that the UL interference is kept at a minimum level. Although this figure shows the typical scenarios of a 3.5G High-Speed Uplink Packet Access (HSUPA) call, the same power control mechanisms and same correlation between retransmission of radio frames and rising RTT can be found in LTE.

4.5.5   Packet Jitter

By definition, packet jitter is the average of the deviations from the network mean latency. It is an important QoS parameter for real-time services using UDP transport due to their delay sensitivity. For non-real-time services like web browsing or e-mail, the jitter has no impact on the user plane QoS.

The impact of jitter can probably best be understood if a VoIP packet stream is imagined in which a constant data stream of thousands of small voice packets is expected to be received with high reliability and continuity for reassembly and playing back the original audio signal smoothly without any error that could be heard by the receiver. Now, if the entity that reassembles and transforms the spoken word from the packet stream must wait longer for some individual packets – even if they are received error-free – this will cause a delay in the reassembler/transformer that in the worst case can be heard as a gap in the audio signal that comes out of the speaker. Packet jitter is a simple measurement that allows one to estimate the risk of having the experience of such gaps in the received audio signal.

There are various UDP jitter measurement definitions. The most needed UDP jitter analysis is a "loose" UDP jitter based on the arrival timestamp of subsequently received UDP packets at the measurement point. For the UDP jitter calculation, the formula outlined in Equation 4.17 is used. In this formula, the individual latency measurement samples for two subsequently received packets of the same stream are defined as D_i and D_J and the difference of these two latency measurement results is defined as $J(j–i)$:

$$j_1 = 0, J_2 = (2, 1), J_i = J_{i-1} + \frac{(|J(i, i-1)| - J_{i-1})}{16} \begin{vmatrix} i = 3 \\ i = \infty \end{vmatrix} \tag{4.17}$$

The factor 16 that is found in the denominator of Equation 4.17 is a smoothing factor proposed in RFC 1889, the standard specification document for RTP.

4.5.6 Packet Delay and Packet Loss on a Hop-to-Hop Basis

When packet loss or abnormal packet delay is measured at a particular measurement point, the question will be raised about which element or part of the network caused such problems. For this kind of root cause analysis, it is necessary to measure packet loss and packet delay on a hop-to-hop basis, for example, between Uu and S1-U, between S1-U and S5, and so on.

Figure 4.51 shows the results for the following measurements:

- *Packet delay UL:* The time the eNB needs to forward a packet received from the Uu to the S1-U interface.

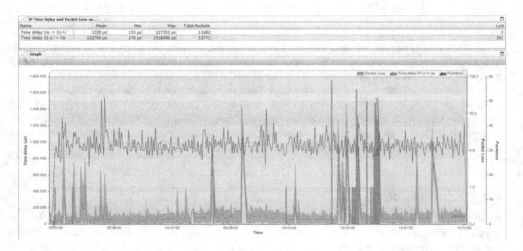

Figure 4.51 Packet delay and lost packet caused by eNodeB

- *Packet delay DL:* The time the eNB needs to forward a packet received from the S1-U interface to the Uu interface.
- *Lost packets:* The total number of lost packets and the time distribution of packet losses according to the tracking results after call trace.

The packet delay and packet loss for DL transmissions from S1-U to Uu are shown in the graph; UL delay measurement results are only displayed in tabular format as a max/min/mean value triple. Note that the graphs in Figure 4.51 correspond to the UDP throughput shown in Figure 4.43. A comparison of both figures reveals the loss of packets in eNB as the root cause of the throughput degradation shown in Figure 4.47.

4.6   KPI Visualization using Geographical Maps (Geolocation)

During the last years, it became popular to visualize KPIs on geographical maps after geolocating events and measurement reports sent by the UE.

The principles of geolocation are simple. The Global Positioning System (GPS) as known from car navigation systems and the geolocation based on the so-called crowd-sourced databases that have stored longitude/latitude information of WLAN SSIDs and mobile network base stations are the methods most known in public.

There are a couple of applications on the market that make use of this positioning method as follows:

- The UE sends error/measurement reports to the app developer's database via user plane connection.
- The error/measurement repots are aggregated into another application that allow to display statistics on a GIS map.
- This application with GIS functionality is bundled with the aggregated error/measurement report data received from subscribers of a particular network operator, and this map including the statistics pattern is offered to be sold to this network operator.

The main benefit of such data is that network operators get a good overview of user quality of experience. However, the data is limited to certain applications and often these kind of specific measurement reports are triggered only in case of problems occurring and the application KPIs cannot be mapped to radio measurements and network alarms by itself.

When it comes to geolocation of signaling information, for example, to build a coverage map based on all RRC measurement reports monitored in the network, the GPS and "crowd-sourced" databases are typically not available. Thus, a different methodology to estimate the geographical position of the UE needs to be found.

The most basic approach is the triangulation method as shown in Figure 4.52. Here, the prerequisite is that radio signals of at least three different cells are reported by the UE. In addition, the timing advance propagation path – this is the one-way distance between UE and cell's antenna – of the serving cell (in the figure this is cell 3) is often known.

For each radio signal sent by measured cell, the distance between UE and the cell's antenna can be calculated based on path loss of the signal as shown in Figure 4.53, where an R is certain distance step between UE and base station and one can see that the path loss leads to declining absolute signal level over distance. The less strong the received signal is, the more far away the UE must be.

While it is difficult to estimate the possible position of the UE when only signal strength of two neighbor cells is reported, the likelihood of the UE's position becomes more accurate if the pathloss distance or timing advance of a 3rd cell is known in addition. Timing advance is only available for the serving cell. Pathloss distance can be calculated for any cell reported.

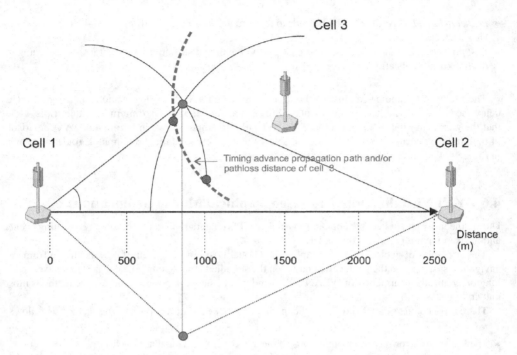

Figure 4.52   Triangulation of an UE position

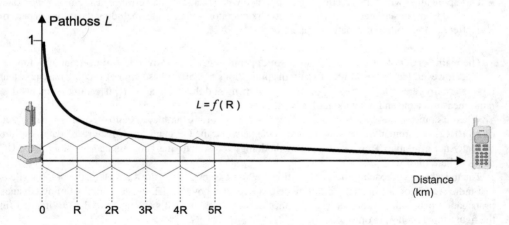

Figure 4.53   Path loss of the cell's radio signal as a function of distance between UE and cell's antenna

Taking now the timing advance of the serving cell and the distance between UE and antenna of cell 3 calculated on behalf of the path loss, the reliable possible geolocation of the UE can be determined as seen in Figure 4.52.

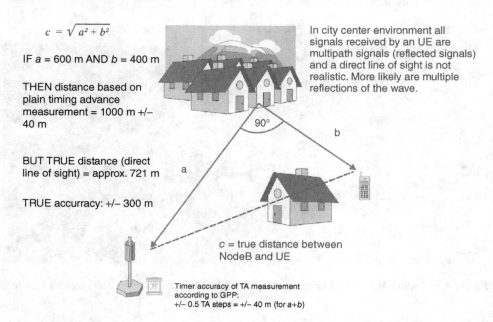

$$c = \sqrt{a^2 + b^2}$$

IF a = 600 m AND b = 400 m

THEN distance based on plain timing advance measurement = 1000 m +/– 40 m

BUT TRUE distance (direct line of sight) = approx. 721 m

TRUE accurracy: +/– 300 m

a

90°

b

In city center environment all signals received by an UE are multipath signals (reflected signals) and a direct line of sight is not realistic. More likely are multiple reflections of the wave.

c = true distance between NodeB and UE

Timer accuracy of TA measurement according to GPP:
+/– 0.5 TA steps = +/– 40 m (for a+b)

Figure 4.54   Distance by timing advance versus true distance in direct line of sight

As a general rule, it can be said that the more neighbor cells are reported the better the geolocation accuracy will be in such cases. However, a fair geolocation accuracy using this methodology will not exceed a median error of 100–150 meters in dense urban areas. This is due to the nature of wave propagation including reflection of waves on building walls and other obstacles as shown in Figure 4.54. In other words, the distance required for the wave to go is often not the distance in the direct line of sight between base station and the UE as is assumed in the triangulation approach. Even taking antenna patterns and human moving patterns into account, such differences cannot be completely eliminated.

Despite the limited accuracy, there are some clear benefits of the RAN signaling-based geolocation solutions:

- Provision of coverage and interference maps based on true RRC measurement reports of all UEs in the network as shown in Figure 4.55 (for reasons of data privacy, the street map is not shown in this figure, but for sure part of the solution).
- Identification of traffic hotspots including any traffic (signaling, payload) of all user applications.
- Identification of hot spots with radio connectivity problems that have direct impact on customer experience such as call drops and call setup failure (blocked calls) as shown in Figure 4.56.

Thus, the RAN signaling-based geolocation solutions are an important input for radio network troubleshooting, optimization, and capacity planning.

4.6.1   The Minimize Drive Test Feature Set of 3GPP

With Release 10, the 3GPP has introduced a feature set that is known as "Minimize Drive Test" (MDT), which is described in 3GPP 36.805 and 37.320. The purpose of this feature set is to enable the UE to

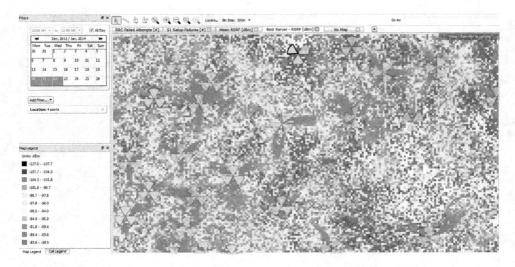

Figure 4.55   RSRP Best Server Coverage Map based on RRC measurement reports of all UEs in the network

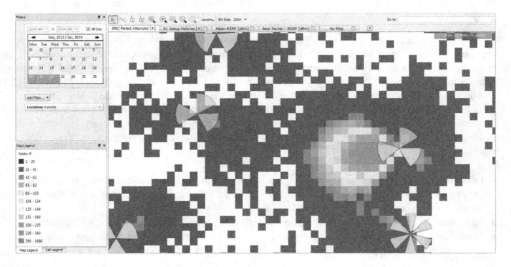

Figure 4.56   Geographical map showing hot spots of accessibility problems (blocked calls = call setup failures)

send specific error logs in case of network problems, for example, call drop. In ideal case, these error logs are tagged with GPS coordinates so that they can be visualized on a map easily and with highest possible accuracy.

The standards distinguish between logged and immediate MDT data.

The logged MDT measurements are used in case of RRC connection setup failure and call drops – when the UE is in RRC IDLE, CELL_PCH, or URA_PCH state, are not able to send an error report immediately. The logged MDT events (can be longer series) will be sent to the eNodeB during next successfully established RRC connection.

The immediate MDT measurements are defined as a mix of radio-related measurements (RSRP, RSRQ, UE power headroom) and physical layer/L2 eNodeB measurements as described in Sections 4.2.3 and 4.3.2. As a result, a particular immediate MDT report may contain the UL/DL data volume transmitted during the measurement period correlated with RSRP/RSRQ, UE power headroom and tagged with GPS coordinates of the UE.

The logged or immediate MDT measurement sent by the UE will be transmitted to the eNodeB using the RRC protocol (3GPP 36.331) that is enhanced with a set of messages and parameters for MDT configuration, activation, and reporting starting from Release 10.

However, while the benefits of the Minimize Drive Test feature set in the area of network performance measurement, troubleshooting, and optimization are clear, there are also disadvantages that need to be taken into account.

On one hand, UE chipsets must be MDT-ready and this will be an extra effort for RF chipset and handset manufacturers while subscribers will not pay for these extra features. On the other hand, there are major concerns regarding subscriber privacy, because as a side effect the MDT feature allows discrete and highly accurate tracking of subscriber activities at any time.

3GPP has specified that MDT shall be tied to the trace functionality of the networks operation and maintenance entities and subscribers shall sign a clause in their contract to permit enabling of MDT measurements. The future must prove if this procedure will be seen as sufficient by those who criticize the network operators already today for storing too much subscriber-specific connection information.

Acronyms

2.5G	2.5 Generation
3G	3rd Generation
3GPP	3rd Generation Partnership Project
4G	4th Generation
8PSK	8 Phase Shift Keying
ACK	Acknowledgment
AM	Acknowledged mode
AMBR	Average Aggregate Maximum Bit Rate
APN	Access Point Name
ARFCN	Absolute Radio Frequency Channel Number
ARQ	Automatic Repeat Request
ASCII	American Standard Code for Information Interchange
ASME	Access Security Management Entity
AUTN	Authentication Token
BCC	Base Station Color Code
BCCH	Broadcast Control Channel
BSIC	Base Station Identity Code
BSS	Base Station Subsystem
CDMA	Code Division Multiple Access
CGI	Cell Global Identification
CI	Cell Identity
CQI	Channel Quality Indicator
CRC	Cyclic Redundancy Check
CS	Circuit Switched
CWR	Congestion Window Reduced (IPv6)
DCH	Dedicated Channel
DECT	Digital Enhanced Cordless Telecommunications
DHCP	Dynamic Host Configuration Protocol
DL	Downlink
DL-SCH	Downlink Shared Channel
DNS	Domain Name System

LTE Signaling, Troubleshooting and Performance Measurement, Second Edition. Ralf Kreher and Karsten Gaenger.
© 2016 John Wiley & Sons, Ltd. Published 2016 by John Wiley & Sons, Ltd.

DRB	Dedicated Radio Bearer
eARFCN	E-UTRAN Absolute Radio Frequency Channel Number
EBI	EPS Bearer ID
ECM	EPS Connection Management
E-DCH	Enhanced DCH
E-DCH	Enhanced Dedicated Channel
EDE	ECN Echo (IPv6)
EDGE	Enhanced Data Rates for GSM Evolution
EGPRS	Enhanced GPRS
EMM	EPS Mobility Management
EMR	Error Modulation Ratio
eNB	eNodeB
EPS	Enhanced Packet System
E-UTRAN	Enhance UMTS Terrestrial Network
E-UTRAN	Enhanced UMTS Terrestrial Network
EVM	Error Vector Magnitude
FACH	Forward Access Channel
F-TEID	Fully Qualified Tunnel Endpoint Identifier
FTP	File Transfer Protocol
GBR	guaranteed bit rate
GERAN	GSM/EDGE Radio Access Network
GGSN	Gateway GPRS Support Node
GMM	GPRS Mobility Management
GMSK	Gaussian Minimum Shift Keying
GPRS	General Packet Radio Service
GSM	Global System for Mobile Communication
GTP	GPRS Tunneling Protocol
GUMMEI	Globally Unique MME Identifier
GUTI	Globally Unique Temporary UE Identity
HARQ	Hybrid Automatic Repeat Request
HLR	Home Location Register
HRPD	High Rate Packet Data
HSDPA	High Speed Downlink Packet Access
HS-DSCH	High Speed Downlink Shared Channel
HSPA	High Speed Packet Access
HS-PSCH	High Speed Downlink Physical Shared Channel
HSS	Home Subscriber Server
HSUPA	High Speed Uplink Packet Access
HTTP	Hyper Text Transfer Protocol
ICMP	Internet Control Message Protocol
IDNNS	Intra Domain NAS Node Selector
IHL	Internet Header Length
IMEI	International Mobile Equipment Identity
IMS	IP Multimedia Subsystem
IMSI	International Mobile Subscriber Identity
IP	Internet Protocol
IPsec	Secure IP
ISO	International Standard Organization
KASME	Key (from) ASME

kbps	Kilobit per second
kmph	kilometer per hour
L1	Layer 1
L2	Layer 2
LAC	Location Area Code
LAI	Location Area Identity
LCID	Logical Channel ID
LLC	logical link control
LMSI	Local Mobile Subscriber Identity
LOS	Line of Sight
LTE	Long Term Evolution
LTE-A	Long Term Evolution – Advanced
MAC	Message Authentication Code
MAC	Medium Access Control
Mbps	Megabit per second
MCC	Mobile Country Code
MF	Multiple Frames (IP fragmentation flag)
MIB	Master Information Block
MIMO	Multiple Input Multiple Output
MME	Mobility Management Entity
MMEC	MME Code
MNC	Mobile Network Code
MS	Mobile Station
NAS	Non-Access-Stratum
NCC	Network Color Code
NMSI	National Mobile Subscriber Identity
NRI	Network Resource Identifier
OFDM	Orthogonal Frequency Division Multiplex
OSI	Open Systems Interconnection (Reference Model)
PAR	Peak-to-Average Ratio
PBCH	Physical Broadcast Channel
PDCCH	Physical Downlink Control Channel
PDP	Packet Data Protocol
PDSCH	Physical Downlink Shared Channel
PDSCH	Physical Downlink Shared Channel
PDU	Protocol Data Unit
PLMN	Public Land Mobile Network
POP3	Post Office Protocol (version 3)
PRACH	Physical Random Access Channel
PRB	Physical Resource Block
PS	Packet Switched
P-SCH	Primary Synchronization Channel
PSH	Push (IPv6)
PUSCH	Physical Uplink Shared Channel
PUUCH	Physical Uplink Control Channel
QAM	Quadrature Amplitude Modulation
QCI	Quality of Service Class Indicator
QoS	Quality of Service
RAB	Radio Access Bearer

RAC	Routing Area Code
RACH	Random Access Channel
RAI	Routing Area Identity
RANAP	Radio Access Network Application Part
RAND	Random Number
RAT	Radio Access Technology
RES	Response (Number)
RFC	Request for Comments
RLC	radio link control
RNC	Radio Network Controller
RNS	Radio Network Subsystem
RNTI	Radio Network Temporary Identity
RNTP	Relative Narrowband Transmission Power
RSRP	Received Signal Reference Power
RSRQ	Received Signal Reference Quality
RSSI	Received Signal Strength Indicator
RST	Reset (IPv6)
RTP	Real-Time Transport Protocol
S1AP	S1 Application Part
SAC	Service Area Code
SAE	System Architecture Evolution
SAI	Service Area Identity
SC-FDMA	Single Carrier Frequency Division Multiple Access
SCTP	Stream Control Transmission Protocol
SDU	Service Data Unit
SGSN	Serving GPRS Support Node
SI	System Information
SINR	Signal to Interference and Noise Ratio, equivalent to SNIR
SIB	System Information Block
SIP	Session Initiation Protocol
SM	Session Management
SMTP	Simple Message Transfer Protocol
SN	Sequence Number (PDCP)
SNIR	Signal to Noise and Interference Ratio, equivalent to SINR
SNR	Serial Number (part of IMEI)
SON	Self Optimizing Network
SRB	Signaling Radio Bearer
SRNC	Serving Radio Network Controller
SRS	Sounded Reference Symbols
SS7	Signaling System Number 7
S-SCH	Secondary Synchronization Channel
SYN	Synchronize (IPv6, TCP)
SYN-ACK	Synchronize Acknowledgment (IPv6, TCP)
TA	Tracking Area
TAC	Type Approval Code
TAC	Tracking Area Code
TAI	Tracking Area Identity
TBF	temporary block flow
TCP	Transmission Control Protocol

TD-SCDMA	Time Division Synchronous Code Division Multiple Access
TEID	Tunnel Endpoint Identifier
TFI	Temporary Flow Identifier
TFT	Traffic Flow Template
TLLI	Temporary Logical Link Identity
TMSI	Temporary Mobile Subscriber Identity
TTI	Time Transmission Interval
uARFCN	UTRAN Absolute Radio Frequency Channel Number
UDP	User Datagram Protocol
UL	Uplink
UL-SCH	Uplink Shared Channel
UM	unacknowledged mode
Urgent (flag)	URG (IPv6)
URI	Uniform Resource Identifier
USIM	UMTS Subscriber Identity Module
VANC	VoLGA Access Network Controller
VoLTE	Voice over LTE
WCDMA	Wideband CDMA
XRES	Expected Response (Number)

Bibliography

Badach, A., *VoIP – Die Technik, 4th edition, Hanser*, Germany, 2010.

Holma, H. and Toskala, A. *LTE for UMTS – OFDMA and SC-FDMA Based Radio Access, 2nd edition*, Wiley, UK, 2011.

Kreher, R. *UMTS Performance Measurement: A Practical Guide to KPIs for the UTRAN Environment*, Wiley, UK, 2006.

Kreher, R. and Ruedebusch, T., *UMTS Signaling: UMTS Interfaces, Protocols, Message Flows and Procedures Analyzed and Explained*, 2nd edition, Wiley, UK, 2007.

Lescuyer, P. and Lucidarme, T. *Evolved Packet System (EPS): The LTE and SAE Evolution of 3G UMTS*, Wiley, UK, 2008.

Sesia, S., Toufik, I., and Baker, M. *LTE, The UMTS Long Term Evolution: From Theory to Practice, 2nd edition*, Wiley, UK, 2011.

Wikipedia, the free encyclopedia: www.wikipedia.org.

3GPP 23.003. *Numbering, Addressing and Identification*.

3GPP 23.007. *Restoration Procedures*.

3GPP 23.203. *Policy and Charging Control Architecture*.

3GPP 23.401. *General Packet Radio Service (GPRS) enhancements for Evolved Universal Terrestrial Radio Access Network (E-UTRAN) Access*.

3GPP 24.008. *Mobile Radio Interface Layer 3 Specification; Core Network Protocols*.

3GPP 24.301. *Non-Access-Stratum (NAS) protocol for Evolved Packet System (EPS)*.

3GPP 29.002. *Mobile Application Part (MAP), Specification*.

3GPP 29.272. *Evolved Packet System (EPS); Mobility Management Entity (MME) and Serving GPRS Support Node (SGSN) Related Interfaces Based on Diameter Protocol*.

3GPP 29.274. *3GPP Evolved Packet System (EPS); Evolved General Packet Radio Service (GPRS) Tunnelling Protocol for Control plane (GTPv2-C)*.

3GPP 32.421. *Telecommunication Management; Subscriber and Equipment Trace: Trace Concepts and Requirements*.

3GPP 32.423. *Telecommunication Management; Subscriber and Equipment Trace: Trace Data Definition and Management*.

3GPP 32.425. *Telecommunication management; Performance Management (PM); Performance measurements, Evolved Universal Terrestrial Radio Access Network (E-UTRAN)*.

3GPP 32.450. *Key Performance Indicators (KPI) for E-UTRAN: Definitions*.

3GPP 33.401. *3GPP System Architecture Evolution (SAE): Security Architecture*.

3GPP 36.101. *Evolved Universal Terrestrial Radio Access (E-UTRA); User Equipment (UE) Radio Transmission and Reception*.

3GPP 36.104. *Evolved Universal Terrestrial Radio Access (E-UTRA); Base Station (BS) Radio Transmission and Reception*.

LTE Signaling, Troubleshooting and Performance Measurement, Second Edition. Ralf Kreher and Karsten Gaenger.
© 2016 John Wiley & Sons, Ltd. Published 2016 by John Wiley & Sons, Ltd.

3GPP 36.133. *Evolved Universal Terrestrial Radio Access (E-UTRA); Requirements for Support of Radio Resource MANAGEMENT.*

3GPP 36.211. *Evolved Universal Terrestrial Radio Access (E-UTRA); Physical Channels and Modulation.*

3GPP 36.212. *Evolved Universal Terrestrial Radio Access (E-UTRA); Multiplexing and Channel Coding Channel Coding.*

3GPP 36.213. *Evolved Universal Terrestrial Radio Access (E-UTRA); Physical layer Procedures.*

3GPP 36.214. *Evolved Universal Terrestrial Radio Access (E-UTRA); Physical layer – Measurements.*

3GPP 36.300. *Evolved Universal Terrestrial Radio Access (E-UTRA) and Evolved Universal Terrestrial Radio Access Network (E-UTRAN), Overall description.*

3GPP 36.304. *Evolved Universal Terrestrial Radio Access (E-UTRA); User Equipment (UE) Procedures in Idle Mode.*

3GPP 36.306. *Evolved Universal Terrestrial Radio Access (E-UTRA); User Equipment (UE) radio access capabilities.*

3GPP 36.314. *Evolved Universal Terrestrial Radio Access (E-UTRA); Layer 2 – Measurements.*

3GPP 36.321. *Evolved Universal Terrestrial Radio Access (E-UTRA); Medium Access Control (MAC) protocol specification.*

3GPP 36.322. *Evolved Universal Terrestrial Radio Access (E-UTRA); Radio Link Control (RLC) Protocol Specification.*

3GPP 36.323. *Evolved Universal Terrestrial Radio Access (E-UTRA); Packet Data Convergence Protocol (PDCP) Specification.*

3GPP 36.331. *Evolved Universal Terrestrial Radio Access (E-UTRA); Radio Resource Control (RRC); Protocol Specification.*

3GPP 36.413. *Evolved Universal Terrestrial Radio Access Network (E-UTRAN); S1 Application Protocol (S1AP).*

3GPP 36.423. Evolved Universal Terrestrial Radio Access Network (E-UTRAN); X2 Application Protocol (X2AP).

3GPP 36.805. Study on Minimization of drive-tests in Next Generation Networks.

3GPP 37.320. *Radio measurement collection for Minimization of Drive Tests (MDT); Overall Description.*

ETSI TS 102.250. *QoS Aspects for Popular Services in GSM and 3G Networks; Part 2: Definition of Quality of Service Parameters and their Computation.*

IETF RFC 2396. *Uniform Resource Identifiers (URI): Generic Syntax.*

IETF RFC 3261. *SIP: Session Initiation Protocol.*

IETF RFC 3588. *Diameter Base Protocol.*

IETF RFC 4960. *Stream Control Transmission Protocol.*

IETF RFC 791. *Internet Protocol.*

IETF RFC 793. *Transmission Control Protocol.*

IETF RFC 2018. *TCP Selective Acknowledgment Options.*

IETF RFC 768. *User Datagram Protocol.*

IETF RFC 2225. *Classical IP and ARP over ATM.*

IETF RFC 2460. *Internet Protocol, Version 6 (IPv6) Specification.*

If not stated otherwise all 3GPP specifications refer to the Release 11 standards and features. Latest versions of these documents can be downloaded from http://www.3gpp.org/ftp/Specs/.

Latest RFC versions of TCP/IP world protocols can typically be found quickly when following the appropriate hyperlinks from www.wikipedia.org.

Index

LTE Signaling, Troubleshooting and Performance Measurement, Second Edition. Ralf Kreher and Karsten Gaenger.
© 2016 John Wiley & Sons, Ltd. Published 2016 by John Wiley & Sons, Ltd.

Printed in the United States
By Bookmasters